First Interregional Conference “Sustainable Development of Eurasian Mining Regions” (SDEMR-2019)

E3S Web of Conferences Volume 134 (2019)

Kemerovo, Russia
25 - 27 November 2019

Editors:

Alexey Khoreshok
Victor Atrushkevich
Stefan Voth
Demirel Nuray
Juraj Janocko
Dean Yunliang Tan
Daniela Marasova
Pawel Stefanek
Mariana Petrova

ISBN: 978-1-5108-9934-6

Printed from e-media with permission by:

Curran Associates, Inc.
57 Morehouse Lane
Red Hook, NY 12571

Some format issues inherent in the e-media version may also appear in this print version.

No changes have been made to the content of these proceedings. There may be changes to pagination and minor adjustments for aesthetics.

Printed with permission by Curran Associates, Inc. (2020)

For additional information, please contact EDP Sciences – Web of Conferences
at the address below.

EDP Sciences – Web of Conferences
17, Avenue du Hoggar
Parc d'Activité de Courtabœuf
BP 112
F-91944 Les Ulis Cedex A
France

Phone: +33 (0) 1 69 18 75 75
Fax: +33 (0) 1 69 28 84 91

contact-edps@webofconferences.org

Additional copies of this publication are available from:

Curran Associates, Inc.
57 Morehouse Lane
Red Hook, NY 12571 USA
Phone: 845-758-0400
Fax: 845-758-2633
Email: curran@proceedings.com
Web: www.proceedings.com

TABLE OF CONTENTS

THE RATIONALE FOR THE DEVELOPMENT OF A METHOD FOR ASSESSING THE ROOF CONDITIONS OF MINE WORKINGS IN THE CONDUCT OF ACTUAL MINING USING THE METHOD OF GEOINFORMATION ANALYSIS 1
Abramovich Alexander, Stepanov Yuri, Kretschmann Jurgen

APPLICATION OF HYDRAULIC PRESS FOR TESTING POWERED ROOF PROP SUPPORTS 7
Kipczak Piotr, Wladzielczyk Krzysztof

BENCH TESTS OF A SUPPORT SYSTEM OF A POWERED ROOF SUPPORT'S HYDRAULIC LEG AIMED AT MINIMIZING CONSEQUENCES OF LEAKS 15
Szurgacz Dawid, Brodny Jaroslaw, Brzózka Marcin, Diederichs Ryszard

THE NONCONTACT DETERMINATION TECHNIQUE RESEARCH OF DANGEROUSLY LOADED ZONES IN THE UNDERGROUND MINING 23
Bizyaev Aleksey, Voronkina Natalia, Savchenko Andrey, Cupov Michail

PROBLEMS OF OPERATIONAL SAFETY OF DUMP TRUCKS IN OPEN PIT MINE 30
Gerike Boris, Zacharov Aleksandr, Gerike Pavel, Drozdenko Yuriy

ENHANCEMENT OF OPEN-PIT MINING EFFICIENCY BY MEANS OF QUALITY IMPROVEMENT OF OPEN-PIT EXCAVATORS DESIGN AND OPERATING IN SPECIFIED CONDITIONS 36
Zykov Pyotr, Zvarych Evgeny, Karasev Andrei, Weber Dieter

OPTIMIZATION METHODS AND ALGORITHMS FOR CALCULATING THE CONSTRUCTION OF NON-RIGID PAVEMENT FOR TECHNOLOGICAL QUARRIES ROADS 45
Tailakova Anna, Pimonov Alexander

ACCOUNTING FOR THE STEEL LINING INFLUENCE ON THE RESULTS OF SOLID ROCK RESISTIVITY PROBE FROM DEVELOPMENT WORKING BOUNDARY 51
Ivanov Vadim, Dyagileva Anna, Anyona Seroni

DEVELOPMENT OF THE DIGITAL MODEL FOR THE SELECTION OF MINE WORKING SUPPORT IN COMPLEX MINING AND GEOLOGICAL CONDITIONS 56
Kopytov Aleksandr, Pershin Vladimir

RESEACH OLD RUBBER TIRES UTILIZATION TO INCREASE THE LIFE OF SUPPORT IN THE LONGWALL AT QUANG NINH VIETNAM 64
Nguyen Phi Hung, Nguyen Cao Khai, Pham Duc Thang, Ta Van Kien

THE DESIGNING OF THE NEW TYPE OF MINE VENTILATION WALL 71
Nikolaev Alexander, Gazizullin Rafail, Nesterov Valery

COMPUTER MODEL OF ELECTRIC DRIVE OF 1L100K CONVEYING UNIT 77
Zolkin Denis, Petrov Vadim

RECOMMENDATIONS FOR DETERMINING THE THICKNESS OF THE ROAD PAVEMENT OF TECHNOLOGICAL ROADS BUILT FROM COAL-BEARING ROCKS 82
Shalamanov Victor A., Shabaev Sergey N., Alama Faruz

THE RESEARCH OF OPERATIONAL TEMPERATURES OF DUMP TRUCKS TIRES 86
Goryunov Sergey, Khoreshok Aleksey, Grigoryeva Natalya, Preis Elena, Alitkina Oksana

OPEN PIT MINE TRUCK ROBOTIZATION AT THE STAGE OF ROCK TRANSPORTATION 93
Zhdanov Vyacheslav, Kosolapov Andrey, Ibragimova Flyura

SUBSTANTIATION OF OPTIMAL DESIGN PARAMETERS FOR NEW SYSTEM ANGLE HEADFRAMES IN STATE-OFTHE ART MINES 98
Kassikhina Elena, Pershin Vladimir, Tan Yunliang

INCREASING SAFETY AT ROCK PREPARATION BY BLASTING IN OPEN PITS 105
Katanov Igor, Kondratyev Sergey, Sysoyev Andrey

CRITERIA FOR THE DIFFICULTY OF WORKING OUT COAL-BEARING ZONES OF QUARRY FIELDS 110
Selyukov Alexei, Demirel Nuray

THE RESEARCH OF ENVIRONMENTAL-AND-ECONOMIC RISKS OF THE COAL MINING ENTERPRISE IMPACT ON WATER RESOURCES 115
Mikhailov Vladimir, Kudrevatykh Natalia, Tyuleneva Tatiana

IMPROVING THE QUALITY OF PROFESSIONAL DRIVERS' SELECTION IN HAULAGE COMPANIES TRANSPORTING MINERALS 121
Semenov Yuri, Semenova Olga, Marasová Daniela

APPLICATION POTENTIAL ANALYSIS OF HEAT ACCUMULATOR FOR MINING MECHANISMS ... 127
Rybár Radim, Beer Martin

ALLOCATION OF THE GEO-DYNAMICALLY HAZARDOUS ZONES DURING INTENSIVE MINING OF FLAT-LYING COAL SEAMS IN THE MINES OF SUEK-KUZBASS JSC ... 137
Yutiaev Evgeny, Meshkov Anatoly, Popov Anton, Shabarov Arcady

MODELING OF PROCESSES AND TECHNOLOGIES OF THE CONCENTRATING RASPADSKAYA FACTORY FOR ASSESSING THE IMPACT OF DILUTION OF RAW COAL ON THE TECHNOLOGICAL PERFORMANCE OF ENRICHMENT PRODUCTS ... 146
Udovitsky Vladimir, Kandinsky Vladimir, Begunov Artyom, Plotnikova Larisa

DESIGN OPTIMIZATION OF TUMBLING MEDIA FOR PROCESSING MINING MACHINERY PARTS ... 153
Lyukshin Vladimir, Shatko Dmitry, Strelnikov Pavel

AIR POLLUTION BY COAL DUST AS A FACTOR OF ECOLOGICAL COMPATIBILITY FOR COAL MINING ENTERPRISES ... 158
Korchagina Tatiana, Bykov Anatoly, Schastlivcev Evgeny

REVITALIZATION OF COAL BROWNFIELDS IN SOLVING ENVIRONMENTAL PROBLEMS AND STRUCTURAL DEVELOPMENT OF KUZBASS ECONOMY ... 165
Zhironkin Vitaly, Janocko Juraj

DEVELOPMENT AND DEBUGGING OF MPI PROGRAMS ON THE ENVIRONMENTAL RESEARCH COMPUTING PORTAL IN THE MINING REGION ... 175
Vlasenko Andrey, Sotnikov Igor, Prokopenko Eugeniya, Demidov Anton

ECOLOGIZATION OF THE DEVELOPMENT MANAGEMENT OF THE MINING REGIONS OF RUSSIA ... 181
Kolechkina Irina, Verchagina Irina, Eltsova Elena, Petrova Mariana

URANIUM ORE MINERALIZATION AT THE WESTERN CARPATHIANS / SLOVAKIA (CONDITIONS AND SOCIAL ASPECTS) ... 186
Jacko Stanislav, Bátorová Kristína, Šèerbáková Barbora, Blištan Peter

THE USE OF DUST SUPPRESSION ON UNPAVED ROADS AT THE MINING ALLOTMENT OF THE KUZBASS MINE ... 191
Cherkasova Tatyana, Cherkasova Elizaveta, Tihomirova Anastasia, Andryushkov Alexey, Dormeshkin Oleg

RESEARCH OF CHARACTERISTICS OF PJSC "KOKS" COAL AND COKE DUST FUEL BRIQUETTES ... 196
Solodov Vyacheslav, Cherkasova Tatyana, Kolmakov Nikolay, Subbotin Sergey, Radchenko Yuriy

APPLICATION OF THE "LEAN THINKING" CONCEPT TO THE ANALYSIS OF COAL MINING REGION'S SUSTAINABLE DEVELOPMENT OBJECTIVE ... 203
Mikhalchenko Vadim, Rubanik Yuriy

MECHANISMS FOR HARMONIZING THE SUSTAINABLE DEVELOPMENT OF KUZBASS AS A RESOURCE REGION ... 212
Fridman Yuri, Loginova Ekaterina, Rechko Galina

RESOURCE REGIONS AS AN INITIAL BASE FOR TRANSFORMATION OF THE RUSSIAN COMMODITY-BASED RENTAL ECONOMY MODEL ... 217
Kusurgasheva Ludmila, Lubyagina Lydia, Stefanek Pawel, Lapinskas Arunas

ECOLOGICAL CONSCIOUSNESS IN A SHIFT OF SCIENTIFIC PARADIGMS ... 223
Yazevich Mariya, Kalinina Oksana, Zhironkina Olga

THE SOLUTION OF ENVIRONMENTAL PROBLEMS AND THE DYNAMICS OF DEMOGRAPHIC PROCESSES IN INDUSTRIALIZED REGIONS ... 231
Bel'kov Anton, Zolotukhin Vladimir, Zolotukhina Natalia, Sedina Natalia, Kozyreva Marina

EXPERT EVALUATION OF THE RISKS IN COAL-MINING ENTERPRISES BASED ON FUZZY LOGIC ... 236
Raevskaya Elena, Pimonov Alexander, Mihailov Vladimir

THE UNITY OF THE TRAJECTORY OF SUSTAINABLE DEVELOPMENT OF THE MINING REGION AND ENSURING ITS ENVIRONMENTAL SAFETY ... 243
Sheveleva Oksana, Slesarenko Ekaterina, Kudrevatykh Natalia, Mamzina Tatiana

ON THE ISSUE OF FOOD SECURITY OF COAL MINING REGION ... 252
Kosinskiy Piotr, Zaruba Natalia, Egorova Natalia, Kharitonov Aleksey

APPROACHES TO EVALUATION OF ENVIRONMENTAL AND ECONOMIC DAMAGE TO THE KUZBASS AGGLOMERATION CAUSED BY COAL MINING INDUSTRY DEVELOPMENT ... 259
Kosinskiy Piotr, Merkuriev Vladimir, Medvedev Aleksey

DIGITAL ECOSYSTEM OF THE REGION AND THE COMPETITIVENESS OF KUZBASS COAL ENTERPRISES ... 267
Samorodova Lyudmila, Shut'ko Larisa, Yakunina Yulia, Lyubimov Oleg, Kovacs Péter

THE MANAGERIAL ASPECTS OF INTEGRATING THE SUSTAINABLE DEVELOPMENT PRINCIPLES INTO PRACTICES OF MINING COMPANIES ... 275
Lazarenko Yuliia, Garafonova Olga, Marhasova Vyktoriia, Grigashkina Svetlana, Kozureva Olena
DIVERSIFICATION STRATEGY AND SUSTAINABLE DEVELOPMENT OF MINING COMPANIES IN THE EURASIAN MINING REGION ... 283
Voronets Dmytro, Garafonova Olga, Verezomska Irina, Oliinyk Oksana
THE CONVERGENT STRUCTURAL BASE OF SUSTAINABLE DEVELOPMENT IN THE 21ST CENTURY ... 291
Zhironkin Sergey, Cehlar Michal, Zhironkin Vitaly
HUMAN POTENTIAL IN THE SYSTEM OF SUSTAINABLE DEVELOPMENT ... 296
Snegireva Tatyana, Kayachev Gennady, Falaleev Albert, Kurgansky Sergey
EFFECTIVE COST ALLOCATION METHOD IN AN INDUSTRIAL ENTERPRISE ENVIRONMENT IN MINING INDUSTRY ... 302
Janovská Kamila, Vilamová Šárka, Piecha Marian, Kutáè Josef, Kozel Roman, Èitbajová Jana
USE OF MODERN COMPUTER TECHNOLOGIES AND 3D MODELLING IN THE STUDY OF HISTORICAL MONTANISTICS ... 312
Pavolová Henrieta, Šimková Zuzana, Petráš Miloš
MACROECONOMIC INDICATORS OF RAW MATERIAL POLICY IN SLOVAKIA ... 321
Pavolová Henrieta, Šimková Zuzana, Seòová Andrea, Wittenberger Gabriel
SOME FEATURES OF RURAL ECONOMIC DIVERSIFICATION IN MINING REGIONS ... 333
Yakushnina Tatyana
FORMATION OF COMPETITIVE AIC IN CONDITIONS OF MINING REGION SUSTAINABLE DEVELOPMENT (ON THE EXAMPLE OF KUZBASS) ... 338
Shelkovnikov Sergey, Lubkova Elmira, Shilova Anna, Ermolaeva Galina, Shustova Elena
ENERGY-SAVING ELECTRONICS PRODUCTION IN VIET NAM: RISE AND PROBLEMS OF SUSTAINABLE DEVELOPMENT ... 345
Ngoc Tran Thi Bich, Binh Dao Thanh
Author Index

Statement of Peer review

In submitting conference proceedings to *Web of Conferences*, the editors of the proceedings certify to the Publisher that

1. They adhere to its **Policy on Publishing Integrity** in order to safeguard good scientific practice in publishing.
2. All articles have been subjected to peer review administered by the proceedings editors.
3. Reviews have been conducted by expert referees, who have been requested to provide unbiased and constructive comments aimed, whenever possible, at improving the work.
4. Proceedings editors have taken all reasonable steps to ensure the quality of the materials they publish and their decision to accept or reject a paper for publication has been based only on the merits of the work and the relevance to the journal.

Title, date and place of the conference

The first interregional conference "Sustainable Development of Eurasian Mining Regions (SDEMR-2019)", 25-27 November 2019, T.F Gorbachev Kuzbass State Technical University, 650000 28 Vesennya st., Kemerovo, Russian Federation

Proceedings editor(s):

Prof. Director Alexey Khoreshok, DSc, Kuzbass State Technical University
Prof. Victor Atrushkevich, Dr.Sc., College of Mining, NUST "MISiS"
Prof. Stefan Voth, Dr.-Ing, TFH Georg Agricola
Assoc. Prof. Demirel Nuray, PhD, Middle East Technical University
Prof. Juraj Janocko, Dr.Sc., Technical University of Kosice
Prof. Dean Yunliang Tan, PhD, Shandong University of Science and Technology
Prof. Daniela Marasová, PhD, Technical University of Kosice
Ing. Pawel Stefanek, PhD, KGHM Polska Miedź SA
Assoc. Prof. Mariana Petrova, Dr.Sc., St.Cyril and St.Methodius University of Veliko Turnovo

Date and editor's signature

Date: 08.11.2019

The rationale for the development of a method for assessing the roof conditions of mine workings in the conduct of actual mining using the method of geoinformation analysis

Alexander Abramovich[1], *Yuri* Stepanov [2], and *Jurgen* Kretschmann[3]

[1]T.F. Gorbachev Kuzbass State Technical University, 650000, 28 Vesennya str., Kemerovo, Russian Federation
[2]NFI Kemerovo State University, 654000, pr. of metallurgists 19, Novokuznetsk, Kemerovo region, Russian Federation
[3]Technische Hochschule Georg Agricola (THGA), Bochum, Germany

Abstract. This article considers the need of developing the method for evaluating the condition of excavation roof using the method of geoinformation analysis. The article also describes the existing research and roof control methods and analyzes their advantages and disadvantages. On the basis of this analysis the authors of the article establish the requirements for the method developed. The authors suggest using the geoinformation technologies for modeling, analyzing and evaluating excavation roof condition. We compile a list of tasks which are to be accomplished to develop the method. The paper presents the list of information objects, the database and the properties of these objects. The approach to database implementation is chosen. The database structure is presented. The article also describes the technology of geospatial data analysis, various sections data required for the development evaluation method. The analysis and selection of software for 3D modeling are made up, and 3D models of excavation sites are built.

1 Introduction

At the present time coal mining is rapidly increasing. The use of various technologies of coal extraction places increasingly high demands on equipment and connections for reliability, performance and a number of other criteria. Replacement of old mine equipment by modern high-tech equipment allows to solve many production applications, the most important of which is productivity gains taking into account issues of safety of mining operations. Modern equipment incorporates a large number of different sensors that allow analyzing the state of equipment and technological processes, but this is not enough to ensure safe working conditions. In the sphere of mining safety, there are many unsolved problems, one of which is the assessment of stress-strain state of a mountain range. The

[1] Corresponding author: abramovich_sanja@rambler.ru

https://doi.org/10.1051/e3sconf/201913401001

process of coal extraction is a dynamic process, as a result of which the stress-strain state of rocks is redistributed in the districts of coal face. Mine workings require timely surveys and final report on the stage of working. The need for monitoring and assessment of status of stress-strain state of rocks occurs with increasing depth of mining coal seams. On the basis of obtained data geoinformation models of mining are built. Despite the wide choice of mountain geoinformation systems, there is lack of facilities of geoinformation analysis of rocks status [1]. In this regard, it is proposed to use the method of geometry analysis of occurrence of emergency situations for timely executive decision-making. This will allow selecting equipment for miming operations, which in turn will contribute to safe and highly productive extraction of mineral products.

2 Materials and methods

There are plenty of research methods of roof conditions, the most common are: visual monitoring, georadiolocation, et al. [2-6]. Table one presents the comparative analysis of the most common research methods of roof conditions in underground mining workings.

Table 1. Advantages and disadvantages of the existing research methods and monitoring of roof displacement condition of mine workings.

Method	Advantages	Disadvantages
Visual monitoring	Assessment by workers of roof conditions of mine workings at the place of installation of deep bench marks.	The inability of permanent monitoring of roof conditions.
	Ease of use.	The inability to analyze the stress-strain state of roof of mine workings.
	Simplicity of installation.	The great influence of human factors.
		There are no accurate readings of the stress-strain state of roof of mine workings.
Georadiolocation	Visualization of scanned roof on radargram.	The inability to analyze the stress-strain state of roof of mine workings.
	The ability to bind the radar gram to the pickets of the mine.	There are no accurate readings of the stress-strain state of roof.
	High speed of radar scan of roof.	The inability of permanent monitoring of roof conditions.
	The ability to analyze changes of roof conditions on the scanned areas.	There is no database of radar grams of roof conditions of mine workings.
		Low current capacity.
Survey of collared holes by videoimage endoscopy	Identification of decompactification zones and rock jointing of massif.	The inability of permanent monitoring of roof conditions.
	Understanding the composition of massif in the study area.	The inability to analyze the stress-strain state of roof of mine workings.
		The great influence of human factors.

On the basis of data in Table 1, it may be concluded, that all existing methods of analysis have a number of disadvantages, the most important of which are: the inability of permanent monitoring of roof conditions, the inability to analyze the stress-strain state of roof of mine workings, the great influence of human factors.

On the basis of a comparative analysis of existing methods for assessing the roof conditions of mine workings, it can be concluded that the existing methods for monitoring the roof conditions are outdated and their use as the basic monitoring method is not relevant and not rational, but it is possible to use them as additional ones.

In addition to the considered methods for monitoring the status of stress-strain state, there are methods [7-15], often based on the use of geographic information systems. Geo-information technologies that are part of geographic information systems (GIS) are used for the analysis of geospatial data.

On the basis of GIS, such tasks as the collection, storage, presentation of spatial information, geoscience analysis, visualization and modeling of geospatial data are solved [10].

There is no ready-made complex and method of geo-information analysis and evaluation of roof conditions of mine workings which confirms the relevance of the problem under consideration.

Thus, one of the tasks to ensure the safety of mining and improve the performance of mining equipment is a development of method of geo-information analysis and evaluation of roof conditions of mine workings.

The following requirements are imposed on the developed method:

1. The possibility of permanent monitoring and analysis of roof displacements of mine workings.
2. Introduction of numerical methods to eliminate the human factor.
3. The existence of prediction methods of roof displacements.

To develop a method it is necessary to solve the following tasks:

1. Makeup of data objects that are necessary for building 3D models of mine working.
2. Development of a database for storing information about the state of massif.
3. Selection of software for building 3D models of mine workings.
4. The choice of equipment and technology for the organization of monitoring the stress-strain state of rock formation.
5. Development of a methodology for modeling the stress-strain state of rock massif in different geological conditions.
6. Development of methods for geo-information analysis of the results.

The following information is required to build 3D models and to store data on roof displacements of mine workings:

1. Indications and time of their fixation with reference to pickets and installation sites.
2. Data on mine divisions, such as: the name of the division, isohyps, type of violation, the intersection of exploration wells with the seam, the seam thickness, etc.
3. Data from the powered support units.

Modern powered supports have many different built-in sensors, one of which allows evaluating the stress-strain state of massif. In the future, it is planned to investigate the analysis of the stress-strain state of massif on the basis of data from the powered support units.

3 Realization

It is necessary to use databases using the object-oriented approach of Entity Framework, built on the basis of the platform .Net. The database structure is shown in Figure 1.

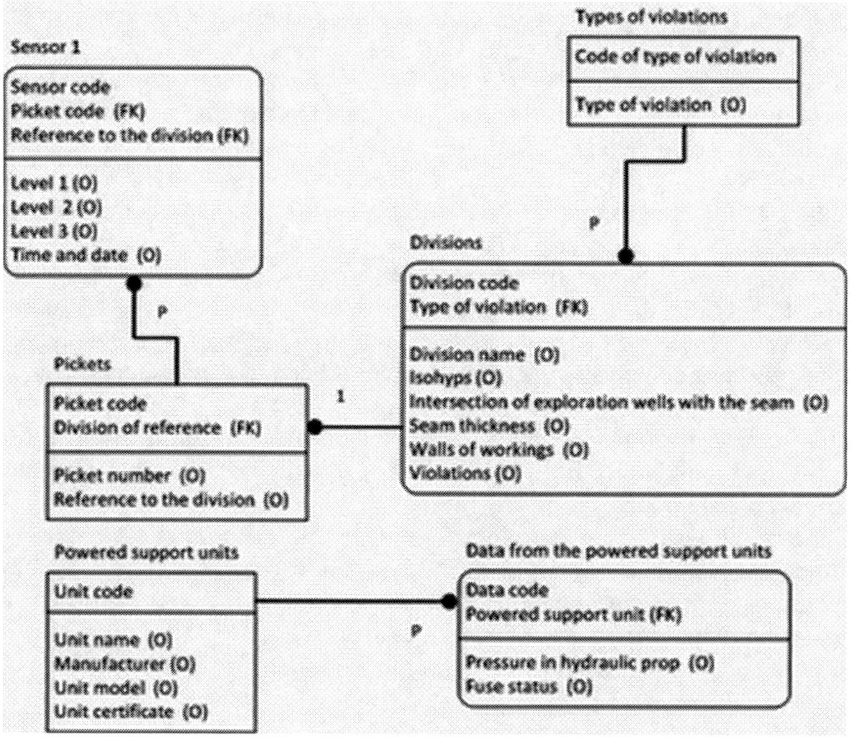

Fig.1. The database structure.

Data warehouses are formed on the base of the collected and calculated data. As a method of analysis, it is advisable to use OLAP-technologies and their varieties (HOLAP, MOLAP), which allow, based on the collected data (raw data) and slicing at selected points, to obtain new knowledge about the conditions of the massif. Through this, it is possible to obtain new dependencies characteristic of this block.

To build 3D models of mine workings, you can use various software applications with built-in 3D modeling tools. Based on the specificity of the visualization of geospatial data, the following requirements are imposed on the 3D modeling program:

1. The existence of the analysis module by the finite element method.
2. Ability to work with triangulated grids.
3. Rendering.
4. Animating a model.

For modeling of the stress-strain state of rock formation in the vicinity of the extraction work, taking into account the changing geological conditions, the software analysis was carried out. It is presented in Table 2.

Table 2. Comparative characteristics of 3D modeling programs.

	AutoCAD	**T-Flex CAD**	**KOMPAS-3D**	**FreeCAD**
License type	Training version, limited	Training version, limited	Training version, limited	Freeware
Rendering	+	+	+	+
Analysis by the finite element method	+	+	+	+
Animating a model	+	+	+	+
Work with triangulated	+	-	-	+

	AutoCAD	T-Flex CAD	KOMPAS-3D	FreeCAD
grids				

Analysis of the software (SW) concluded that to meet the challenges of modeling of massif, AutoCAD or FreeCAD can be used. The work will use FreeCAD, as it meets all the requirements and applies to FreeWare. The triangulated block and the constructed 3D models of the block are presented in Figures 2, 3, 4.

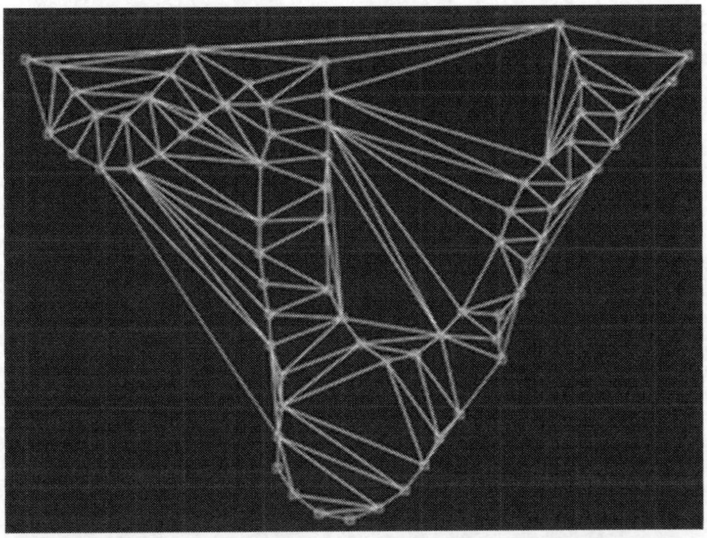

Fig.2. Triangulation of block.

Fig. 3. 3D - model of the block. Isometric view.

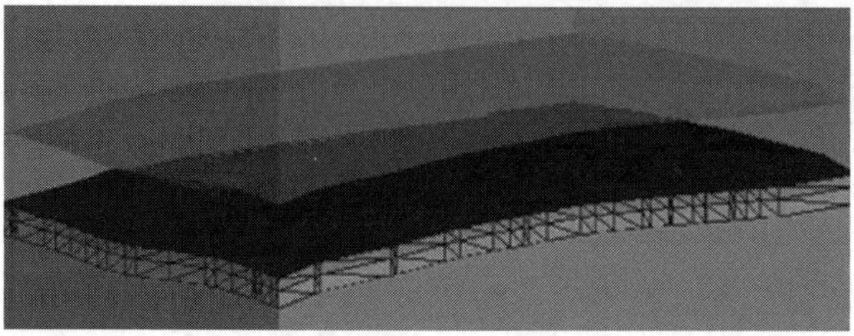

Fig.4. 3D - model of the block. Front view.

4 Conclusion

As a result of this work, prerequisites were created for the development of a method for assessing the roof conditions of mine workings in the conduct of actual mining using the method of geoinformation analysis. A list of tasks that need to be solved for the development of the method was compiled. A database structure has been developed, which is necessary for modeling mine workings, automatisation the construction of 3D - model of blocks and modeling the stress-strain state of rock formation. A 3D modeling program was selected in which models of blocks were constructed.

The development and application of the method for assessing the roof conditions of mine workings in the conduct of actual mining using the method of geoinformation analysis will improve the safety of mining, increase labor productivity and reduce the time for modeling, analysis and assessment of roof conditions of mine workings.

References

1. M. Satya, O. Anurag, M. Sachin, Nat. Conf. OSGIS-2015, **150** (2015)
2. G.L Feofanov, Urgalugol JSC **3**, 18 (2017)
3. C. Zhao, Y. Zhang, G. Zhang, J. Li, S. Ma, Sensors, **17**, 1044 (2017)
4. J. Doherty, A. Hasan, G. Suazo, A. Fury, Can. Geotech. J., **52**, 1901 (2015)
5. E. Lawson, D. Tesarik, M.Larson, H. Abraham, Int. Conf. GCM, **35**, 26 (2016)
6. Y. Zhao, N. Zhang, G. Si, Sensors, **16**,112 (2016)
7. Y. Mustafa, A. Ahmet, O. Muharrem, K. Bayram, Y. Hayati, Int. J. E. Res. Tech., **7**, 199 (2018)
8. M. Larson, D. Tesarik, J.Seymour, R. Rains, Inf. Circ., **151**,253 (2000)
9. Z. Zhang, J. Bai, Y. Chen, S. Yan, Int. J. Rock Mech. Min. Sci., **80**, 1 (2015)
10. J. Blachowski, A. Chrzanowski, A. Szostak-Chrzanowski, Conf. Int. Mine Surv. C. Aachen, **15**, 529 (2013)
11. A. Verma, D. Deb, Int. J. of Geomech., **13**, 181 (2013)
12. Y. Lu, W. Yang, J. of Central South. Univ., **20**, 1397 (2013)
13. A. Chanyshev, I. Abdulin, J. of Mining Sci., **50**, 18 (2014)
14. V. Vaziri, J. Hamidi, A. Sayadi, Environ. Earth Sci., **77**, 1 (2018)
15. A. Iannacchione, T. Bajpayee, L. Prosser, Min. Eng., **59**, 47 (2007)

Application of hydraulic press for testing powered roof prop supports

Piotr Kipczak[1,*], and *Krzysztof* Władzielczyk[1]

[1]AGH University, Department of Machinery Engineering and Transport, al. Mickiewicza 30, 30-059 Krakow, Poland

Abstract. The report presents the possibility of using press forging, used in a smelter in the crushing and forming steel ingots processes, to the operation of compression powered roof supports. Press have been redeveloped which is to construct, to produce and to exchange handles so that they can be mounted in these housing racks in an upright position. Given the capability of measured quantities, measurement procedure and methodology of the study. There were presented some exemplary results of the tests.

1 Introduction

The necessity to eliminate hazards and to secure the performance of works in underground mining excavations are very important scientific and technical problems, requiring innovative and responsible theoretical and practical solutions. The geological complexity of the rock layers, but also the accumulation of the initial energy of the rock gravity deformation caused by the forces of gravity is of great importance for the safety of mining works [6, 8]. This is related to the increasingly deteriorating mining conditions prevailing on the mining walls. The dependence of the rock mass on the mechanized obuszowa during shocks and rock bursts can be explained by the so-called "Rock mass model", the scheme of which is shown in Figure 1 [1, 9] Two layers of rock can be distinguished in the rock massif. The lower one is a direct ceiling, which forms the most densely fractured layers, passing, in case of exploitation for a heart attack, into a chaotic debris in the goaf. The upper part is the main ceiling, composed of heavily cracked layers, preserving some form of geometrical continuity in the goaf. The shape of the relaxed rock mass has a significant impact on the increase of the rock mass pressure in a given mining excavation. The correct operation, adapted to the existing mining and geological conditions of the housing depends mainly on the safety of the crew, and in the further consequence the rhythmicity of the mining process. The problem of housings designed for work in difficult mining and geological conditions, and especially in thawed deposits, is of particular importance [1, 5, 7]. This is a very serious problem because currently about 50% of active hard coal mining walls in Poland are exploited in rock blast beds [6, 8]. The existing situation causes that the hydraulic powered housings used in these conditions should have appropriate safeguards

* Corresponding author: kipczak@agh.edu.pl

allowing for taking over dynamical loads occurring in time of bumps [4, 9]. At present, there are workplace tests simulating the real conditions prevailing in the face, aiming at testing new solutions for the protection of hydraulic racks [2, 3]. Bearing in mind the conditions prevailing in the excavation, hydraulic industrial presses with a pressure force of 25MN [2] were used for testing hydraulic racks of mechanized housings working in rock-blast beds.

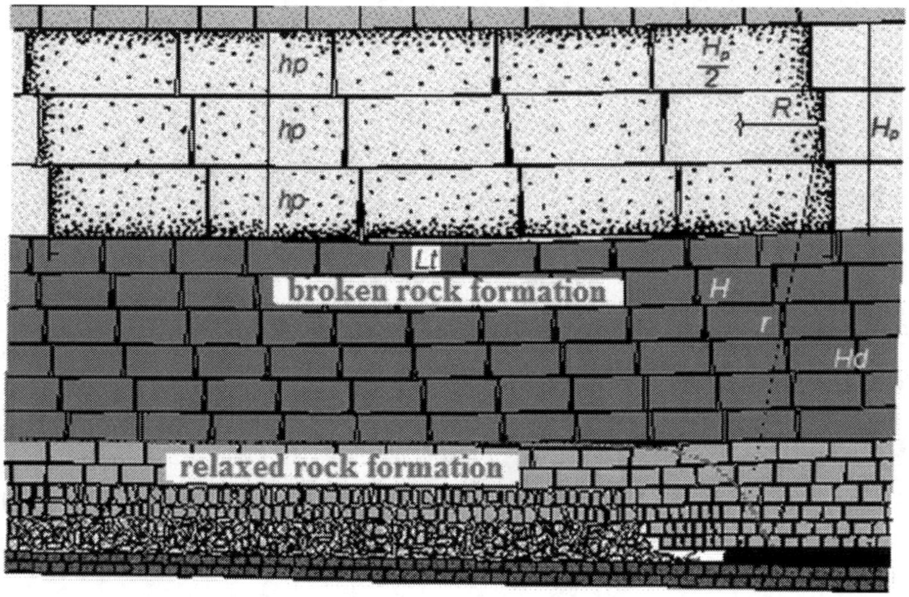

Fig. 1. Schematic disturbed rock wall around the bump at risk [1].

2 Description of the hydraulic press

Hydraulic press has standard equipment for a crushing and forging ingot, so it became necessary to make the corresponding fixture for testing hydraulic props. The basic starting point to create a table and handles the upper anvil, which will implement the process of subsidence of the r. Using the holes and guides technology of the hydraulic press (Fig. 2) mounting rack brackets are designed to serve as a mounting rack in the roof bar and floor. For research purposes carrying large cylinders, based on certain assumptions, have been adapted for hydraulic press. The task of the press was to put pressure on the test cylinder (Fig. 3) to determine the value of opening and closing pressure relief valve. Basic technical assumptions as stated in relation to the test bench, the opportunities for the position:
• the maximum compressive force is 20 [MN]
• research hydraulic props with length within the limits of 1.0 - 3.5 [m]
• research hydraulic props stand within diameters of 0.1-0.6 m
• supply pressure stands from 30 to 50 MPa
• feed pumps actuators working cylinder > 80 l / min

The test stand used for the study consists of rails, fixed beams, sliding beams, a table and working actuators. Supporting structure (four slides) followed by moving sliding beams, using the actuators generate force is a very important arrangement, because of the possibility of adjusting the pressure and velocity of falling working actuators. The sliding and moveable beams of the top move, depending on the desired strength and speed falling, three cylinders Ø 650 mm. An important element of the press is a sliding table, allowing easy assembly of the actuator in the handle and it can slip in under the anvil mounted on the sliding beam. Actuator mounted fixtures can be seen on Figure 2. Upper handle is mounted in an anvil slot, while the bottom bracket is in the sliding table. The task of the handles was to generate pressure in the cylinder axis.

Fig. 2. View mounted handles [2].

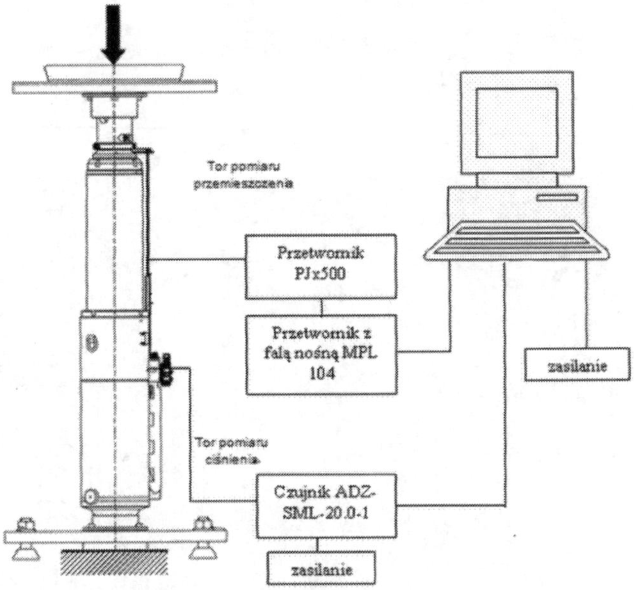

Fig. 3. Schematic of the measurement system used in the study [2].

3 The research place

The research place consisted of the following three systems:
• mechanical system visible on Figure 2; hydraulic press with handles,

• hydraulic power system and venting stands (Fig. 4) - which made the process of venting and power stand in during the experiment. The composition of this system included: hydraulic power unit, consisting of pumps, valves and system pressure relief valve and a switch turns of the electric motor. The switch turns allow you to quickly change the direction of flow of the emulsion, either for Spreading stand or the purpose of plundery. During the operation, the unit being supplied with hydraulic reservoir was disconnected and a pump connected to tank (a barrel) with an emulsion. Collection of emulsions was made after cleaning it through the filter mesh 100[μm]. The unit while adjusting the bypass valve and on the bench is presented in the Fig.4 and Fig.5. In order to vent and control the operation of tested hydraulic cylinders, the powering them up was done through the hydraulic unit shown in Figure 4 with a 1.0 grade gauge to control pressure.

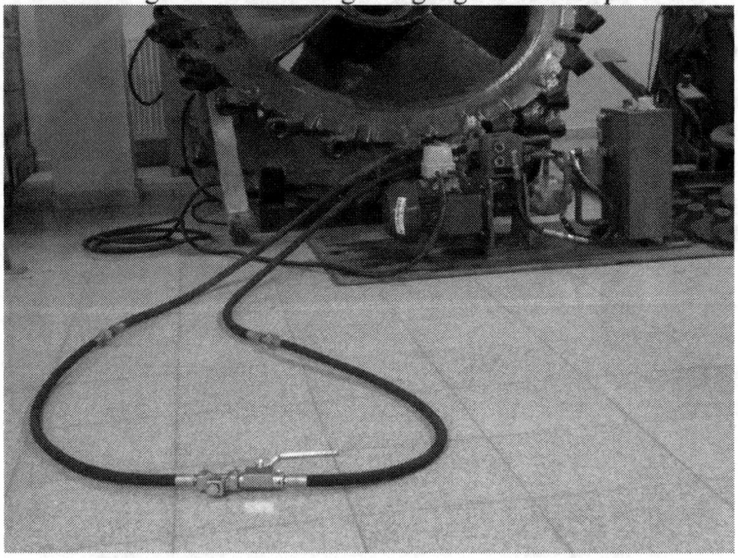

Fig. 4. Adjusting the settings of a hydraulic drive for supplying stand [2]

Fig. 5. Hydraulic drive on the test bench in HSW Stalowa Wola [2]

• measuring system consisted of two parts, ie mechanical and electrical ones. Mechanical part consisted of pre-designed and built equipment for mounting sensors (Fig. 6) and a simple hydraulic components for directing a stream of emulsion to the over or sub-crushing zones, for either prizing apart or plundery of the stand or to shut off the flow allowing the load rack. For the implementation of the procedures was used hydraulic valve connected to a specific constructed shut-off valve. The measurement of electrical parts included:

- Inductive displacement meter with amplifier,
- Inductive pressure transducer with amplifier,
- Voltage divider,
- Measuring computer with sampling time 0.001 [s].

Schema shows the measuring system used Fig.6, and this picture shows a fully armed measuring stand with the mechanical and electrical elements. The stand was used to determine the pressure of opening and closing the safety valves tested at different testing options of the actuator, in order to verify the new approach of a new design valve for useful actuators with large stability. Implementation of the basic assumptions in relation to the research place, were concerning:

• the maximum compressive force is 20 [MN],
• research of hydraulic props with lengths within the limits of 1.0 - 3.5 [m]
• research of hydraulic props diameters within 0.1-0.6 m
• the pressure for stands from 1930 to 1950 MPa
• efficiency of the pumps connected to working actuators> 80 l / min

The measurement system, whose scheme is presented in Fig 3, consisted of a pressure sensor CL-1L, range up to 100 [MPa] and sensitivity of 2.67 [mV / MPa] with a maximum value of 10% overload. Measurement of displacement was made by using an inductive displacement sensor type PJx 500 with a measuring range of ± 500 [mm] and with a maximum nonlinearity of 0.41%.

Fig. 6. Vented and fully armed stand before you try squeezing [2]

4 The procedure of preparing a hydraulic press for testing hydraulic racks

In order to carry out tests on hydraulic stands, it was necessary to create a procedure for adapting the hydraulic press to the tests performed. After the analysis of the structure of the hydraulic press, its technical parameters, the method of controlling the work of the press and the tooling installed on the press, the procedure of preparing the hydraulic press for the individual stages of testing the stands was adopted. The procedure was as follows:

1. disassembly of anvils and working handles of the standard equipment of the press, depending on the construction of the tested rack,
2. assembly of anvils and working handles depending on the construction of the tested stand,
3. checking the correctness of the assembly of the press tooling and the reliability of mounting the tested rack in mounted holders,
4. performing a preliminary test of the anvil slide for specific values of the sliding velocity.

5 Measured quantities and methodology of conducting tests

The measurements measured during the racks were:

• pressure prevailing under the piston of the external core as a function of time,
• displacement of the inner core piston rod as a function of time.

The tests were carried out in such a way that before starting the tests, after ejecting the hydraulic stand and bleeding its hydraulic system, the rack was put under the anvil of the hydraulic press using a sliding table. The next step was to eliminate the misalignment of the actuator axis with the anvil axis and closing the valve that cuts the stand power supply. The next stage of the racks' tests was to measure their sliding speed for certain values of the anvil's fingerprints. After making a series of measurements, the actuator was extended to complete the operating medium in the stand before the next test. Measurements of a given rack were carried out until the predetermined range of rack sliding speeds was exhausted, which predicted lowering the upper press anvil at speeds of 10%, 20%, 30%, 40%, 50%, 60%, 70%, 80%, 90% and 100% of its maximum speed. At the rate of less than 10%, the hydraulic systems of the press were not allowed.

Fig. 7. View ram compression during measurement [2]

The speed of lowering the anvil and thus the speed with which the inner core was pressed into the external core displaced from it through the bleed valve was determined by

 https://doi.org/10.1051/e3sconf/201913401002

setting the overflow valves with certain values controlled from the hydraulic press work station. The set ranges were determined each time for a specific measuring series of a given type of hydraulic stand. After carrying out a full series of tests of the hydraulic rack, it was replaced with another type of stand, for which a further examination was carried out according to the methodology presented above.

6 Summary

Presented in the paper, methodology of testing racks of mechanized housings of high load capacity with the use of hydraulic presses with high nominal pressures allows obtaining reliable results of testing the working parameters of the stands. The implementation of rack tests in their full scope of work requires only minor modifications of the press construction involving the assembly of appropriate hydraulic rack mounts on the press using already existing mounting holes. Thanks to this, it is possible to securely immobilize the hydraulic rack during its testing, as well as easy assembly and disassembly of the stand after the tests, Fig. 6.

It is also worth paying attention to the fact of ensuring full safety of testing the racks on the hydraulic press in question. As it can be seen in Fig. 7, the design of the hydraulic press provides free access to the tested rack, and the considerable space around the press eliminates the dangers resulting from possible breakage of the stand during the tests.

Another advantage of using hydraulic presses to test the roofs of mechanized housings is the possibility of precise adjustment of the pressure force of the press, and thus the load on the tested rack. Thanks to this, it is possible to draw up a very precise working characteristic of the tested stand in the form of an appropriate graph. Examination of the stands on the hydraulic press also allows to determine the operation and opening pressure of quick-return valves mounted in hydraulic racks, intended for mechanized roofs working in excavations threatened by rock-bumps. Verification of the operation of these valves is extremely important from the point of view of work safety of the mining wall. Fig. 8 presents an example of the operation characteristics of the DN12 quick-release valve obtained during the TAGOR N-142 hydraulic rack test.

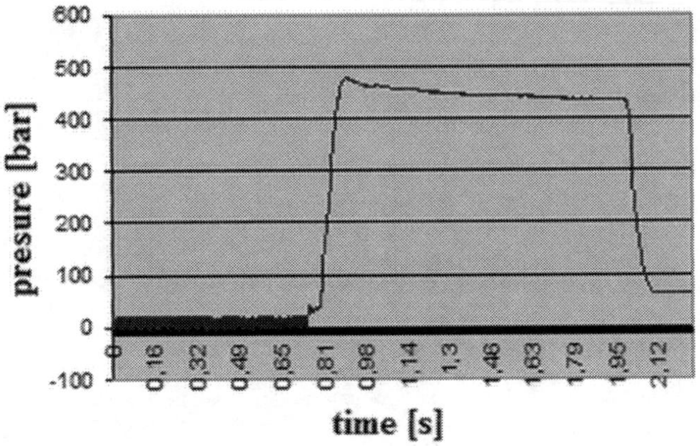

Fig. 8. Measurement of the displacement of 40% [2]

References

1. A. Biliński, T. Kostyk, PG, **3** (2014)
2. *The test stand Forge Stalowa Wola* (Stalowa Wola, Raport, 2014) (not published)

3. J. Marianowski, P. Kipczak, A. PG, **67** (2011)
4. D. Szurgacz, Mining Science, **22 :Special Issue 2,** 171–179 (2015)
5. L. Doległo, J. Gil, K. Stoiński, MG, **4** (2013)
6. K Krauze, PG, **63(11)**, 2007
7. T. Łagoda, Transport, **4:22** (2013)
8. S. Prusek, *Stateczność wyrobisk ścianowych podczas eksploatacji pokładów węgla kamiennego z zawałem skał stropowych* (Główny Instytut Górnictwa, Katowice, 2016)
9. D. Szurgacz, A. Kasprusz, Prace naukowe GIG. Górnictwo i Środowisko, **3**, 89–99 (2011)

E3S Web of Conferences **134**, 01003 (2019) https://doi.org/10.1051/e3sconf/201913401003
SDEMR-2019

Bench tests of a support system of a powered roof support's hydraulic leg aimed at minimizing consequences of leaks

Dawid Szurgacz[1,*], *Jarosław* Brodny[2], *Marcin* Brzózka[3], and *Ryszard* Diederichs[3]

[1]Polska Grupa Górnicza S.A., KWK ROW Ruch Chwałowice, ul. Przewozowa 4, 44-206 Rybnik, Poland
[2]Silesian University of Technology, Faculty of Organization and Management, Roosevelta 26, 41-800 Zabrze, Poland
[3]DOH Hydraulics Center, ul. Konstytucji 147, 41-906 Bytom, Poland

Abstract Effectiveness of a powered roof support significantly impacts on the efficiency and safety of a mining process. The main element of the support is a hydraulic leg which transfers loads imposed by the rock mass. One of the basic issues occurring during the extraction process are internal and external leaks of hydraulic systems. A new support system was designed to limit and later eliminate the consequences of such leaks. Its main component is a valve block equipped with two independent check valves. This solution allows to maintain high pressure in both chambers of a leg despite leaks. The article discusses the developed solution and presents preliminary results of bench tests of a leg equipped with the new block and loaded dynamically. The results are promising, and the tested block fulfils its task by maintaining the working pressure in the leg during load.

1 Introduction

Development of world economy is strongly affected by accesses to cheap energy sources. Despite the dynamic changes taking place in recent years, the basis for its development is still cheap and easily available energy [15]. The share of energy of renewable sources increases [17, 18], however, the energy obtained from conventional sources has the largest share in the global energy balance. Hard coal remains the basic raw material from which this energy is produced. Despite the increasing share of brown coal, crude oil and gas, hard coal production is still very high. The costs of energy generated from coal are the lowest, which means that it is widely used in its production. As a result, coal mining has been increasing in the world in recent years. The main factor that limits its exploitation in some regions of the world, mainly in Europe, concerns the impact it has on the environment. Both in the process of operation and during combustion. Many harmful gases and dust enter the environment both in the process of operation and during combustion. Various technical

* Corresponding author: dawidszurgacz@doh.com.pl

and technological activities are undertaken to limit the impact. Nevertheless, it is indisputable that coal, due to its prevalence and price, will continue to be one of the dominant energy resources in the world.

In the entire process of energy and other products production (e.g. coke) from hard coal, effective and safe operation of the exploitation is essential. This raw material deposits underground and therefore its extraction is based on deep mining.

Currently, the most popular method of underground exploitation of hard coal is a longwall system. It is a highly efficient and relatively safe way of operation, whose stages can be mechanized [4, 16]. This method applies powered roof support systems. The efficiency of the systems depends mainly on the effectiveness of its individual machines [4, 21, 22]. The main machines included in the systems are shearers or plow machines (designed to mine the rock mass), scraper conveyors (transporting the output) and a roof support whose main task is to protect the longwall working and support the whole unit. Effective and trouble-free operation of the support is fundamental as it decides on the entire coal mining process. Generally, it can be assumed that mining support, both for galleries and longwall excavations, is a basic factor impacting on the safety of underground mining operations.

The powered roof support comprises individual sections. They are connected with each other by means of hydraulic and control systems and are a uniform system protecting the area of operation in a longwall working. This is very important in the context of natural hazards occurring in the exploitation process. The most dangerous of these threats are ventilation and dust hazards [19, 23, 24, 25, 26]. Ventilation hazards include mainly methane and fire hazards, the development and extent of which depend on the effectiveness of the support operation and the extent to which it protects the cave-in zone [20, 23, 25, 27].

The wide scope of tasks that the support has to fulfil means that its selection and work are subject to special supervision.

The parameters of hydraulic legs operation are significant, as they are a basic load-bearing element of each section of a powered roof support. Obtaining adequate parameters of operation of hydraulic legs is the foundation for correct operation of the section and the entire unit. Therefore, it is necessary to conduct tests to determine the parameters of operation of these legs. This process is a major challenge for companies producing mining equipment. Mainly because they operate in very different mining and geological conditions. Consequently, they must be highly reliable and available.

One of the basic conditions determining the level of operational safety of a powered roof support section in a longwall excavation is its correct spragging. This requires proper initial load-bearing capacity that would ensure continuity of roof layers and adequate level of stress in the seam and the excavation's roof. The mining process is dynamic and therefore the bearing capacity of the support must be maintained for a set period of time and at required level. The section cannot lose its stability and the individual sections cannot slid down fully or partially during the mining process. For this reason, there are a number of studies that present various results of research and analyses related to the continuity of the mining process [1,5,6,7,8,9].

Leg support diagnostic systems are increasingly used in order to ensure adequate support of the powered support section [11,12]. Unfortunately, they are not applicable everywhere. A drop in load-bearing capacity of legs during operation often occurs. One of the basic reasons are internal and external leaks of hydraulic fluid. This in particular concerns two-stage legs. Here, the unsealing occurs between the first and second stage, in the bottom seal or in the form of external unsealing.

Consequently, the fluid leaks or the pressure in the leg drops unexpectedly which leads to loss of load-bearing capacity. These are very dangerous phenomena that disturb the work of the entire support and cause a threat to the machines of the complex and the safety of employees. Any additional loads imposed by the rock mass are an additional factor that can lead to serious damage or destruction of these machines.

The occurrence of leaks in the hydraulic leg causes that it does not obtain the required support, which in turn may lead to increased convergence of the roof, resulting in the occurrence of roof falls or breaks in the continuity of the roof layers which in turn results in difficulties in the operation.

One of the reasons for such leaks in legs and the hydraulic system of the support section are frequent and difficult to predict static and dynamic overloads of the legs. The most common leakage is small but increases with the passage of time and subsequent cycles of operation of the section. Lack of system for monitoring section load-bearing capacity causes that such a leak may cause that the section will cease to transfer loads imposed by the rock mass.

For this reason, all types of pressure monitoring systems in the legs, which enable the detection of pressure anomalies resulting from leaks, are crucial. A well-functioning monitoring system can detect leaks already at a pressure lower than the initial load-bearing capacity. Therefore, the lack of such a system means that a leg with leaks, and consequently entire sections, operates for a long period of time without proper load-bearing capacity.

Detection of leaks must be followed by quick replacement of a leg that must be conducted each time the leak has been registered. Otherwise, a section with such a leg will pose a great threat to the effective operation of the entire support.

The authors developed and then tested an original system for detecting leaks in a leg and protecting it against consequences of such leaks. The paper discusses the developed system and presents preliminary results of its application in legs that were testes in a testing station. The results are highly promising. The authors believe that it is necessary to conduct further bench tests followed by underground tests to fully assess the system. Practical use of the system should significantly improve the parameters of the operation of the powered roof support's section and the safety of mining operation.

2 Materials and Methods

Currently, the commonly used hydraulic system for controlling the powered roof support' leg is equipped with a single control block with one check valve. This valve closes the fluid outflow from the space under the piston in the first stage of the leg. A diagram of this system with its basic elements marked is shown in Figure 1. I

t is clear that in this system the second stage of the leg is not protected. The hydraulic fluid can flow out of it quite freely. In the event of external or internal leakage, this fluid may flow out through a drainpipe or canal without obtaining the required pressure.

It can be assumed that this system does not protect and does not provide the ability to control the pressure in the under-piston and over-piston part.

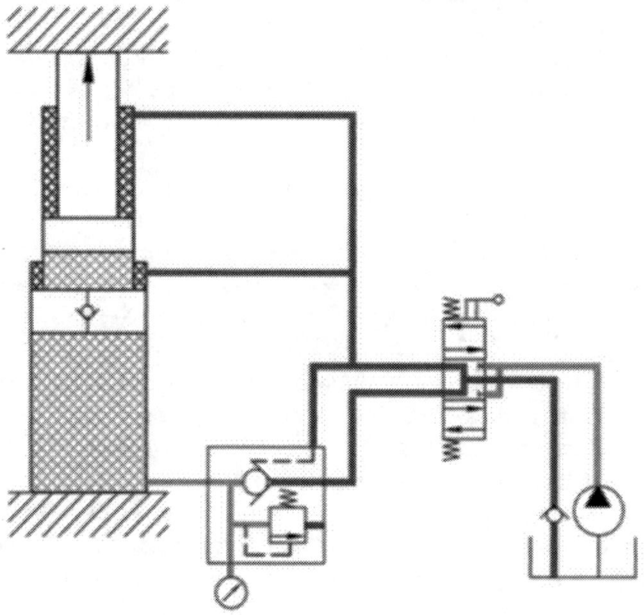

Fig. 1. Standard support system using a block with a single check valve.

The solution that removes some of these inconveniences is presented in Figure 2. This support system is equipped with a valve block which has two check valves.

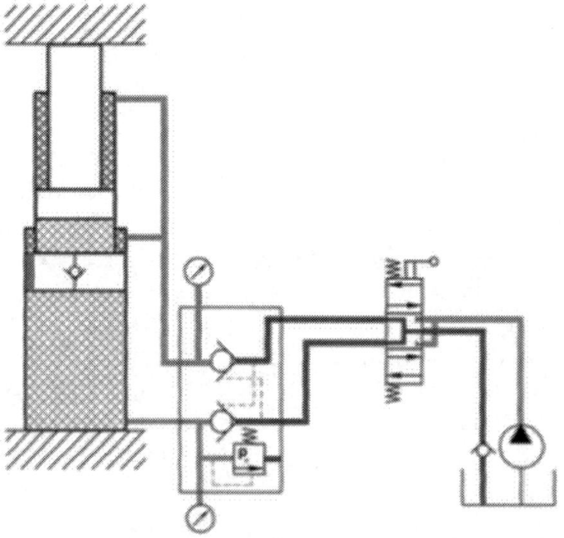

Fig. 2. A design of a support system using a block with a double check valve

The solution presented in Figure 2 is based on the idea that the fluid outflow from the under-piston and over-piston space is closed independently by means of two independent check valves. At the same time, the use of these valves in no way limits the functionality of the entire leg section or the entire support. A leg equipped with such a valve block performs the same function in a powered roof support section as a leg equipped with the hydraulic system used so far.

The undoubted advantage of the system shown in Figure 2 is that it maintains high pressure in the spacer over the piston. In the event of internal leakage of the leg, e.g. due to the pressure of the rock mass, the pressure of the fluid in the piston space increases, causing it to flow through the leaking piston seal to the space above the piston, from which it cannot flow into the main line. The second check valve maintains high fluid pressure in both chambers of the leg.

Pressure gauges were installed in the spaces above and under the piston in order to control the pressure in both chambers of the leg. This allows continuous control of the pressure in these spaces. This, in turn, gives the opportunity to check the technical condition of the leg for leaks. When the system indicates a leakage, the service team is required to either repair the damage or replace the leg. This monitoring enables quick response to occurring disorders and, consequently, effective response to occurring failures in the work of the leg.

The technical parameters of the overflow valves used in the block in the space under the piston are matched to a given leg in the same way as when used in the hydraulic system with a single block. The overflow valve is matched with the space above the piston and should be set to a pressure not greater than the working pressure of the lines connecting the valve with the block and the leg. The double valve block has an additional AM inlet that allows connection of a pressure sensor included in the pressure monitoring system of the legs.

Due to the fact that the outflow of fluid from the over-piston space is cut off by a check valve, no further flow of fluid through the hydraulic distributor to the drain main is possible. Therefore, the pressure in the over-piston space increases, allowing the pressure in the under-piston space to increase up to the value of the operating pressure limited by the setting of the overflow valve. Based on the pressure readings indicated by the p_{pt} and p_{nt} pressure gauges, it is possible to calculate the value of the pressure difference p at which fluid leakage from the leg occurs from the following relationship:

$$p = p_{pt} - p_{nt}, \tag{1}$$

Fluid pressure in the space over the piston reduces the initial and working bearing capacity of the leg. The reduction in the working bearing capacity of the leg P_r can be calculated knowing the hydraulic ratio of the leg k (the ratio of the space over the piston to the space under the piston):

$$_{D}P_r = P_r\, x\, k \tag{2}$$

3 Results and Discussion

The biggest threat to all types of mining supports is the dynamic load imposed by the rock mass. Therefore, it is necessary to conduct tests under dynamic load both in case of gallery (yielding) [2,3] and hydraulic supports. This necessity also results from the fact that currently the frequency of such loads constantly increases. Deteriorating mining and geological conditions mean that the supports operate in increasingly extreme conditions. In particular, this applies to longwall excavations where there is a large accumulation of all types of hazards, which were already mentioned in the introduction of the article.

Therefore, the research team had decided to performed bench tests of the leg with dual hydraulic valve block in order to obtain sufficient assessment of the new solution (Fig. 2). The tested leg, equipped with a new support system, was dynamically loaded with free-falling impact mass.

The scheme of the testing station is shown in Figure 3a and the hydraulic leg mounted on the station is presented in Figure 3b. The tests were carried out on a specialized dynamic testing station located at TLO Opava in the Czech Republic. A free-falling mass was dropped on a cross bar placed on a leg. The results presented in the article show that the mass of the cross bar and the impact mass were constant. The change in the impact energy occurred as a result of the change in height from which the impact mass was dropped.

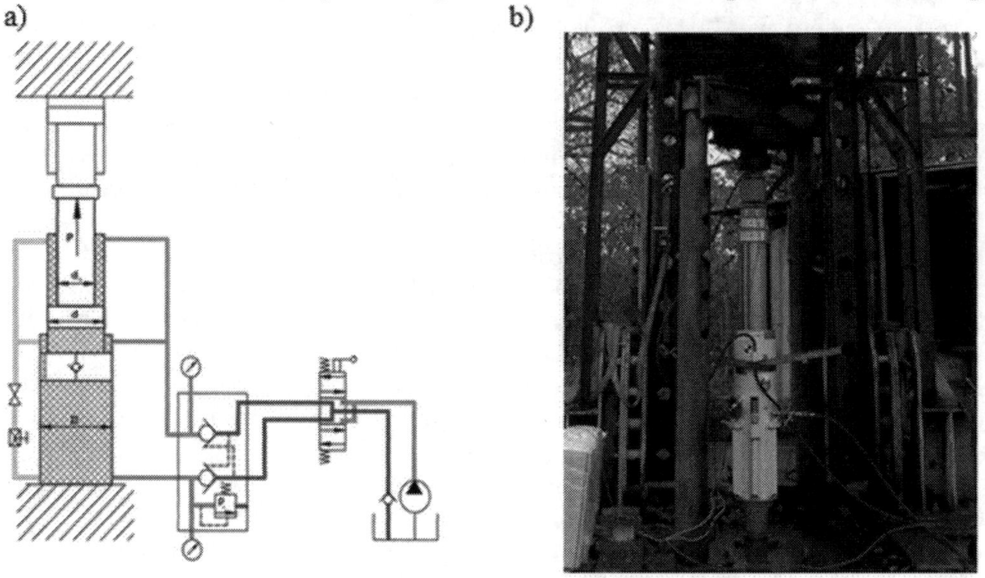

Fig. 3. Bench testing: a) a layout diagram of the system, b) the testing station

The tested leg was properly prepared so that the parameters necessary to determine internal leaks in its system could be registered. The conducted research provided a number of interesting results. Figures 4 and 5 show examples of pressure change patterns in the space under and over the piston.

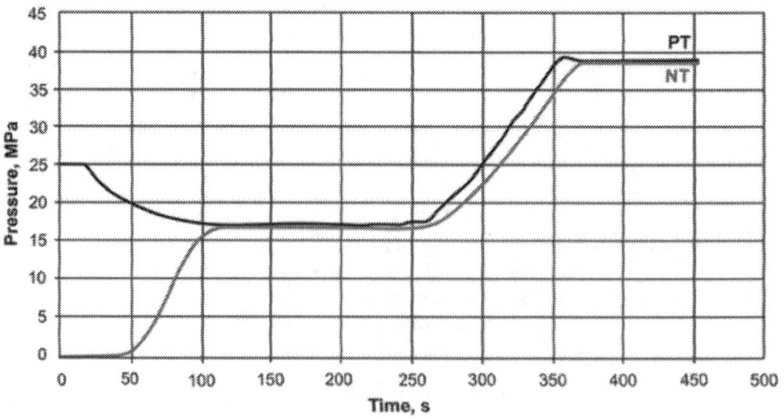

Fig. 4. The course of pressure changes in the space under and over the piston with a valve block at working pressure $P_r = 25$ MPa

The results indicate that as a result of a dropped impact there is a sudden increase in pressure in the space under and over the piston. The tested support system, using a double block, maintains pressure, preventing loss of support and stability of the leg. The recorded pressure waveforms (in Figure 4) indicate that the leg was slid out at a pressure of 25 MPa in the space under the piston. The time after which the internal leak was detected was about 0.25 s, and it took place at a pressure of about 18 MPa. The pressure in both chambers of the leg increased to approximately 38 MPa as a result of dynamic loading. The system used did not allow the pressure to drop and maintained it almost at the same level. Thanks to the system, the leg carried external loads all the time, despite the leak.

4. Conclusion

The issues discussed in this paper have a significant impact of the mining process. Leaks in the hydraulic system are a very unfavourable phenomenon which is indicated by practical uses and tests in testing stations. They disturb the work of the leg and prevents the implementation of the basic functions of the section and the entire mining support. The problem concerns both the powered roof support section and individual legs used to secure mining excavations.

Tests were carried out using the developed valve block to minimize the effects of common internal leaks in hydraulic legs that. As provided by world recommendations in this area, the tests were carried out under dynamic loading using a drop weight tester [10, 13,14]. The results clearly indicate that the proposed solution is very effective and allows minimizing the effects of leaks in a two-stage hydraulic leg. The application of the developed solution can significantly reduce any negative effects of leaks. It seems reasonable to replace valve blocks with a single check valve with blocks with two check valves. The authors believe that the presented solution will operate properly in underground excavations.

References

1. M. Blaschuk, A. Dronov, A. Koperchuk, R. Chernukhin, V. Litvienko, E3S Web of Conferences, **15**, 03003 (2017)
2. J. Brodny, Arch. Min. Sci., Vol. **55**:4, 733–746 (2010)
3. J. Brodny, Arch. Min. Sci., Vol. **56**:2, 303–318 (2011)
4. J. Brodny, M. Tutak, Energies, **12**, 2505, doi:10.3390/en12132505 (2019)
5. G. D. Buyalich, K. G. Buyalich, V. V. Voyevodin, IOP Conference Series: Materials Science and Engineering, **91**, 012087 (2015)
6. G. D. Buyalich, K. G. Buyalich, V. V. Voyevodin, IOP Conf. Series: Materials Science and Engineering. – IOP Publishing, **127**, 012034 (2016)
7. G. Buyalich, K. Buyalich, M. Byakov, E3S Web Conf., **21**, (2017)
8. G. Buyalich, M. Byakov, K. Buyalich, E3S Web Conf., **41**, (2018)
9. G. Buyalich, M. Byakov, K. Buyalich, E. Shtenin, E3S Web Conf., **105**, 03025 (2019)
10. V. I. Klishin, T.M. Tarasik, Jour. of Min. Sci., **37**, 77–84 (2001)
11. D. Szurgacz, J. Sust. Min., **14:4**, 157-163 (2015) DOI: 10.1016/j.jsm.2015.12.001
12. D. Szurgacz, J. Brodny, SGEM, **17:13**, 781-788 DOI: 10.5593/sgem2017/13/S03.099 (2017)
13. D. Szurgacz, J. Brodny, Sustainability, **11**:9, 2570 (2019) doi.org/10.3390/su11092570
14. K. Stoiński, M. Mika, Jour. of Min. Sci., **39**, 72–77 (2003)
15. S. Zhironkin, S. Demchenko, G. Kayachev, E. Taran, O. Zhironkina, E3S Web Conf., **105**, 03008 (2019)
16. I. Yermakova, V. Fedusov, E3S Web Conf., **105,** 01049 (2019)
17. M. Tutak, J. Brodny, 17th Intern. Multidisci. Sci. GeoConf. SGEM, Vienna GREEN Conference Proceedings, **17**:43, 35-642; (2017) DOI: 10.5593/sgem2017H/43/S29.080

18. K. Stecuła, J. Brodny, 17th Intern. Multidisci. Sci. GeoConf. SGEM2017, Vienna GREEN Conference Proceedings, **17**:43, 749-756 DOI: 10.5593/sgem2017H/43/S29.094 (2017)
19. M. Tutak, SGEM2017 Vienna GREEN Conference Proceeding, **17**:43, 301-308 DOI: 10.5593/sgem2017H/43/S19.038 (2017)
20. M. Tutak, 17th Intern. Multidisci. Sci. GeoConf. SGEM, Vienna GREEN Conference Proceedings, **17**:15, 53-60 DOI: 10.5593/sgem2017H/15/S06.007 (2017)
21. J. Brodny, S. Alszer, J. Krystek, M. Tutak, Arch. of Cont.Sci. Vol. **27:LXIII:2,** 197–209 (2017) DOI : 10.1515/acsc-2017-0012
22. J. Brodny, M. Tutak, E3S Web Conf. **29** (2018) doi.org/10.1051/e3sconf/20182900014
23. J. Brodny, M. Tutak, Jour.l of Appl. Flu. Mech., **11**:3, 545-553 (2018) DOI: 10.18869/acadpub.jafm.73.246.27240
24. J. Brodny, M Tutak, Intern. Jour. of Envir. Rese. and Pub. Hea. **15**:9, 1-16 (2018) 10.3390/ijerph15091846
25. M. Tutak, J. Brodny, Energies, **11**(11), 3076, pp. 1-28. doi.org/10.3390/en11113076 (2018)
26. J. Brodny, M. Tutak, 16th Intern. Multid. Sci. GeoConf. SGEM, 1:**2**, pp. 75-82, (2016)
27. M. Tutak, J. Brodny, IOP Conf. Ser. Earth Environ. Sci., **95**, 042025 (2017)

The noncontact determination technique research of dangerously loaded zones in the underground mining

Aleksey Bizyaev[1], *Natalia* Voronkina[2], *Andrey* Savchenko[3], and *Michail* Cupov[3]

[1]Novosibirsk State Technical University, Russia
[2] Siberian Coal Energy Company, Head of the Department for Control of Dynamic Phenomena and Monitoring of Rock Mass, Russia
[3]Chinakal Institute of Mining SB RAS, Novosibirsk, Russia

Abstract. The electromagnetic radiation signals associated with the destruction of rock samples were studied under field and laboratory conditions. The parameters of the signal were found that preceded the violation of the integrity of the rock in laboratory studies. It is shown that the highest efficiency of the method of electromagnetic radiation is achieved when predicting the dynamic manifestations of rock pressure in the form of rock shock.

1 Introduction

In the works [1-8] it was shown that the destruction of a rock is accompanied by electromagnetic radiation. The nature of this radiation is not fully understood. The radiation of rock during destruction is used in predicting the dynamic manifestations of rock pressure. In the works [1-4,7-8] it was shown that the discontinuity of the rock is accompanied by an increase in the flow of impulses. The equipment Angel-M, which is used in the mines of JSC SUEK-Kuzbass, is built on this observation. In the works [5-8] the process of destruction of rocks, accompanied by four stages of destruction, is considered. From the first to the third stage, the process is considered as pre-destruction. At the third stage of destruction, a discontinuity of the rock and an avalanche-like destruction are observed. This paper presents a laboratory analysis of the electromagnetic radiation (EMI) signals of the host rock from the S.M. Kirov mine. The results of profiling the mine work site in a place prone to discontinuity are presented. The recommendations for the development of the architecture of the registration and diagnostic complex REMI, which includes software and an electromagnetic radiation recorder with a graphic display, are given.

2 Laboratory studies of mining samples

2.1 Description of the laboratory stand

The stand includes a press, an EMP sensor, a load measurement sensor, a displacement sensor, an amplifier system, a recording system, and a system for storing, displaying, and analyzing data. The press consists of two metal plates with dimensions of 300x180x45mm, fixedly mounted on four steel columns with a diameter of 35 mm. Uniaxial loading of the samples was carried out until they were completely destroyed, which was determined by a sharp drop in load.

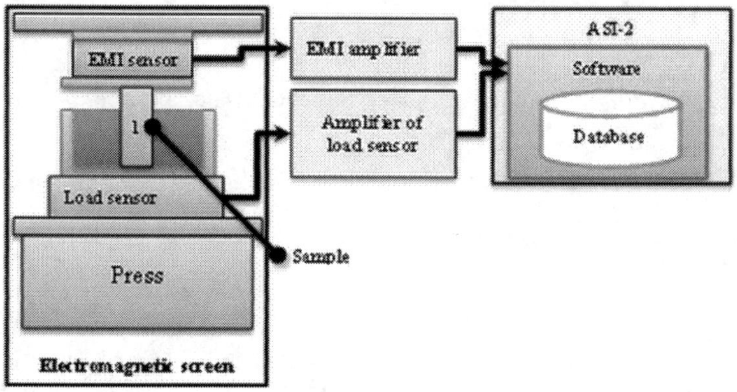

Fig. 1. Structural scheme of the laboratory stand ASI-2.

When testing the host rock mine samples from the S.M. Kirov mine were taken in size 25x25x40 mm, then subjected to uniaxial compression until complete destruction on the ASI-2 laboratory stand. During the destruction, the signals of electromagnetic radiation and loads were synchronously recorded.

2.2 Description of the laboratory experiment

During the registration, a synchronous oscillogram of the destruction process was obtained (Fig. 2). Roman numerals show the stages of destruction, obtained according to [8], by the change in the flow of pulses.

From the oscillogram it is seen that in the process of destruction, the sample is not able to accumulate mechanical load and is destroyed when the load is not more than 1 kN. The first mechanical load unloading occurred in the interval 15800-20000 ms, with the load corresponding to 3 kN. The second failure occurred in the interval 27500-30000 ms, with the accumulation of load.

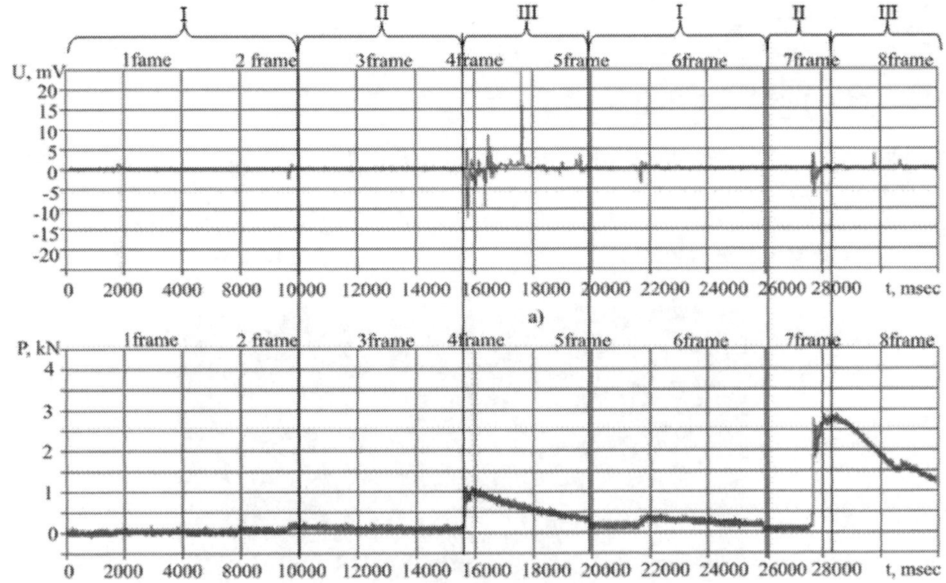

Fig. 2. EMI (a) and loads (b) waveforms recorded during the destruction of a rock specimen.

Fig.3. Fragments of a destroyed host rock specimen.

From fig. 2-3 it can be seen that the rock is not able to accumulate mechanical stresses and collapses under load of no more than 5 kN. Breed crumbles with minor loads. This shows that in an array with such a rock, the stress-strain state will manifest itself in the form of desquamation, which is observed in the S.M. Kirov mine. However, if such a peeling is manifested in the mine cavity or not timely implementation of preventive measures to replace the lining can lead to significant collapses.

3 Natural Studies

3.1 Description of the equipment

Field studies were carried out using the Angel-M mine equipment [10] and the REMI-3 registration and diagnostic complex [4].

Mine automated complex "Angel-M" is presented in Fig. 4 and consists of a registration unit with an indicator, an electromagnetic radiation sensor, and a data reception program [10].

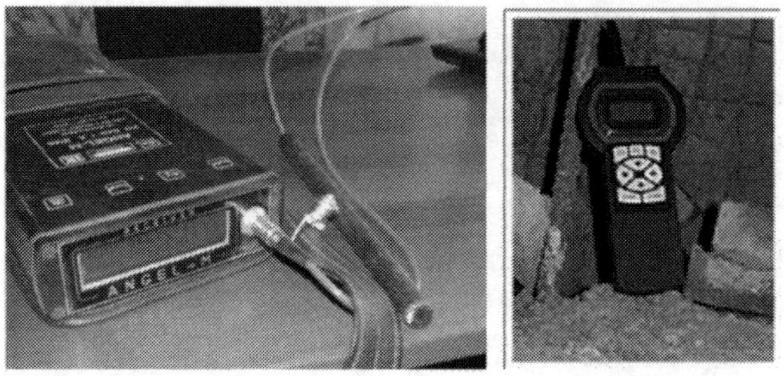

a) b)

Fig. 4. Detector of Electromagnetic Radiation Angel-M (a) and REMI-3 (b).

Monitoring of the electromagnetic environment is carried out in four stages: reception and selection of signals from the sensor during a specified time interval; data processing with determining the number of amplitude components that exceed threshold values per unit time; comparison of the obtained distribution of amplitude components with standards (criteria for assigning a control area to a certain category); displaying the category score on the display and storing the results in non-volatile memory.
REMI-3 recorder is capable of registering a signal at a frequency of 100-500 kHz, with a sampling frequency of 1 MHz. Registration results are recorded in non-volatile memory.

3.2 Taking measurements

The mine of S. M. Kirov, JSC SUEK-KUZBASS, located in city Leninsk-Kuznetsk, Kemerovo region, was chosen as the place of research. The methane mine category is a super category. Work on identifying hazardous areas was carried out in productive strata: the Boldyrevsky stratum — power is 1.8–2.4 m, the bedding angle is from 0 to 10°; Polenovsky stratum - power is 1.4-1.8m, the bedding angle is from 0 to 12°. The absolute gas content of the mine is 181.7m^3/min. The mine has two lava coal mining, four tunneling face. Studies were conducted in the lava 25-101, tunneling slab CMPH 25-03, capital workings, ventilation furnace 25-97 in the zone of geological disturbance and the influence of lava 25-101. To assess the impact of interference from equipment and cables, measurements were made, including in close proximity to areas with a cable network, substations, and mechanisms. The production at various sites was profiled and the comparative characteristics of the registration and diagnostic complex "REMI-3" and "Angel-M" were given.

During the measurements, signals were jointly recorded by the REMI-3 and Angel-M instruments. The measurement results of the device REMI-3 with the integration in the interval of 1 ms were recorded, and the signal component in the interval of 15 seconds was recorded in the memory. Oscillograms at various sites of mine workings are shown in fig. 5 and Fig. 6. The readings of the Angel-M instrument were recorded simultaneously, the readings of which were recorded in the form of two parameters. The measurement results are shown in table 1.

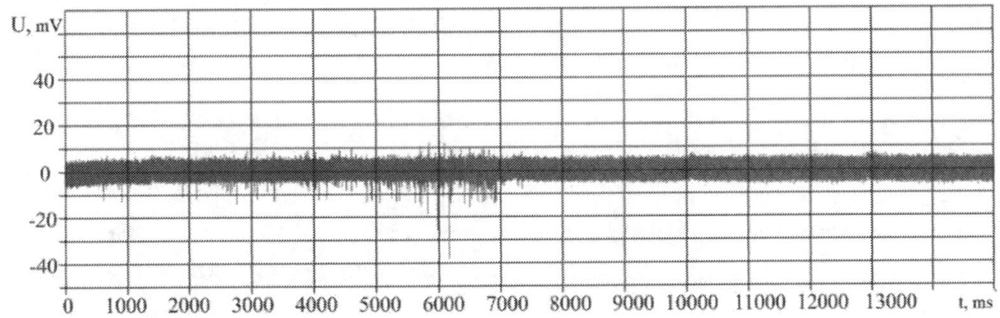

Fig. 5. The results of measurements of the EMP level at PK72

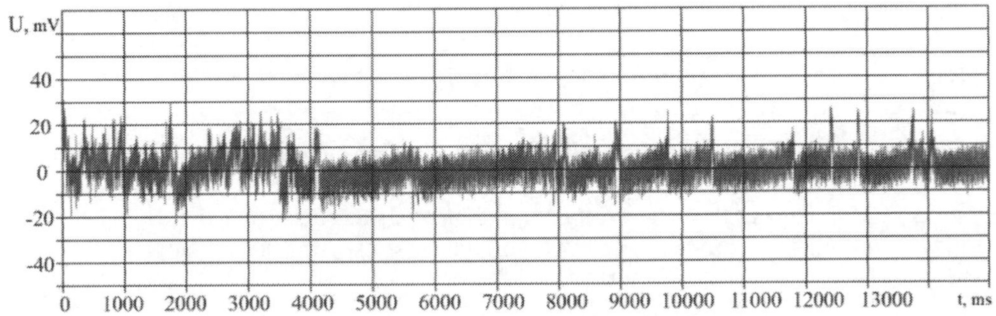

Fig. 6. The results of measurements near the lava 50 meters from KSH.

Table 1. Results of measurements with the device REMI-3 and Angel-M.

Location	"Angel-M"	"REMI-3"
Conjugation with BIS 2503 and CMPH	A=08,5 B=0,143 A=25,5 B=0,173 A=10,8 B=0,128	0-150 50-300 0-150
Fault 24 FPO 2503	A=4,2 B= 0,278 A=4,4 B= 0,279 A=4,6 B=0,251	50-160 20-150 20-180
PK 20 VH Lava	A=4,1 B= 0,281 A=4,3 B= 0,288 A=4,1 B=0,286	15-400 40-350 40-150
PK52	A= 4,8 B= 0,294 A=11,1 B= 0,336 A= 8,6 B= 0,308	70-180 10-120 50-140
Conjugation with lava 25101	A=15,9 B= 0,200 A=15,4 B= 0,366 A=12,0 B=0,242	200-500 230-300 330-400
Lava 50	A=309,7 B=0,026 A=641,4 B= 0,008 A=292,5 B=0,05	350-600 400-750 360-630
Conjugation with KH	A=50,0 B=0,079 A=38,4 B= 0,097 A= 2,8 B=0,127	30-70 40-80 40-80
CMPH 2503	A=4,5 B=0,291 A=4,1 B=0,286 A=4,3 B=0,288	30-60 60-90 80-100

The measurement results show that measurements with the device REMI-3 and Angel-M correlate with a correlation coefficient of 0.73. From the oscillograms recorded by the device REMI-3 it can be seen that in those places where there is increased cracking and collapse, the period of single pulses increases from 5 ms to 100 or more. In places where abundant desquamation is observed, the signal amplitude increases from 5 mV to 20 mV, directly near the source.
In the process of profiling the mine workings with the Angel-M and REMI-3 devices, the pickets were selected on an interval of 50 meters (Fig.7).

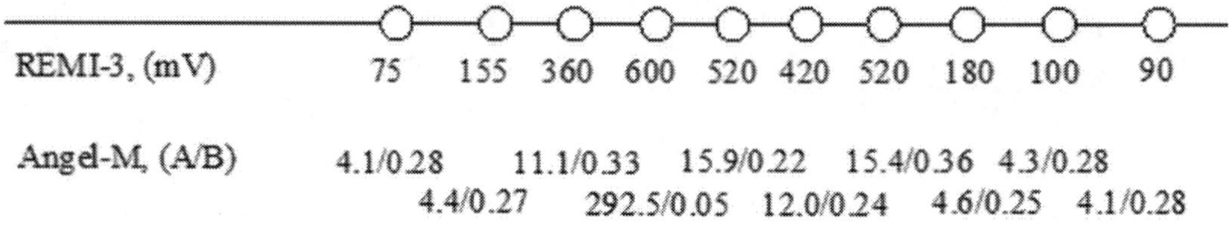

Fig. 7. Measurement results when profiling mine workings at the site of interfacing with lava.

An area was found in which an excess of the background component was observed more than 10 times, at this point the Angel-M device showed the value of the parameter A = 292.5, and B = 0.05. During the measurements, a dynamic manifestation of rock pressure in the form of shooting occurred, after which the instrument values approached the background.

4 Conclusions

1. Conducted laboratory studies of the enclosing rock from the Kirov mine. It is shown that the rock is not able to hold the load, there are dynamic manifestations of rock pressure in the form of desquamation.
2. Conducted field studies of the mine working area, with the help of equipment Angel-M and REMI-3. It is shown that on the site, the most significant from the point of view of forecasting are oscillograms, which show the nature of the boards in contrast to the averaged parameters.
3. In places with a probable manifestation of rock pressure, an excess of the background component is observed by more than 10 times.

References

1. G.E. Yakovickaya, Physical problems of the destruction of rocks, **1,** 98–103 (2003)
2. M.V. Kurlenya, G.I. Kulakov, G. E. Yakovitskaya, Fiziko-tekhnicheskie problemy razrabotki poleznykh iskopaemykh, **1**, 12–21 (1991)
3. V.S. Kuksenko, I.Ye. Inzhevatkin, B.C. Manzhikov, Fiziko-tekhnicheskie problemy razrabotki poleznykh iskopaemykh, **1**, 9–22 (1987)
4. A. G. Vostretsov, G. E. Yakovickaya, A. A. Bizyaev, Reports of the Academy of Higher Education of the Russian Federation, **3:28,** 29–38 (2015)
5. A.G. Vostretcov, A.A. Bizyaev, Fiziko-tekhnicheskie problemy razrabotki poleznykh iskopaemykh, **6,** 204-210 (2017)
6. A.G. Vostrtcov, G.E. Yakovickaya, A.A. Bizyaev, Reports of the Academy of Higher Education of the Russian Federation, **3:40,** 38-48 (2018)

7. A.G. Vostretcov, G.E. Yakovickaya, A.A. Bizyaev, Reports of the USSR Academy of Sciences, **2:21,** 46-54 (2013)
8. A.A. Vorobiev, E.K. Zavodovskaya, V.N. Salnikov, Reports of the USSR Academy of Sciences, **220:1**, 82-85 (1975)
9. A.G. Vostretcov, A.A. Bizyaev, Scientific Bulletin of the Novosibirsk State Technical University, **3:32,** 37-44 (2008)
10. D.V. Yakovlev, S.N. Mulev, Coal, **10**, 14-19 (2014)

 https://doi.org/10.1051/e3sconf/201913401005

Problems of Operational Safety of Dump Trucks in Open Pit Mine

Boris Gerike[1,2], *Aleksandr* Zacharov[1], *Pavel* Gerike[2], and *Yuriy* Drozdenko[1,*]

[1] T.F. Gorbachev Kuzbass State Technical University, 650000, Kemerovo, Russia
[2] Institute of Coal of the Siberian Branch of the RAS, 650065, Kemerovo, Russia

Abstract. One of the risk factors for dump trucks drivers is the over-vibration on various parts and control units. This leads to the formation of high intensity of work. To search for the causes of over-vibration it is proposed to use vibration analysis methods. To search for the causes of over-vibration it is proposed to use vibration analysis methods in time when dump truck is not moving. This allows the escaping the influence of external factors such as like profile of the road, slope, etc. The analysis of the vibrational signal spectrum allows determining the type of defect with a high degree of reliability.

1 Introduction

There is a known problem of occupational morbidity of the drivers of dump trucks [1-3], which is associated with over-vibration in the workplace. This requires a reduction in time of relay of the driver, but this is often impossible due to the organization of labor in the pit mine. Despite the continuous improvement in the design of mining dump trucks and their service systems, there is in the level of occupational morbidity among drivers do not decrease [4, 5].

Vibrodiagnostic [6–9] is the most acceptable for manufacturing and assembling mining equipment most suitable for searching and identifying various types of mining equipment (including heavy mining dump trucks) [6–9], which allows 90% confidence [8, 10-13] to identify and diagnose developing defects at an early stage.

2 Methods of examination

Vibrodiagnostics refers to functional methods for assessing the technical condition of equipment [9, 11], therefore, in principle, it can be used for working loads of dump trucks when driving them on technological roads, during loading and unloading of a transported rock mass. However, when a diagnostic examination of the trucks is to search for defective equipment is more appropriate Parking test, when you set various operation modes of the engine (idling, the nominal rpm maximum allowable rpm when the Parking test), then the disconnect is attached equipment (pump, drive shaft and fan wheel).

* Corresponding author duv.gmik@kuzstu.ru

The values of vibration velocity V, vibration acceleration a and vibration displacement S measured in the frequency range (3 - 4000 Hz) were chosen as diagnostic features.

Table 1 presents comparative estimates of classes of working conditions with the maximum and permissible levels of vibration acceleration.

Table 1. Comparison of European and Russian classes of working conditions

Directive 2002/44/EC - vibration [14]		**Russian classification of working conditions** (general and local vibration) [15-17]		
Invalid level $a >$ 1,15 m/s^2	Red	4	$a_z > 8{,}9$ m/s^2; $a_{y,x} > 6{,}3$ m/s^2	Extreme
		3.4	$a_z = 4{,}46$-$8{,}9$ m/s^2; $a_{y,x} = 3{,}2$-$6{,}3$ m/s^2	Harmful
		3.3	$a_z = 2{,}23$-$4{,}46$ m/s^2; $a_{y,x} = 1{,}6$-$3{,}2$ m/s^2	
Limit level $a =$ 1,15 m/s^2	Yellow	3.2	$a_z = 1{,}12$-$2{,}23$ m/s^2; $a_{y,x} = 0{,}79$-$1{,}6$ m/s^2	
		3.1	$a_z = 0{,}56$-$1{,}12$ m/s^2; $a_{y,x} = 0{,}4$-$0{,}79$ m/s^2	
Acceptable level $a \leq 1{,}15$ m/s^2	Green	2	$a_z < 0{,}56$ m/s^2; $a_{y,x} < 0{,}4$ m/s^2	Accepted
		1	$a_z = 0$ m/s^2; $a_{y,x} = 0$ m/s^2	Optimum

Figure 1 shows a diagram of the control points for measuring local vibration in the driver's cab.

Fig. 1 The scheme of vibration measurement on the driver's seat, driver's cab, and steering wheel

To search for a source of over-vibration was conducted examination units that generate the vibration (diesel engine and generator).

Figure 2 shows a control points scheme of measurement of local vibration on the left side of the diesel engine. A similar vibration measurement scheme was used on the right side of the engine.

Figure 3 shows a control points scheme of measurement of local vibration on the electric generator.

https://doi.org/10.1051/e3sconf/201913401005

3 Measurement result

The object of the study was chosen dump truck BelAZ-75306, which is operated on the pit mine "Vinogradovsky" PJSC "Kuzbasskaya Toplivnaya Company". During operation, drivers had a subjective feeling of over-vibration. Measurement of vibration level was carried out in accordance with the accepted measurement procedure with the sequential shutdown of the components of the power unit.

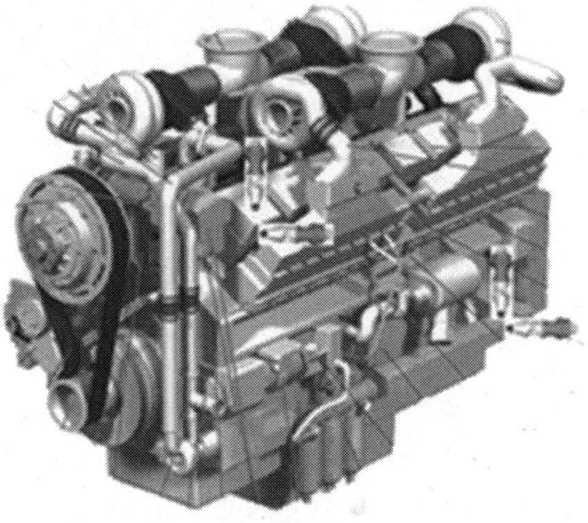

Fig. 2 The scheme of vibration measurement on a diesel engine

Fig. 3 The scheme of vibration measurement on an electric generator.

 https://doi.org/10.1051/e3sconf/201913401005

From the analysis of the vibration that affects the driver of the dump truck, it follows that the maximum acceleration is formed on the cab floor in the vertical component amax = 0.77 m/s^2 at idle, which exceeds the maximum permissible value [a] = 0.5 m/s^2 [14-17]. As follows from the spectral analysis of vibration acceleration (figure 4) in some cases, vibration isolation of the cabin passes the mechanical vibrations generated by the engine to the floor of the driver's cab.

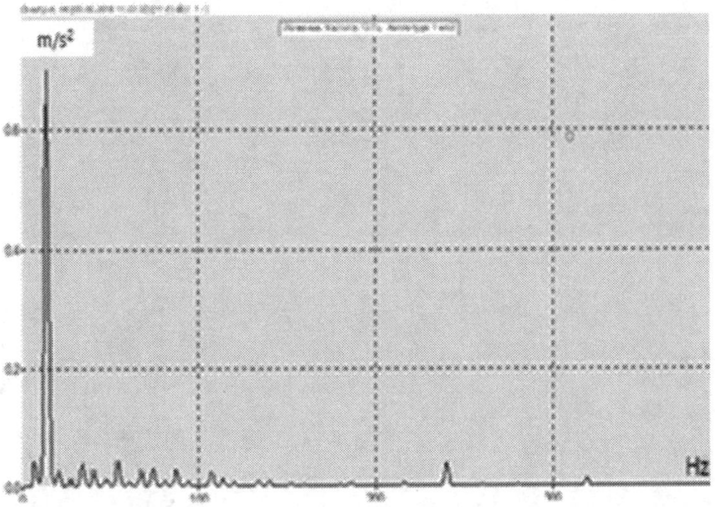

Fig. 4 Spectrum of the vertical component of vibration generated by the engine to the floor of the driver's cab n=800 min^{-1}, a=0,77 m/s^2

From the analysis of vibration acting on the driver of the dump truck, it follows that the maximum value of vibration acceleration acting on the hands on the steering wheel (local vibration) reaches the value of a_{max} = 0.40 m/s^2, which does not exceed the maximum permissible value [a] = 2.5 m/s^2 [14-17].

Spectral analysis of the results of vibration measurement on the steering wheel indicates poor engine performance (figure 5) – there is the appearance of the second and higher harmonics, there is a harmonic at half the crankshaft speed. All this testifies to the unsatisfactory work of the power supply system of the engine.

Analysis of the test results shows that the main source of mechanical vibrations is a diesel drive dump truck. At the same time, the maximum value of vibration acceleration amax=27 m/s2, which is less than the maximum permissible value [a] = 4g, and the value of vibration velocity, by which the technical condition is estimated, V = 6,2 < [V] = 7,2 mm/s, which indicates a satisfactory technical condition of the drive system.

Increasing the engine speed from 800 min^{-1} (idle run) to 1200 min^{-1} (maximum speed during testing) almost always leads to a decrease in the level of vibration, but the value of vibration acceleration is almost always increasing.

Disconnection of attachments (pump, cardan shaft and fan wheel) has virtually no effect on the overall and local vibration level, although the vibration process pattern is slightly changed.

 https://doi.org/10.1051/e3sconf/201913401005

a)

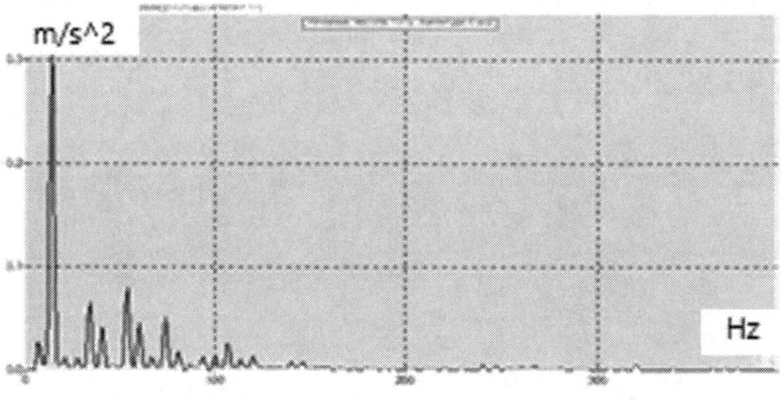

b)

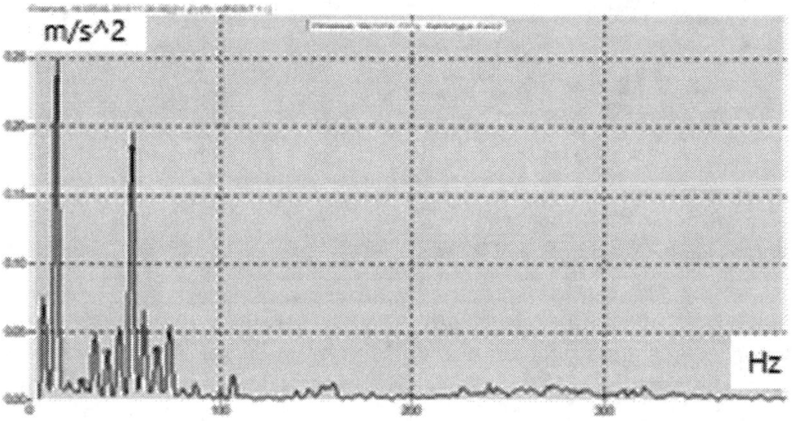

c)

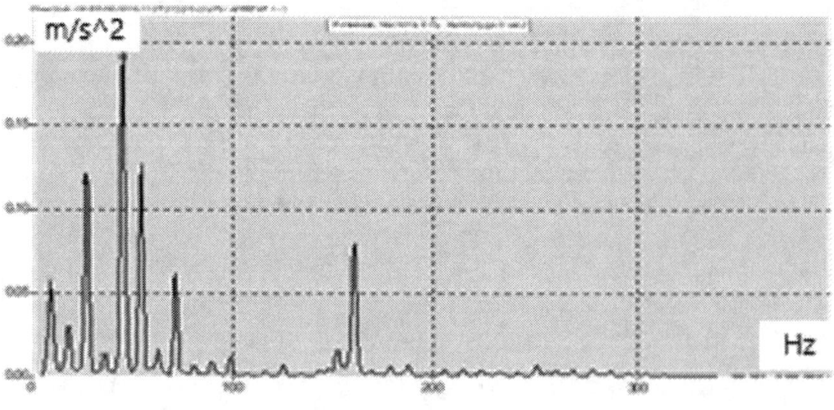

Fig. 5 Spectra of the vibration on the steering wheel: a - vertical component (n=800 min^{-1}; a=0,36 m/s^{-2}); b - quadrature component (n=800 min^{-1}; a=0,38 m/s^{-2}); c - vertical component (n=1000 min^{-1}; a=0,31 m/s^{-2})

4 Conclusion

Summarizing the above, we can say that:

- the search for sources of increased vibration is possible only using the method of functional diagnostics, namely vibration diagnostics;
- when carrying out diagnostic measurements, it is necessary to exclude the influence of external production factors for the identification of defects in the manufacture, assembly, and operation of dump trucks;
- the spectral composition of vibration processes makes it possible to determine the type of defect with a high degree of reliability (probability p=0.9) and, when using the Russian classification of working conditions, to determine the degree of danger of the detected defect.

References

1. L. Saidi, J. Ben Ali, F. Fnaiech, ISA Transactions, **53**, 1650-1660 (2014)
2. J. Tuma, *10.1002/9781118797563.ch3* (Dekker mechanical engineering, Berlin, 2014)
3. J. Xiaohang, C. Fangzhou, P. Yayu, Q.Wei & Qu, Liyan, IEEE Industry Applications Magazine, **10.1109/MIAS.2017.2740470** (2018)
4. L.Jing, M. Zhao, P. Li, X. Xu, Measurement, **111**, 1 (2017)
5. E. Kuzin, B. Gerike, Y. Drozdenko et al. IOP Conf. Series: Materials Science and Engineering, **253**, 012013 (2017)
6. B. Gerike, I. Panachev, E. Kuzin, E3S Web of Conf., **15**, 03008 (2017)
7. G. E. Morales-Espejel, Evolution, **2**, 25 (2010)
8. A. Mohamed, S. Sassi, M. Paurobally, Shock and Vibration, **2018**, 1913289 (2018)
9. W. Yu, Y. Shao, and C. Mechefske, Engineering Failure Analysis, **54**, 103 (2015)
10. K. Heidarbeigi, H. Ahmadi, M. Omid and A. Tabatabaeefar, Journal of Agricultural Technology, **5**, 1 (2009)
11. Kelly S. Graham, *Advanced vibration analysis* (Dekker mechanical engineering, Berlin, 2013)
12. Anil Rana,. Journal of Quality in Maintenance Engineering, **22:2**, 180 (2016)
13. B. Gerike, A. Mokrushev, The 9th China-Russia Symposium Coal in the 21st Centuri: Mining, Intelligent Equimpent and Environmental Protection, **213** (2018)
14. *Directive 2002/44/EC of the European Parlament and of the Ccouncil of 25 June 2002* (2002)
15. J. Obuchowski et al, Appl. Mech. Mater., **683**, 171 (2014)
16. W. Bartelmus, Key Engineering Materials, **588**, 184 (2014)
17. Z. Ming, J. Xiaodong, Mechanical Systems and Signal Processing, **94**, 129 (2017)

 https://doi.org/10.1051/e3sconf/201913401006

Enhancement of Open-Pit Mining Efficiency by Means of Quality Improvement of Open-Pit Excavators Design and Operating In Specified Conditions

Pyotr Zykov[1,*], *Evgeny* Zvarych[1], *Andrei* Karasev[1], and *Dieter* Weber[2]

[1]T.F. Gorbachev Kuzbass State Technical University, Novokuznetsk branch, 654000 Novokuznetsk, Russian Federation
[2]Czech Technical University in Prague, 166 36 Zikova 1903/4, Praha, Czech Republic

Abstract. The objective set in the research is the following: to develop the methodology and application tools for assessment of the open-pit excavators engineering level when selecting and designing them for specific mining-geological and technical service conditions. To achieve the objective, the methods of statistical data analysis, methods of regression models and of mathematical experimentation building, analysis and synthesis, methods of comprehensive assessment of the engineering level indicators, as well as mathematical modeling and appliance of linear algorithm were used. As a result, there was developed the methodology of comprehensive assessment of engineering level and parameters optimization of open-pit shovel excavators, there was also developed a software package, based on the methodology mentioned above, that allows science-based selection of excavators that possess optimum parameters for the specified conditions.

1 Introduction

Open-pit shovel excavators are the prime extracting equipment used for open-pit works and they perform over 80% of the total work content. In this regard, quality of their performing functions determines to a greater extent the efficient operation of the whole mining enterprise [1, 2].

Currently there is a process of technical re-equipping of the out-of-date excavator fleet by the new generation of extraction-and-loading machines implementation, as a rule such new machines have a hydraulic drive. Mining specialists' selection of excavators is mainly based on the operating history of the previous machines, their price and operation costs, and, what is more, no quality assessment of the machines in specific conditions of the enterprise is performed. Such approach has lots of disadvantages the main one is that it is impossible to «bind» the parameters of the selected machine and the parameters of the environment. As a result such approach does not lead to significant improvement of

* Corresponding author: zykovpetr@yandex.ru

excavator performance quality. When selecting and designing machinery with necessary relevant parameters it is necessary to use systems approach that will help improve the quality of open-pit excavation equipment performance.

Appliance of science-based range of methodologies that allow selecting or developing machines possessing optimum performance according to the quality criteria for specific mining conditions will make it possible to improve open-pit mining efficiency, to reduce costs of mining process, to increase excavator-car complex productivity. Development of software package, based on such range of methodologies, will make it possible to simplify and computerize the problem of excavator model selecting according to the specific conditions.

2 Methods

2.1 The methodology of excavator performance assessment

The assurance of required indicators of excavator performance in specific conditions according to engineering specifications, as well as production and operational requirements greatly depend on the quality of the excavation process performed by the excavator. The excavator quality can be displayed only during the process of performing its function according to its purpose, that is, during excavation of mined rock in specified operating conditions [1, 5-7]. The excavator mustn't be considered apart from the elements of the whole digging system «excavator-mined rock-human being-outside environment».

The main purpose of quality assessment of an excavator - developed mined rock – human being - outside environment interaction, as the elements of the whole digging system, is to determine how efficiently each element functions in specified mining-geological and engineering conditions [8, 9]. Quality assessment makes it possible to solve the problem of open-pit excavators selecting and designing according to specific mining plant conditions.

The conducted analysis of existing quality assessment techniques showed that the most acceptable analogue for the assessment of open-pit shovel excavators engineering level is G. I. Solod's methodology which he proposed in 1991.The methodology is based on the fundamental principles of qualimetry and allows conducting quantitative assessment of quality and engineering level of operationally non-varying machines of different types, dimensions types and structural designs.

2.2 Methodology of excavator parameters optimization while designing

Currently there are no sufficient normative-technical documents that specify such important direction of product quality improvement as its optimum design.

Each object of design has its own designation and like any other system it is characterized by its structure and parameters. The structure that defines the components of the object of design and their links, must guarantee the proper product operating and achieving the goals set according to its designation. Each component of the object of design is also characterized by its own parameters that are one way or other linked and are relevant to the object of design parameters. All these characteristics (the structure of the designed object, its parameters, the parameters of its components and their linkage) present the raw information necessary for the optimum design of the object.

For the purpose of the object parameters optimization it is necessary to create a formalized mathematical model. Mathematical model of optimization (MMO) consists of an objective function, restrictions and boundary conditions. The objective function presents mathematical description of reliance of the designed object application purpose on the parameters being optimized and in general term can be expressed as follows:

$$P = f(P_j) \rightarrow max(min), \tag{1}$$

where Pj – are the parameters being optimized (j = 1, 2,….., m, where m is the number of the parameters being optimized).

Boundary conditions indicate maximum allowable values of variables. In general case boundary conditions are two-sided.

$$b_j \geq P_j \geq a_j, \tag{2}$$

But often there is no upper boundary bj- ($b_j \rightarrow \infty$), and the lower boundary aj is equal to zero. Such boundary condition requires only so-called non-negative quantity for the parameter being optimized, that is $P_j \geq 0$.

Restrictions in MMO may present the descriptions of connections between parameters being optimized, formalization of expressions according to correspondence of actual indicators of designed object to the specified ones as well as formalization of additional purposes that were left beyond the objective function. Besides, the restrictions are introduced into MMO to reduce the dimension of optimization problem which is done to simplify problem statement and solving. The restrictions are expressed either in terms of relations that describe links between parameters being optimized:

$$c_j = f_c(P_j); j = 1,2,..., m, \tag{3}$$

or in terms of inequalities that describe different structural, operating and regulatory restrictions:

$$H_k = f_н(P_j); k = 1,2,..., l. \tag{4}$$

In such a way, from the informative point of view the process of design of any object includes acquisition of raw information about the object of design, its parameters optimization performed on the ground of the objects purpose as well as making a decision based on the results optimization and presentation by the form of normative-technical documents. The documents serve the foundation for project translation into end products. It should be mentioned that under free market conditions state regulatory documents, that were widely used earlier, today lose their importance at large extent, so the normative-technical documents mentioned above are developed by each customer with regard to their specific conditions and can be presented in the form of technical design specification.

3 The Results

3.1 Development of the methodology of engineering level and optimization of open-pit shovel excavators parameters comprehensive assessment

The basis of the methodology is the functional criterion of the machine that determines its main purpose. G. I. Solod proposed determining the criterion in the form of product of mining machine production capacity while performing its function in specific conditions and specific energy of the process of machine performing its function in such conditions.

Thereunder, the functional criterion of the open-pit excavator quality assessment can be presented by the form of the following *(kJ/h)*:

$$\lambda = P \cdot W, \tag{5}$$

where P – is estimated-theoretical production capacity of the excavator, m^3/h; W – specific energy of the mining rock excavation, *(kJ/ m^3)*.

The expression for functional criterion calculation for shovel excavator of the «front shovel» type in the specific conditions (kW/m^3) will take the following form:

$$\lambda' = \frac{3600 \cdot k_н \cdot k_в \cdot k_{кв}}{t_ц \cdot k_р}\left[k_f + \frac{k_н}{k_р}\rho_п \cdot g\left(\frac{h_к}{2} + \frac{\pi^2}{900 \cdot \alpha \cdot g}(n \cdot R_в)^2\right)\right], \tag{6}$$

for the open-pit shovel excavators of «back-hoe» type:

$$\lambda' = \frac{3600 \cdot k_н \cdot k_в \cdot k_{кв}}{t_ц \cdot k_р}\left[k_f + \frac{k_н}{k_р}\rho_п \cdot g\left(\left(\frac{h_к}{2} + h_в\right) + \frac{\pi^2}{900 \cdot \alpha \cdot g}(n \cdot R_в)^2\right)\right], \tag{7}$$

where α is a rotational angle of the unloading process, degrees; h_B– maximum height of the unloading, m; k_H – coefficient of bucket (shovel) full; k_P – mining rock fragmentation coefficient; k_B – excavator operation time coefficient; k_{KB} – shovelman qualification and handling quality coefficient; k_f – specific rock resistance to digging, kPa; ρ_n – mining rock density, t/m^3; h_K – maximum excavator digging height, m; n – swing speed, rad/s; R_B – maximum unloading radius, m; t_C – cycle duration, s; g – gravitational acceleration, m/s^2.

Excavator quality indicators are taken from the USSR's State Standard 4.377–85 and what is more, the indicators of purpose, reliability, technological effectiveness are of the greatest importance and other groups of indicators are not so important, this fact is connected with the chosen criterion. To reduce the number of values, some parameters may be translated into specific type (the advantage of any specific indicators is the following: they combine several absolute indicators) that completely corresponds to the principles of qualimetry.

By way of the indicators that determine engineering level of open-pit shovel excavators the following are suggested: specific energy density (E_y); specific metal content (k_M); specific excavator digging force (k_f^F); specific maximum digging height and radius (k_p^{hK} and k_p^{RK}); excavator service life (T_P).

As a result, we have obtained a complex consisting of six indicators that determine the engineering level of open-pit shovel excavators (fig. 1).

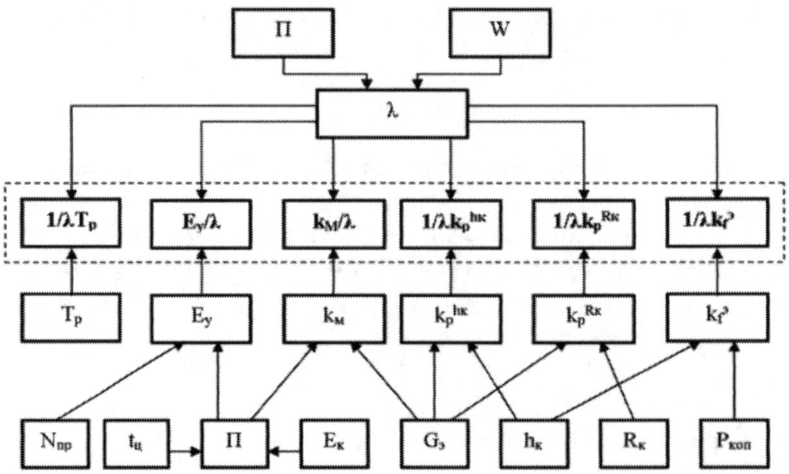

Fig. 1. The complex of indicators that determine engineering level of open-pit shovel excavators: λ – functional criterion, *kJ/h*; k_f^E – specific excavator digging force, MPa; E_Y– specific energy density, kW/(m³/h); k_M – specific metal content, t/(m³/h); $k_p^{h\kappa}$ – specific maximum digging height; $k_p^{R\kappa}$ – specific maximum digging radius; T_p – service life, h; P_{kon} – specific digging force, kH; N_{np} – drive power, kW; P – productive capacity, m³/h; G_E – operation weight of the excavator together with operating equipment, t; h_K – maximum digging height, m; R_K – maximum digging radius, m; t_C – theoretical cycle duration, s; E_K – shovel capacity, m³; W – specific energy of mining rock excavation, kJ/m³.

Composite indicator of excavator quality is determined in accordance to the algorithm methodology presented in figure 2.

3.2 Development of the methodology of excavator parameters optimization while designing

On the basis of the systemic approach there was developed a mathematical model of open-pit shovel excavator parameters optimization (MMO). In the way of optimization criterion there were used generalized estimators of engineering level, and in the way of parameters being optimized there were used its simple indicators.

$$\begin{cases} k = a_0 + a_1 k_{\text{м}} + a_2 N^{y}_{\text{пр}} + a_3 \lambda + a_4 T_p \rightarrow \max \\ a_{10} + a_{11} k_{\text{м}} + a_{12} N^{y}_{\text{пр}} + a_{13} \lambda + a_{14} T_p \le K^{0}_{\text{п}} + S_0^{K\text{п}} \\ a_{10} + a_{11} k_{\text{м}} + a_{12} N^{y}_{\text{пр}} + a_{13} \lambda + a_{14} T_p \ge K^{0}_{\text{п}} - S_0^{K\text{п}} \\ a_{20} + a_{21} k_{\text{м}} + a_{22} N^{y}_{\text{пр}} + a_{23} \lambda + a_{24} T_p \le K^{0}_{\text{м}} + S_0^{K\text{м}} \\ a_{20} + a_{21} k_{\text{м}} + a_{22} N^{y}_{\text{пр}} + a_{23} \lambda + a_{24} T_p \ge K^{0}_{\text{м}} - S_0^{K\text{м}} \\ a_{30} + a_{31} k_{\text{м}} + a_{32} N^{y}_{\text{пр}} + a_{33} \lambda + a_{34} T_p \le K^{0}_{\text{э}} + S_0^{K\text{э}} \\ a_{30} + a_{31} k_{\text{м}} + a_{32} N^{y}_{\text{пр}} + a_{33} \lambda + a_{34} T_p \ge K^{0}_{\text{э}} - S_0^{K\text{э}} \\ \lambda_{max} \ge \lambda \ge \lambda_{min} \\ k_{\text{м}} \ge 0; N^{y}_{\text{пр}} \ge 0; \lambda \ge 0; T_p \ge 0 \end{cases} \tag{8}$$

where k is a composite indicator of engineering level (optimization criterion); T_P – excavator service life, h; k_M – specific metal content of the excavator, t/m³; λ – functional criterion of open-pit shovel excavators quality, kJ/h; N_{np}^{Y} – specific drive power, kW/m³; K_M^0, K_E^0, K_n^0 – specified values of specific weight restrictions, as well as specific power consumption and indicator of productivity restrictions; S_0^{KN}, S_0^{KE}, S_0^{KM} – values of standard deviations while finding out the regressional reliance of outcome performance data on the parameters being optimized.

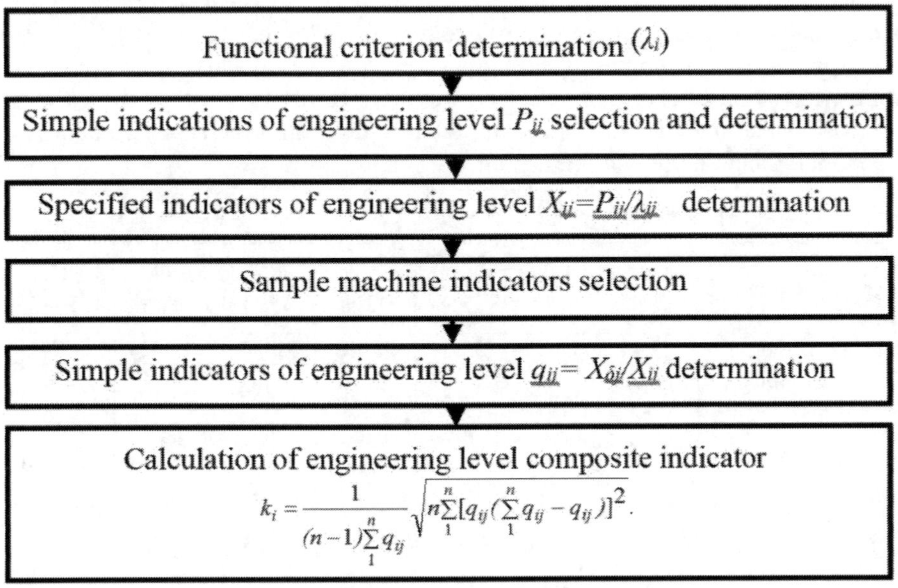

Fig. 2. The calculation algorithm of excavator composite quality indicator.

Constants of goal function models and restrictions are determined as a result of mathematical experiment realization that is understood to be calculation of composite engineering level and outcome performance data (K_M^0, K_E^0, K_n^0) subjected to specified combinations of variables according to the experiment planning matrix.

After the optimization of composite quality indicators, it is possible to synthesize the main parameters of open-pit excavators technical characteristics.

4 Discussion

4.1 Comprehensive assessment of engineering level of the existing open-pit hydraulic excavators fleet

On the grounds of the developed range of methodologies there was conducted a comprehensive assessment of engineering level of existing open-pit hydraulic «front shovel» type excavators fleet according to proposed nomenclature of simple indicators, using specific for open-pit excavators functional criterion.

Assessment of the engineering level was conducted under the following conditions: out of the whole variety of the models and modifications of open-pit excavators there were selected machines of all dimension types with shovel capacity equal to 5.2–45 m^3. Almost all the excavators under discussion are produced in lots but some models were taken out of production. However, they have also been included into the group of machines under the discussion. It has been done to allow taking into consideration the previous experience of open-pit excavating equipment design as fully as it is possible.

To compare the engineering level of domestic production excavators (OMZ) with foreign equipment there were examined sufficiently proven and widely used open-pit excavators manufactured by the leading in this field companies: American «Caterpillar», Japanese «Komatsu», «Hitachi», German «Liebherr» and Swiss «Terex O&K».

An overall analysis of calculation results of the open-pit hydraulic excavators engineering level composite indicator shows that all these machines have this indicator higher than 0.5. This fact confirms that the technical level of excavators is high enough. The average technical level of the group of excavators under discussion is $Y_{CP} = 0.764$.

Table 1 presents the following: the higher positions in the ranked list are occupied by excavators with small bucket (shovel) - capacity. This can be explained by the fact that while increasing the machine size and complicating its design the scaling factor as well as system factors begin to come out increasingly.

Table 1. Composite indicator of open-pit excavators engineering level.

The excavator model	**385C FS**	**EG-150**	**365C FS**	**EX 1200**	**EX 1900**	**R994 Litronic**	**PC 1250-7**	**R984 Litronic**
Functional criterion. kJ/h	23.04	24.22	22.69	22.46	22.30	21.93	21.57	20.88
Composite quality indicator	0.913	0.894	0.877	0.850	0.830	0.816	0.815	0.769
The excavator model	**EG - 550**	**RH 40-E**	**PC 3000-6**	**EG - 350**	**PC 4000**	**PC 8000-6**	**RH 90-C**	**RH 120-E**
Functional criterion. kJ/h	22.54	20.65	20.88	22.07	21.24	20.75	20.75	20.63
Composite quality indicator	0.768	0.766	0.759	0.757	0.751	0.738	0.738	0.738
The excavator model	**R996 Litronic**	**RH 400**	**EX 5500**	**R995 Litronic**	**EX 3600**	**EG - 110**	**RH 200-E**	**EG - 5.5**
Functional criterion. kJ/h	21.54	21.68	21.24	20.68	20.47	21.50	19.85	19.58
Composite quality indicator	0.736	0.730	0.728	0.708	0.704	0.700	0.647	0.616

If to examine the engineering level of the machines produced by foreign manufacturers, the companies Caterpillar and Hitachi should be put emphasis on because the machines produced by them have higher than average engineering level. The machines manufactured by Terex have the lowest indicator of engineering level. The average engineering level of domestic machines is slightly below the level of excavators produced by foreign companies. The indicator value Y expressing the averaged values of the engineering level composite indicator for foreign machines under discussion is $Y_{foreign} = 0.769$, that is 3% higher than for excavators produced by domestic company OMZ, for which $Y_{domestic} = 0.747$. However, domestic machines have Y-indicator 3.2% higher than the machines produced by Terex Company. This means that from the point of view of construction arrangement perfection and of new ideas and solutions existence, excavators produced by Russian manufactures almost keep a pace with foreign machines. It should be also mentioned that domestic excavators EG150 and EG550 occupy the 2 and 9 lines in the ranking list correspondingly (table 1). However, other models are at the bottom of the list that dramatically reduces the average value of the indicator.

High quality of the excavators manufactured by foreign companies is mainly achieved by a longer service life (more than 90 thousand hours), higher maintenance and operation efficiency, which, in its turn, is caused by the appliance of up-to-date methods and design tools, extensive use of advanced production technologies and construction materials, modern control and diagnostics systems as well the latest achievements of ergonomics and design [6, 12-14].

The main directions of improving the open-pit hydraulic excavators engineering level are the following: increasing the specific digging force, reducing the specific metal content and energy density. The main directions of domestic machines improvement are the following: increasing service life and reliability of excavators by the hydraulic drive design improvement, as well as widening the hydraulic excavators range by means of manufacturing bigger dimension-types [15-18].

4.2 Open-pit excavator parameters optimization at the stage of developing its technical design specifications

Having the optimum values of optimized indicators λ^*, N_{np}^{y*}, k_M^* и T_p^* as well as the engineering level composite indicator k^* that are determined by the relation (8), it is possible to determine the basic parameters of the open-pit excavator's technical characteristics.

The results of optimal parameters synthesis for excavator with bucket (shovel) capacity equal to 15 m^3 and technical characteristics of its analogue are presented in table 2.

Table 2. The results of optimal parameters synthesis.

Parameters	KOMATSU PC-3000-6	Parameter optimum value
Excavator bucket capacity (E_K), m^3	15	15
Operating weight (G_E), t	253	224.3
Drive power (N_{np}), kW	940	949.5
Estimated cycle time (t_c), s	26	23.4
Travel speed (v_E), km/h	2.4	2.7
Maximum digging radius (R_K), m	13.5	14.6
Maximum unloading radius (R_B), m	10.94	12.4
Maximum digging height (h_K), m	14.7	15.1
Maximum unloading height (h_B), m	10.2	11.1
Maximum digging force (P_{Kon}), kN	1100	993.3
Service life (T_p), thousand hours	90	84.5
Swing speed (n), rpm	4.6	4.8
Bucket raise speed (V_r), m/s	0.78	0.91
Composite indicator of engineering level k	*0.759*	*0.981*

5 Conclusion

The research work contains a new solution of the recent scientific problem of parameters optimization for open-pit shovel excavators with the purpose to select and create machines of higher engineering level which is essential for the development of mining machines design and operation theory and practice.

The proposed methodology of the optimal selection and parameters optimization of open-pit shovel excavator models will allow performing science-based selection of excavators possessing optimum for the specified conditions parameters as well as will make it possible to install and optimally coordinate their parameters upon specified criteria at the design stage.

The was performed a comprehensive assessment of the engineering level of open-pit hydraulic «front shovel» type excavators existing fleet that use the software package, the engineering level of the open-pit excavators existing fleet was evaluated, the ways to increase the engineering level and methods of their realization at the stage of design were found out.

As a result of studying optimization mathematical model there were determined the optimal values of composite quality indicators out of which the basic parameters of the technical characteristics of open-pit excavators were synthesized. For example, there were examined the parameters of the excavator KOMATSU PC-3000-6 and of an excavator with optimal parameters (table 2) and what is more, the composite quality indicator of the latter is 30% higher.

The range of methodologies and software package can be used by mining enterprises as a basis for selecting the most suitable for their conditions machine models and development of technical specifications for their design. Selection of an optimum excavator model will make it possible to reduce expenses that are caused by production loss in case of its unreasonable selection.

References

1. S. Er, M. Yilmaz, A. Ertin, A. Tugrul, N. Tokgöz, J. Geol. Soc. India, **89**, 5 (2017)
2. S. Fiscor, Eng. and Mining J., **210**, 3 (2009)
3. J. Yan, B. Li, G. Guo, Y. Zeng, M. Zhang, Chin. J. Mech. Eng., **26**, 6 (2013)
4. S. Frimpong, Y. Hu, H. Inyang, Int. J. Geomech., **8**, 1 (2008)
5. K. Awuah-Offei, S. Frimpong, Mech. and Mach. Theory, **42**, 8 (2007)
6. T. Yoshida, T. Koizumi, N. Tsujiuchi, K. Chen, Y. Nakamoto, Trans. Jap. Soc. Mech. Eng. C., **78**, 787 (2012)
7. H. Wang, M. Sun, Z. He, S. Huang, F. Rong, X. Yuan, J. Hunan Univ. Natur. Sci., **45**, 2 (2018)
8. Y.-g. Wang, Q. Fang, Coal Mine Mach., **27**, 11 (2006)
9. O. J. Abdel-baqi, A. Nasiri, P. J. Miller, IEEE Trans. Ind. Appl., **52**, 6 (2016)

I. Mitsuhashi, H. Tsukamoto, N. Sakai, Sumitomo Heavy Ind. Techn. Rev., **165**, 7 (2007)

10. H. F. Bilgin, M. Ermis, K. N. Kose, I. Cadrici, A. Acik, T. Demirci, A. Terciyanli, C. Kocak, M. Yorukoglu, IEEE Trans. Ind. Appl., **43**, 1 (2007)
11. J. E. Fernández, R. Vijande, R. Tucho, J. Rodríguez, A. Martín, Wear, **250**, 1 (2001)
12. C. Dogruoz, N. Bolukbasi, Bull. Eng. Geol. and Environ, **73,** 3 (2014)
13. Y. Cao, Y. Xie, Nonlinear Dyn., **91**, 1 (2018)
14. O. Litvin, M. Tyulenev, S. Zhironkin, S. Prokopenko, Acta Montanistica Slovaca, **22:2**, 146-152 (2017)
15. E.A. Tyuleneva, Yu.V. Lesin, Ya.O. Litvin, Journal of Mining and Geotechnical Engineering, **1(4)**, 35-49 (2019). DOI: 10.26730/2618-7434-2019-1-35-49
16. M. Tyulenev, O. Litvin, M. Cehlár, S. Zhironkin, M. Gasanov, Acta Montanistica Slovaca, **22:3**, 296-302 (2017)
17. M. Tyulenev, O. Litvin, S. Zhironkin, M. Gasanov, Acta Montanistica Slovaca, **24(2),** 88-97 (2019)
18. M.A. Tyulenev, S.A. Zhironkin, E.A. Garina, Int. J. Min. Miner. Eng., **7:4**, 363-370 (2016)

E3S Web of Conferences **134**, 01007 (2019) https://doi.org/10.1051/e3sconf/201913401007
SDEMR-2019

Optimization Methods and Algorithms for Calculating the Construction of Non-Rigid Pavement for Technological Quarries Roads

Anna Tailakova[1,*], and *Alexander* Pimonov[1,2]

[1]T.F. Gorbachev Kuzbass State Technical University, Kemerovo, 650000, Russian Federation
[2]Institute of Economics and Industrial Engineering, Siberian Branch of the Russian Academy of Sciences, Novosibirsk, 630090, Russian Federation

Abstract. Previously developed by the authors of the optimization model for calculating the construction of non-rigid pavement of public roads is proposed to be used for the calculation of the construction of pavement technological career roads. The article describes the optimization methods and algorithms for calculating the construction of pavement. Possibilities of using methods of coordinate-wise descent, multi-start, dynamic programming for the selection of the optimal construction of pavement are presented. Application of genetic algorithms for the decision of an optimization problem of calculation of a construction of pavements on the basis of comparison of their efficiency with efficiency of search methods is proved. Described results of computational experiment of selection of genetic algorithm operators to reduce the volume of calculations and ensure the stability of the results.

1 Introduction

Quarry roads located on the territory of mining enterprises are constructed for the movement of specialized vehicles and other means of ensuring the technological process of mining. The cost of construction and maintenance of technological quarries roads make up a significant part of the cost of transporting the coal mass. The efficiency of mining transport is also directly dependent on the quality and durability of the haul road and the main element of the pavement. Road clothes of internal roads of industrial enterprises should be constructed in accordance with the basic provisions of the system of normative documents in construction. Typical solutions can significantly reduce the range of search for the most rational construction of pavement, but they solve this problem only partially, because each material of construction has its own cost, and these costs vary depending on the area of construction of the road, the location of bases, quarries, warehouses, the distance of delivery of materials, etc. Ground conditions are also different even with the same structures. It is impossible to take all this into account in standard solutions. The best solutions to minimize the cost of construction of technological quarries roads can be obtained by calculating the construction of pavement on the basis of the optimization models proposed by the authors. These optimization models make it possible to calculate numerically (to determine the thickness of the layers and the volume of the required materials) the minimum cost of materials of construction of non-rigid pavement. Limitations of models are requirements for strength and frost resistance of the calculated

*Corresponding author: taa.vtit@kuzstu.ru

construction in accordance with instructions of normative documents on the construction and calculation of road pavement. Optimal construction includes construction, calculation, technical and economic analysis and selection of the most rational construction solution. The software developed by the authors can significantly reduce the time spent on the construction calculation and consider a greater number of options. The most effective is the automated construction of road pavement using the proposed optimization models, methods and algorithms to obtain optimal solutions to the problem (to minimize the cost of construction) under given soil conditions and strength constraints.

Quarries technological roads are natural and technical objects with a layered structure. The task of calculating the construction of the pavement is reduced to the selection of the thickness of their layers in the options outlined in the construction, or to the choice of materials with appropriate deformation and strength characteristics at a given layer thickness and taking into account the requirements of strength. Thus, it is proposed to use two optimization models to solve the problem of selection of road pavement construction:

1) selection of construction by varying the thickness of the structural layers;
2) selection of construction by varying the materials of the structural layers.

Parameters of the problem are characteristics of structural layers, and limitations are strength requirements in accordance with the normative documentation. In this case, the problem of finding the optimal construction of pavements for technological roads is a problem of multidimensional optimization containing nonlinear constraints, and is reduced to minimizing the cost of construction work in terms of the consumption of materials of structural layers. In this case, the objective function is a multi-extremal function, since it can contain both global and several local extrema.

2 Application of non-linear programming methods of searching the optimal pavement construction

Due to nonlinear constraints in proposed optimization models, it is impossible to use analytical methods to solve the problem. By discretizing the intervals, the optimization problem can be reduced to a discrete one and solved by a brute force method. The brute force method ensures that the global extremum is found. A complete search of solutions can be implemented using the method of coordinate descent.

Coordinate descent algorithms is based on the universal approach to algorithmic, numerical optimization: solving an optimization problem by solving a sequence of simpler subproblems. Each iterate is found by fixing most of components of the variable vector at their current values and approximately minimizing the objective function with remaining chosen components. Thus, the calculation of all possible variants of the road pavement construction can be made with successive variations in parameters of each layer of the structure. But this approach raises the problem of inefficiency of brute force, associated with its complexity. During the computational experiment it was found that the time of such a method to solve the problem increases exponentially with respect to the size of the input data. So when calculating the construction of the pavement of 3 layers on the basis of the proposed model, it is necessary to calculate more than 1000 variants of possible constructions (depending on the value of the minimum allowable thickness of the structural layers and step), the construction of 5 layers – more than 100,000 [1–2]. Alternatively, the timeliness of coordinate descent for large-scale problems can be also improved by parallelizing the coordinate updates. The multi-start method involves a parallel run of the coordinate descent method from several points of the set of valid solutions. The best result obtained when running parallel processes is chosen as a solution to the problem. Approximate speedup (ideally) is proportional to the number of processors. [3–4].

3 Application of dynamic programming methods for search the optimal construction of pavement

The solution of the problem of inefficiency of a full search can be the use by dynamic programming methods, so that it is possible to search the optimal construction that meets the requirements of strength. The dynamic programming approach is solving each subproblem only once, thereby reducing the number of computations. This is especially useful in cases where the number of repetitive subtasks increases exponentially. The method of dynamic programming with profile can be used for the task.

With regard to the problem to be solved, this method involves the allocation of the entire space of feasible solutions to several subsets (in accordance with the number of parameters of the problem), fixing all the variable parameters, except one. Thus, the problem of finding the optimal variant of the construction is divided into subtasks of finding the optimal variant of the construction for each of subsets. Then there is a choice of the best option from received. But in the course of the algorithm in several subsets may have same options. Then the caching method is used to exclude the repeated calculation of construction variants.

Depending on the location of the global minimum point of the objective function, the amount of calculations can be significantly reduced. In the worst case, the operation time of the algorithm will be comparable to the time of the coordinate descent method.

Dynamic programming is a well-proven method used when the problem can be solved recursively, which leads to an algorithm with exponential time complexity. If the subproblems are overlapped, then the dynamic programming approach can reduce unnecessary computation, so each subproblem is solved not more than once, and the time complexity of the algorithm is reduced to polynomial. On the other hand, algorithms based on dynamic programming are usually difficult to parallelize, since the subproblems are interdependent, that means each subproblem requires the results of solving the previous subproblem [5–8].

4 Application of heuristic methods of finding the optimal construction of pavement

To reduce the volume of calculations, heuristic methods can be used to obtain an acceptable solution to the problem, using numerical methods focused on finding the best solutions from a variety of alternatives, while avoiding a complete search and comparison of all options. The best known evolutionary algorithm is the genetic algorithm. Mathematically, a genetic algorithm can be defined as a stochastic optimization method for discrete optimization problems.

Genetic algorithms (GA) are adaptive search methods used to solve optimization problems. They use both an analogue of the mechanism of genetic inheritance and an analogue of natural selection. At the same time, the biological terminology in a simplified form and the basic concepts of linear algebra are preserved. Main operators of the genetic algorithm are crossing, mutation, selection of parents and selection (selection of chromosomes in a new population). The type of operator plays an important role in the implementation and efficiency of the GA. There are basic forms of operators whose clean use or modernization leads to a GA suitable for a particular task.

During the computational experiment (table. 1) the following GA parameters were set: in the evolutionary model, each variable corresponds to a gene in the chromosome; to represent chromosomes, the decimal coding is used; the fitness function completely repeats the target function of the proposed mathematical model; the population size is 40 chromosomes; the maximum number of generations is 100. Different combinations of GA operators were used to adjust GA. For each case, the HA was run 50 times for each of the 5 test constructions of non-rigid pavement. An indicator of the stability of the genetic algorithm is the ratio of the number of successful runs to the total number of runs of the algorithm, in addition, the number of generations in which the algorithm was stopped is estimated.

The work of the canonical GA involves the generation of the initial population from a given number of chromosomes, with the help of selection, an intermediate array is formed, from which 2 parents are randomly selected. Next, a single-point crossover is made, and created two descendants mutate (single-point mutation) with a given probability. Mutated descendants take the place of their parents. The process continues until the completion criterion of the algorithm is reached. Computational experiment has shown (tab. 1) that this algorithm demonstrates low stability, slowly converges and is not effective in time due to the large number of calls to the objective function to solve the problem of selecting the optimal construction of road pavement.

Table 1. Results of the computational experiment.

	Canonical genetic algorithm	**Elitism**	**Rank based selection**	**Tournament selection**	**Sampling with partial replacement of the population**	**Hybrid algorithm**
Stability	**0.015**	**0.8**	**0.82**	**0.4**	**0.87**	**0.7**
Average operating time	8.05	15.37	13.58	2.11	16.96	3.9
Average number of evaluation construction options	2784	3578.82	4000	691.2	4000	997.6
Average number of calls to the target function	955.89	1221.83	3535.62	86.45	3535.8	47.9
Average generation of the result GA	69.6	89.635	100	17.28	100	12.47

In the canonical genetic algorithm, the fittest individuals do not always pass on to the next generation. This means that the new population does not always contain the chromosome with the highest fitness function value from the previous population. Elite strategy is used to prevent the loss of such individuals. This individual is guaranteed to be included in the new population. The use of an elitist strategy (table. 1) improves the stability of the algorithm for the task of optimizing the pavement construction.

In the canonical genetic algorithm, the selection of chromosomes in the parent pool is random, replacing the method of random selection by ranked selection (table. 1) increases the stability of the algorithm, making the search for the best solution directional. In this selection, individuals of the population are ranked by the values of their fitness function. This can be represented as a sorted list of individuals, ordered in the direction from the most adapted to the least adapted, in which each individual is assigned a number that determines its place in the list and called the rank. The number of copies of each individual introduced into the parent population is calculated by a priori given function depending on the rank of the individual.

In addition to the ranked selection, tournament selection is possible, where all individuals of the population are divided into subgroups, followed by the choice of individuals with the best fitness in each of them. The population is divided into subgroups of 2 individuals each. Computational experiment has shown (tab. 1) that tournament selection allows to reduce the amount of calculations, but the stability of the algorithm is reduced.

As a result of the limitations of the problem a large number of chromosomes is unsuitable, i.e. such constructions do not meet the criteria of strength according to the normative documentation. To prevent premature degeneration of the population, a genetic algorithm with partial replacement of the population, otherwise called a genetic algorithm with a fixed state was used. This algorithm is characterized by the fact that some of the best chromosomes in the population goes into the next generation without any changes. The reliability of the algorithm increases, but the amount of calculations increases in comparison with the elite strategy [9–10].

One of the problems of using evolutionary computations to optimize multiextremal functions is to get into a local extremum. To solve this problem, it is possible to use hybrid genetic algorithms [11–13]. The idea of hybrid algorithms is to combine a genetic algorithm with some other classical search method suitable for solving this problem. In each generation, all the generated descendants are optimized by the selected method and then entered into a new population. Thus, it turns out that each individual in the population reaches a local optimum, near which it is located. Further being manufactured for normal GA operations: selection of pairs of parents, crossover and mutation. In practice, hybrid algorithms are very successful. This is due to the fact that the probability of getting one of the individuals in the area of global extrema is usually high. After optimization, such an individual will be the solution to the problem. This approach combines the ability to consider multiple solutions with the accuracy of iterative methods to achieve faster convergence, as well as improving the reliability of the algorithm. Thus, hybrid algorithms show a sufficiently high reliability, allowing quickly to find the optimal or close to the optimal solution of the problem [14].

5 Conclusion

The use of optimization models and software that implement optimization methods and algorithms in the calculation of pavement allows specialists to eliminate possible errors that arise in the traditional method of construction. Due to the consideration of a large number of options, it is possible to optimize the construction of pavements according to the specified criteria.

Methods of mathematical modeling are a powerful research tool that allows to conduct computational experiments of any complexity, to search for optimal parameters of the system, to identify patterns of behavior of the object under study. In the course of the computational experiment methods of coordinate descent, multi-start method, dynamic programming method were considered. It is proved the efficiency of application of evolutionary computation in combination with exhaustive search methods to optimize the construction of pavements for technological quarries roads, according to the criterion the total value of the materials.

References

1. R. Impagliazzo. *A personal view of average-case complexity* (Gazzone, Rome, 1995)

2. Michael Shi Hao-Jun, Tu. Shenyinying, Xu. Yangyang, Yin. Wotao, UCLA CAM, **16,** 67 (2016)
3. R. Marti, C. Ribeiro Celso, Mauricio G. C. Resende, Multi-start methods for combinatorial optimization, **226:1**, 8 (2013)
4. R. Marti, R. Aceves, M.T. Leon, J.M. Moreno-Vega, and A. Duarte. Intelligent Multi-Start Methods, **27,** 2 (2018)
5. Moshe Sniedovich, *Dynamic Programming Foundations and Principles Second Editio* (University of Melbourne Melbourne, Australia, 2011)
6. A. Lew, H. Mauch, Studies in Computational Intelligence, **3,** 8 (2007)
7. D. P. Bertsekas, *Dynamic Programming & Optimal Control,* (Massachusetts Institute of Technology, Boston, 1995)
8. A. Laaksonen, *Competitive Programmer's Handbook* (Massachusetts Institute of Technology, Boston, 2018)
9. D. Bednárek, M. Brabec, M. Kruliš, Improving matrix-based dynamic programming on massively parallel accelerators, **64,** 175 (2017)
10. K. Oliver, Studies in Computational Intelligence, **67,** 9 (2017)
11. E.V. Vasileva, V.S. Doroganov, A.B. Piletskaya, Coke and Chemistry, **60:9,** 356 (2017)
12. E.V. Vasileva, V.S. Doroganov, A.B. Piletskaya, Coke and Chemistry, **62:2,** 47 (2019)
13. E.V. Vasileva, V.S. Doroganov, A.B. Piletskaya, Coke and Chemistry, **62:2,** 40 (2019)
14. S. Ganjefar, M. Tofighi, Engineering Applications of Artificial Intelligence, **65,** 346 (2017)

Accounting for the Steel Lining Influence on the Results of Solid Rock Resistivity Probe from Development Working Boundary

Vadim Ivanov[1], *Anna* Dyagileva[2], and *Seroni* Anyona[3]

[1]AO "Scientific Center VostNII", Kemerovo, Russian Federation
[2]T.F. Gorbachev Kuzbass State Technical University, 650000, 28 Vesennya st., Kemerovo, Russian Federation
[3]Jomo Kenyatta University of Agriculture & Technology, Mining, Materials & Petroleum Engineering Department, P.O. Box 62000 – 00200, Nairobi, Kenya

Abstract. When solving problems of predicting the hazards of solid rock bump and outburst in development and stope workings, the resistivity probe methods are often used, in particular, the four-probe method, in which electrodes are placed on the roof or the sill of the mine. The presence of steel lining in the mine significantly affects the measurement results until the complete collapse of electric field, so the electrodes are embedded into the rock. To obtain the true value of rock resistivity, it is necessary to make corrections due to the metal lining influence. This paper is devoted to the determination of these correction factors.

1 Introduction

Currently, the development of mining is associated not only with the improvement of mining equipment [1-3] and geotechnology [4-9], including mineral enrichment [10-12], but also with the introduction of information technologies [13-17]. This includes complex modeling of sustainability of rock arrays.

2 Materials and Methods

Consider a three-layer environment (Fig. 1), in which the layers are numerated as 1, 2, 3. The layer 3 is the rock layer enclosing the coal seam, the h-thickness layer 2 is a steel lining, the layer 1 is the roadway and the host rocks.

Let the current sources A and *B*, as well as the receiving electrodes *M* and *N* be located on the same line and be separated from the boundary of the layer 2 at a distance of z_1 (see Fig. 1). The probe arrangement is symmetric (see Fig. 1). Select the beginning of the cylindrical coordinate system in the current source *A* and direct the *z* axis in the direction of the layer 1. As is known [18], the solution of the Laplace equation in cylindrical coordinate system is expressed as follows

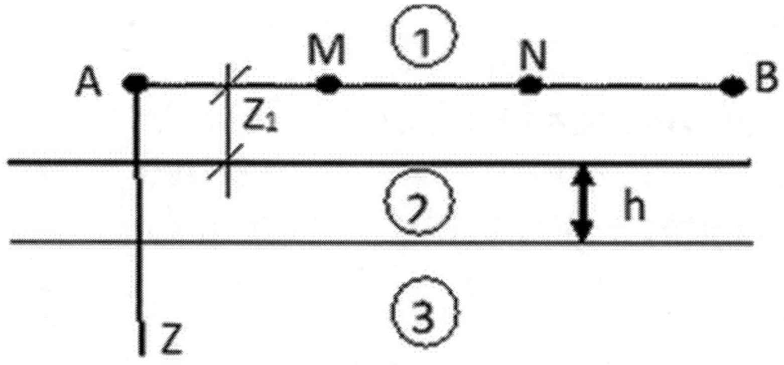

Fig. 1. Scheme of the calculation of the correction factor to the apparent specific resistivity when probing a solid rock from a steel-lined development working.

$$\begin{aligned}
u_1^{(3)} &= \frac{\rho_1 I}{4\pi}\int_0^\infty A_1 \exp(-m\cdot z) J_0(m\cdot r)\,dm; \\
u_2^{(3)} &= \frac{\rho_2 I}{4\pi}\left[\int_0^\infty A_2 \exp(-m\cdot z) J_0(m\cdot r)\,dm + \int_0^\infty B_2 \exp(m\cdot z) J_0(m\cdot r)\,dm\right]; \\
u_3^{(3)} &= \frac{\rho_3 I}{4\pi}\left[\int_0^\infty \exp(-m\cdot |z|) J_0(m\cdot r)\,dm + \int_0^\infty B_3 \exp(m\cdot z) J_0(m\cdot r)\,dm\right],
\end{aligned} \tag{1}$$

where number 3 in the potential's superscripts means the layer's number, in which the current source A is located, the subscripts correspond to the potentials of the corresponding layers, $\rho_{1,2,3}$ – specific resistivity of the layers; I – the source current; $J_0(m\cdot r)$ - the zero-order Bessel function; $A_{1,2}$, $B_{2,3}$ - constant coefficients to be determined from the boundary conditions of the problem.

$$u_3^{(3)}\Big|_{r\to 0, z\to 0} = \frac{\rho_3 I}{4\pi\sqrt{r^2+z^2}}; \quad u_3^{(3)}\Big|_{z\to -\infty} = 0; u_1^{(3)}\Big|_{z\to \infty} = 0; \tag{2}$$

$$u_3^{(3)}\Big|_{z=z_1} = u_2^{(3)}\Big|_{z=z_1}; u_1^{(3)}\Big|_{z=z_1+h} = u_2^{(3)}\Big|_{z=z_1+h};$$

$$\frac{1}{\rho_3}\frac{\partial u_3^{(3)}}{\partial z}\Big|_{z=z_1} = \frac{1}{\rho_2}\frac{\partial u_2^{(3)}}{\partial z}\Big|_{z=z_1}; \frac{1}{\rho_2}\frac{\partial u_2^{(3)}}{\partial z}\Big|_{z=z_1+h} = \frac{1}{\rho_1}\frac{\partial u_1^{(3)}}{\partial z}\Big|_{z=z_1+h}. \tag{3}$$

3 Results and Discussion

The first three boundary conditions automatically hold by virtue of writing the solution in the form (1). Substituting equations (1) into the remaining boundary conditions (2.3), we obtain a system of equations for arbitrary constants $A_{1,2}$, $B_{2,3}$:

$$\begin{aligned} &1 + B_3 \exp(2\,\mathrm{mz}_1) = \frac{\rho_2}{\rho_3}\left[A_2 + B_2 \exp(2\,\mathrm{mz}_1)\right]; \\ &A_1 = \frac{\rho_2}{\rho_1}\left[A_2 + B_2 \exp(2\,\mathrm{m}(\mathrm{z}_1 + \mathrm{h}))\right]; \\ &A_1 = A_2 - B_2 \exp(2\,\mathrm{m}(\mathrm{z}_1 + \mathrm{h})); \\ &\mathrm{B}_2 \exp(2\,\mathrm{mz}_1) - \mathrm{A}_2 = B_3 \exp(2\,\mathrm{mz}_1) - 1. \end{aligned} \tag{4}$$

Introduce the layers' reflection coefficients by the formulas

$$k_1 = \frac{\rho_2 - \rho_1}{\rho_2 + \rho_1}, k_3 = \frac{\rho_3 - \rho_2}{\rho_3 + \rho_2}. \tag{5}$$

Then the solution of the system of equations (4) is given by

$$\begin{aligned} &A_2 = \frac{2\exp(2\,\mathrm{mh})}{k_1 k_3 + \exp(2\,\mathrm{mh})}; B_2 = -\frac{2k_1 \exp(-2\,\mathrm{mz}_1)}{k_1 k_3 + \exp(2\,\mathrm{mh})}; \\ &B_3 = \frac{(\mathrm{k}_1\mathrm{k}_3 - 2\,\mathrm{k}_1)\exp(-2\,\mathrm{mz}_1) - \exp(2\,\mathrm{m}(\mathrm{h} - \mathrm{z}_1))}{k_1 k_3 + \exp(2\,\mathrm{mh})}; \\ &A_1 = \frac{2(1 + \mathrm{k}_1)\exp(2\,\mathrm{mh})}{k_1 k_3 + \exp(2\,\mathrm{mh})}. \end{aligned} \tag{6}$$

Since the specific resistivity of the layer 2 (steel lining) is zero, then

$k_1 = -1; k_3 = 1.$

It follows that $B_3 = -\exp(-2\,\mathrm{mz}_1)$.

Substituting this coefficient into the solution for the potential of the third layer gives

$$u_3^{(3)} = \frac{\rho_3 I}{4\pi}\left[\int_0^\infty \exp(-\mathrm{m}|z|)\mathrm{J}_0(\mathrm{m}\,\mathrm{r})\mathrm{dm} - \int_0^\infty \exp(-\mathrm{m}\,\mathrm{z}_1)\mathrm{J}_0(\mathrm{m}\,\mathrm{r})\mathrm{dm}\right], \tag{7}$$

From where it is obvious that the potential of the third layer does not depend on the thickness of the second layer, i.e. does not depend on the thickness of steel lining

Both integrals in equation (7) are Weber integrals and are easily evaluated. Suppose that the solid rock (layer 3) is probed practically from the border of layers 2 and 3, the depth of the electrodes (Fig. 1) in layer 3 is equal to z_1, the distance from current sources A, B to measurement points M , N is much larger than the coordinate Z , that is $r = z$. Then the solution (7) can be written as follows

$$u_3^{(3)} = \frac{\rho_3 I}{4\pi r}\left[1 - 1/\sqrt{1 + (2z_1/r)^2}\right]. \tag{8}$$

This solution shows that when the depth of the electrodes is equal to zero, i.e. $z_1 = 0$, the field completely collapses; therefore, to eliminate the effect of metal lining, it is necessary to embed the electrodes in the roof (sill) of the roadway.

Let's now calculate the apparent resistivity of the layer 3 when it is probed by the symmetric four-electrode arrangement shown in Fig. 1, we obtain the potential difference between the electrodes M , N

$$\Delta u_3^{(3)} = u_{3M}^{(3)} - u_{3N}^{(3)} = \frac{\rho_3 I}{4\pi}\begin{bmatrix} 1/AM - 1/BM - 1/AN + 1/BN - 1/\sqrt{AM^2 + 4z_1^2} + \\ 1/\sqrt{BM^2 + 4z_1^2} + 1/\sqrt{AN^2 + 4z_1^2} - 1/\sqrt{BN^2 + 4z_1^2} \end{bmatrix}. \tag{9}$$

Taking into account that

$AM = BN, AN = BM, AN - AM = BM - BN = MN,$ will get

$$\Delta u_3^{(3)} = \frac{\rho_3 I}{2\pi}\left[1/AM - 1/AN - 1/\sqrt{AM^2 + 4z_1^2} + 1/\sqrt{AN^2 + 4z_1^2}\right]. \tag{10}$$

According to the well-known resistivity prospecting formula [1], we consider the apparent specific resistivity of the layer 3

$$\rho_k = \frac{2\pi\Delta u_3^{(3)}}{I(1/AM - 1/AN)} = \rho_3\left[1 - AN/MN\sqrt{1 + (2z_1 AM)^2} + AM/MN\sqrt{1 + (2z_1 AN)^2}\right] = C_n\rho_3.$$

4 Conclusion

Thus, in order to obtain the true value of the specific resistivity of the host rocks, it is necessary to divide the resulting apparent specific resistivity by the correction factor C_n

$$C_n = \left[1 - AN/MN\sqrt{1 + (2z_1 AM)^2} + AM/MN\sqrt{1 + (2z_1 AN)^2}\right]. \tag{11}$$

The last formula includes the distances between the electrodes of a four-probe symmetric arrangement and the depth magnitude of the electrodes (see Fig. 1).

References

1. S. Zhironkin, S. Demchenko, G. Kayachev, E. Taran, O. Zhironkina, E3S Web of Conferences, **105**, 03008 (2019)

2. S. Zhironkin, M. Gasanov, G. Barysheva, K. Kolotov, O. Zhironkina, E3S Web Conf., **15**, 03012 (2017)
3. S.A. Zhironkin, K.A. Kolotov, A.E. Genin, F.V. Agafonov, S.A. Kovalevsky, IOP Conf. Ser.: Earth Environ. Sci., **50:1**, 012011 (2017)
4. T. Gvozdkova, M. Tyulenev, S. Zhironkin, V. A. Trifonov, Yu. M. Osipov, IOP Conf. Ser.: Earth Environ. Sci., **50:1**, 012010 (2017)
5. M. Tyulenev, S. Zhironkin, E. Tyuleneva, S. Anyona, M. Hellmer, Coal International, **265:3**, 30-34 (2017)
6. O. Litvin, M. Tyulenev, S. Zhironkin, S. Prokopenko, Acta Montanistica Slovaca, **22:2**, 146-152 (2017)
7. E.A. Razumov, Journal of Mining and Geotechnical Engineering, **2(5)**, 56-77 (2019). DOI: 10.26730/2618-7434-2019-2-56-77
8. M. Tyulenev, O. Litvin, S. Zhironkin, M. Gasanov, Acta Montanistica Slovaca, **24(2),** 88-97 (2019)
9. M. Tyulenev, S. Markov, M. Cehlar, S. Zhironkin, M. Gasanov, Acta Montanistica Slovaca, **23:4**, 368-377 (2018)
10. Ya.V. Kuskova, N.A. Lipnitsky, E3S Web of Conferences, **41**, 01005 (2018)
11. L.A. Vaisberg, Ya.V. Kuskova, Obogashchenie Rud, **4**, 54-60 (2017)
12. V.Yu. Bazhin, V.B. Kuskov, Ya.V. Kuskova, Ugol**, 4**, 50-54 (2019)
13. M.A. Tyulenev, S.O. Markov, M.A. Gasanov, S.A. Zhironkin, Geotech. Geol. Eng., **36:5**, 2789-2797 (2018)
14. F. Abu-Abed, N. Borisov, E3S Web of Conferences, **21**, 01019 (2018)
15. F.N. Abu-Abed, D.V. Martynov, A.V. Ivanova, R.V. Dopira, R.Y. Kordyukov, ARPN Journal of Engineering and Applied Sciences, **11:16**, 9636-9645 (2016)
16. F. Abu-Abed, A. Khabarov, E3S Web of Conferences, 41, 01028 (2018)
17. F. Abu-Abed, E3S Web of Conferences, 41, 01025 (2018)
18. V.N. *Dakhnov Interpretation of the results of geophysical measurements of well logs* (Nedra, Moscow, 2015)

Development of the Digital Model for the Selection of Mine Working Support in Complex Mining and Geological Conditions

Aleksandr Kopytov[1,*], and *Vladimir* Pershin[1]

[1] T.F. Gorbachev Kuzbass State Technical University, 650000 28 Vesennya st., Kemerovo, Russian Federation

Abstract. The methodological basis for the development of a multi-factor digital model for selecting efficient and safe mine working support for given mining and geological conditions on the basis of the assessment of rock mass stability conditions depending on its fracturing and stress state when mining iron ore deposits of Gornaya Shoria is presented. The need to solve fundamentally new problems related to the application of various types of support systems is due to the changeover of mines to ore mining at great depths and classifying deposits as liable to rock-bumps. It is a very important factor, because a decrease in the stability of workings is associated with a high level of stresses in the deep levels being worked out, increased by bearing pressure, which ultimately leads to an increase in the cost of maintaining them. The developed multi-factor digital model "EvrazrudaKrep", which is based on the results of the analysis of guidelines and instructions of the institutes such as All-Russian Research Institute of Mining Geomechanics and Survey, East Research Institute of Ore Mining, Mining Institute of Siberian Branch of Russian Academy of Science, T.F. Gorbachev Kuzbass State Technical University, as well as the mine working support experience in Russia and abroad, allows to quickly solve the problem of choosing a support and improving mining safety in complex mining and geological conditions.

1 Introduction

There are Temir-Telbes (Kaz, Kemerovo Region, Western Siberia, Russia) and Kondoma groups of iron ore deposits in Gornaya Shoria. The most promising are Kondoma group deposits - Tashtagol and Sheregesh (Fig. 1) [1].

In Tashtagol, stoping operations have reached a depth of 700 m, development works have reached 900 m and are carried out on five levels (–70)÷(–350) m, sinking operations are carried out on six levels of the mine. In the course of the year, more than 1,000 m of permanent workings, more than 2,000 m of development workings and about 3,000 m of entries are sunk, 2.5-3 million tonnes of ore are mined.

Mining operations at Sheregesh deposit are carried out at six levels of the mine (+525) – (+115) m. Stoping operations have reached a depth of 470 m, development works - 600 m.

* Corresponding author: L01bdv@yandex.ru

During the year, more than 2,000 m of development workings and more than 12,500 entries are sunk in the deposit. Up to 4.6 million tons of ore are mined [2, 3].

Deposits are classified as liable to rock-bumps [4].

It was found by the research of East Research Institute of Ore Mining that the maximum main stresses in Tashtagol and Sheregesh deposits have north-west direction and are 2.5-3 times greater than the roof strata weight γH.

In the course of the year, there are 3-4 bulk blasts with seismic energy of 10^8-10^9 J in the mines [2, 5].

Under these conditions, the choice of an effective type of mine working support during the sinking cycle, taking into account its subsequent operation, is the most important task for ensuring the required safety of geotechnology for mining iron ore deposits in Gornaya Shoria [6, 7].

2 Development of a digital model for the selection of mine working support

Long-term practice shows that the basic variables affecting the choice of a support are: the stability of rock and ore masses, the hardness factor *f* according to the Protodyakonov scale of hardness, the structural weakness rate K_{sw}, the average distance between fractures L_{av}, the size of the zone of inelastic strain of rock mass around a mine working $l_в$.

The stability of rocks and ores is determined by the strength characteristics of rock types, the intensity of fracturing and the original stress state of rock mass [8].

The original parameters are determined by the geological survey. The main changes in rock and ore fracturing, the propagation of joint systems, and their association with certain rock types are established. The degree of stability in disjunctive fault lines is due to the zones of crushing, shear, and foliation. Fault zones with poor case-hardening (clay material, carbonates, iron hydroxides, chlorite) are particularly hazardous (unstable).

When documenting by the geological service, it is necessary to identify areas of rock mass weakening associated with actively manifested superimposed carbonization processes and low-temperature hydrothermal metamorphism (carbonatization, chloritization, seritization, hematitization, and iron hydroxiding). It should be noted that silicification does not reduce the stability of rocks.

The stability of rock mass around a mine working is directly dependent on fracturing of rocks and ores. The main types of joint systems are continuous, discontinuous and chaotic. All fractures, dividing rock into blocks, ranging in size from ten centimeters to several meters, should be taken into account. Tine fractures within the block are not taken into account. To assess fracturing, it is necessary to determine the parameters of the main joint systems — width, type of filling (if any), angle of incidence [8]. By type of filling fractures are divided into three types:

– tight and welded fractures, fractures with quartz and quartz carbonate filling;

– fillings are products of low-temperature hydrothermal dynamo-metamorphism (chlorite, sericite, carbonate, hematite, iron hydroxides);

– fractures in fault lines with clay gauge; fractures in crushing zones with poor case-hardening (clay material, especially the watered one) and fractures in shear, and intensive foliation zones with milonitization and polish faults.

The effect of fracturing on the stability of rock mass around a mine working is taken into account by means of the structural weakness rate K_{sw} and the average distance between fractures L_{av}.

Several (at least 2-3) joint systems are found almost throughout the mine-take. The number of joint systems is determined in the face in three different segments horizontally and vertically in 2-3 m long segments, depending on the cross-section of working. The average distance between fractures is determined by the formula

$$L_{av} = \frac{1}{6}\left(\frac{L_1^{\mathrm{h}}}{n_1^{\mathrm{h}}} + \frac{L_2^{\mathrm{h}}}{n_2^{\mathrm{h}}} + \frac{L_3^{\mathrm{h}}}{n_3^{\mathrm{h}}} + \frac{L_1^{\mathrm{v}}}{n_1^{\mathrm{v}}} + \frac{L_2^{\mathrm{v}}}{n_2^{\mathrm{v}}} + \frac{L_3^{\mathrm{v}}}{n_3^{\mathrm{v}}}\right) \quad (1)$$

where L_1^{h} and L_1^{v} – the distance between the extreme fractures in horizontal and vertical segments, m; n_1^{h} and n_1^{v} – the number of fractures in the horizontal and vertical segments.

As a rule, the average distance between fractures is taken according to the most intense fracturing.

Fracturing measurements are taken for all rock and ore types with a distance between areas of no more than 30-40 m along strike. Additional measurements are made in various structural units. The results of measurements are recorded in the book in the following form (Table 1).

Table 1. The measurement recording book.

Segment referencing	**Rock, ore character-istic**	**The results of measurements**											
		L_1^{h}	n_1^{h}	L_2^{h}	n_2^{h}	L_3^{h}	n_3^{h}	L_1^{v}	n_1^{v}	L_2^{v}	n_2^{v}	L_3^{v}	n_3^{v}

For example, ores and rocks of the Sheregesh deposit are characterized by the following fracture intensity parameters (Table 2).

Table 2. Rock and ore fracturing classification.

Mass fracture intensity	***L_{av}*, m**	***K_{sw}***
Intensive fracturing	до 0,1	≤0,1
High fracturing	0,10–0,20	0,1–0,2
Medium fracturing	0,20–0,50	02–04
Below-medium fracturing, mostly tight fractures	0,50–0,60	0,5–0,7
Low fracturing, sparse tight fractures or lack thereof	>0,60	0,8–1,0

The changeover of the mines of Gornaya Shoria to mining ore bodies at great depths and classifying of parts of the deposits as burst-prone and liable to rock bumps caused the need to solve fundamentally new problems of the conditions of using different support types and systems [9, 10].

According to the deposits, there are workings that are outside and within the zone of stoping influence. Depending on the location of workings relatively to the direction of action of the main components, the stress field and the stability of workings will differ.

The original stress state of rock mass in quantitative terms is characterized by the values of vertical and horizontal stresses - along strike and transverse.

The deposits are characterized by gravitational and tectonic stress fields, in which the values of the horizontal stress components are greater than those of the vertical ones. These stresses are established in iron ore deposits (Table 3). In addition, each field has its own azimuth of the main horizontal stress component $A_{\sigma.max}$ [11].

Table 3. Parameters of gravitational and tectonic stress field in the deposits of Gornaya Shoria

Deposit	Outside the zone of stoping influence				Within the zone of stoping influence			
	σ_1	$\sigma_2 = \sigma_{max}$	σ_3	$A_{\sigma.max}$, degrees	σ_1	$\sigma_2 = \sigma_{max}$	σ_3	$A_{\sigma.max}$, degrees
Kaz	0,4 γH	2,4 γH	γH	85	2,5 γH	3,7 γH	γH	85
Tashtagol – to a depth of 890 m below 890 m	1,3 γH 1,8 γH	2,5 γH 3,0 γH	γH γH	320 350	3,5 γH	5 γH	γH	50
Sheregesh	1,4 γH	2,6 γH	γH	330	3,8 γH	5,2 γH	γH	48

The strength of enclosing rocks surrounding the mine workings is determined by effective tangential shearing stress.

Tangential shearing stress in the walls and the roof of workings can be found from the expression [11]:

$$\tau_{av} = 0,5\left(\sigma_{//} - \sigma_{\perp} + 2\frac{a_w}{h_h}\sigma_{//}\right) \tag{2}$$

where $\sigma_{//}$, $\sigma_{\perp}$ – the value of the horizontal component of stress in rock mass, acting along and perpendicular to the axis of a working; a_w – width of working, m; h_h – height of a working, m.

In the case when the direction of a working coincides with the direction of the maximum horizontal component of stress, tangential stresses in the roof and the walls of a working are calculated respectively from the expressions

$$\tau^r_{av} = 0,5\left(\sigma_1 - \sigma_2 + 2\frac{h_h}{a_w}\sigma_1\right) \tag{3}$$

$$\tau^w_{av} = 0,5\left(\sigma_2 - \sigma_3 + 2\frac{a_w}{h_h}\sigma_3\right) \tag{4}$$

When the direction of a working coincides with the direction of the minimum horizontal component σ_1, tangential stresses in the roof and the walls of a working are calculated respectively from the expressions

$$\tau^r_{av} = 0,5\left(\sigma_2 - \sigma_3 + 2\frac{h_h}{a_w}\sigma_2\right) \tag{5}$$

$$\tau^w_{av} = 0,5\left(\sigma_2 - \sigma_3 + 2\frac{a_w}{h_h}\sigma_3\right) \tag{6}$$

where σ_1 – the minimum horizontal component of stress; σ_2 – the maximum horizontal component of stress, σ_3 – vertical component of stress; $\sigma_3 = \gamma H$.

Tangential shearing stress in the roof and the walls of workings located at an angle to the maximum horizontal component of stress is obtained by the formula

$$\tau_{av} = \sqrt{\frac{\tau^2_{\min.av}}{1 - \varepsilon^2_э \cos^2}} \tag{7}$$

where $\tau_{min.av}$ – the minimum tangential shearing stress in the roof and in the walls of working; ε – eccentricity – geometrical locus; θ – the angle between the maximum horizontal component of stress and the axis of a working, measured from 0 to 90°.

Eccentricity is determined separately for the roof and the walls of a working from the ratio

$$\varepsilon^2 = 1 - \frac{\tau^2_{\min.av}}{\tau^2_{\max.av}} \tag{8}$$

where $\tau_{max.av}$ – tangential shearing stress in the roof and in the walls.

In the case when the width of a working is equal to its height, the minimum effective shearing stress for workings located at an angle to the minimum horizontal component of stress is defined as

$$\tau_{av} = \tau_{\max.av} \cos(\theta) \tag{9}$$

The stability factor of mine workings is determined:
for the roof by

$$K_{stab} = \sqrt{\frac{[\tau_{av}]}{\tau^r_{av}}} \tag{10}$$

for the walls by

$$K_{stab} = \sqrt{\frac{[\tau_{av}]}{\tau^w_{av}}} \tag{11}$$

where $[\tau_{av}]$ – the ultimate shearing strength of enclosing rock, $[\tau_{av}] = K_{sw}\,[\sigma]\,\mu$; τ^w_{av} – the effective shearing stress in the roof of a working; τ^w_{av} – the effective shearing stress in the walls of a working; $[\sigma] = f \times 10$, MPa – the uniaxial compression strength of enclosing rock; μ – Poison's ratio; K_{sw} – the structural weakness rate. The K_{sw} value is determined by the distance between fractures (Table 4).

Table 4. The dependence of the structural weakness rate K_{sw} on rock (ore) fracturing L_{av}.

The distance between fractures L_{av}, m	The structural weakness rate K_{sw}
>1,5	0,9
1,5–1,0	0,8
1,0–0,5	0,6
0,5–0,2	0,4

After calculating the rock (ore) stability factor, the category of rock stability (Table 5), the design and parameters of mine support are determined.

Table 5. The category of rock (ore) stability.

Rock stability factor	Stability degree	Category of stability
>1	strongly stable	I
1,0–0,65	stable	II
0,65–0,45	medium stable	III
0,45–0,25	unstable	IV
>0,25	strongly unstable	V

When choosing a support, it is necessary to consider the category of rock-bump hazard in the deposit. According to the level of rock-bump hazard, the deposits or rock mass areas around workings are divided into two categories: "Hazard" and "No hazard". The category "Hazard" corresponds to the stress state of rock mass around a mine working, at which a rock bump may occur. The category "No hazard" corresponds to a non-hazardous state and does not require measures against rock bump.

3 Results

Based on many years of research and methodological analysis of the interrelationships of parameters during sinking and operation of mine workings under conditions of stress-strain state and rock bump hazard of iron ore deposits in Gornaya Shoria, the multi-factor digital model "EvrazrudaKrep" was developed to select the type of support and its possible detachment depending on the deposit depth, the categories of rock mass stability and the cross-section of a working [12] (Fig. 1)

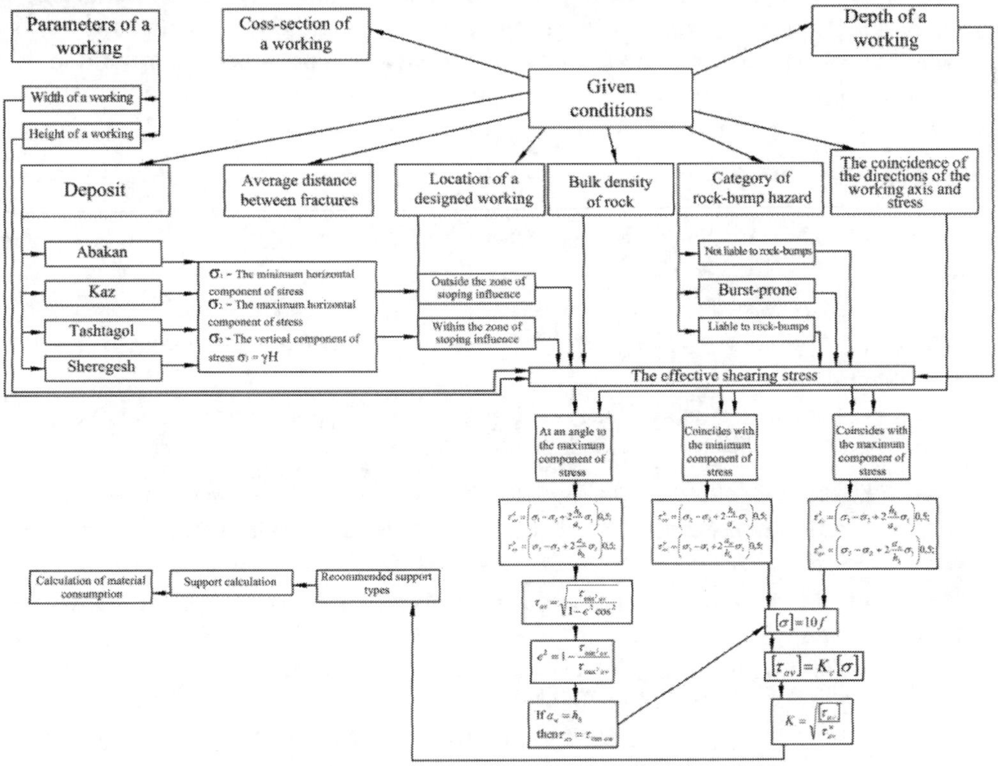

Fig. 1. The multi-factor digital model "EvrazrudaKrep".

The multi-factor digital model "EvrazrudaKrep" is based on the results of the analysis of recommendations and guidelines of the institutes such as All-Russian Research Institute of Mining Geomechanics and Survey, East Research Institute of Ore Mining, Mining Institute of Siberian Branch of Russian Academy of Science, T.F. Gorbachev Kuzbass State Technical University, as well as the experience of using different support types in the Gornaya Shoria mines and other iron ore deposits, including foreign ones. The stability of rocks and ore is determined by the method developed by Siberian Branch of Russian Academy of Science, as it takes into account the stress state of rock mass of a particular deposit. This is a very important factor, because a decrease in the stability of workings is associated with a high level of effective stresses in deep levels being worked out, increased by bearing pressure, increased rock bump hazard of rock mass, which ultimately leads to an increase in the cost of maintaining them.

4 Conclusions

For automated designing of the support construction, electronic models of fastening of mine working supports, mine support systems (cross-section and longitudinal sections, lining units, materials and consumption of materials) were developed. The electronic layout of the support construction design includes: a title page, a sheet - a list of persons informed of and acknowledged; mining geotechnical and engineering data and a report on the rock bump hazard of rock mass, necessary for the automated search for a rational support option; printout results of processing the specified conditions by the "EvrazrudaKrep" program, graphic material.

Using this program, "Methodological guidelines for mine working support and support state monitoring at the "Evrazruda" mine were developed and put into effect by order No. 305 of April 25, 2013 [11]. The manual allows to quickly address the issues of choosing the mine support system in changing mining geological and geodynamic conditions when driving a working, as well as to improve the safety of sinking operations when mining rock-bump hazardous iron ore deposits of Gornaya Shoria.

References

1. V.P. Orlov, B.M. Aleshin, V.M. Alikberov, *Iron ore base of Russia* (Nedra, Moscow, 2007)
2. A.A. Eremenko, V.A. Eremenko, A.P. Gaidin, *Mining, geological and geo-mechanical conditions of iron ore mining in the Altai-Sayan folded area deposits* (Science, Novosibirsk)
3. A.A. Eremenko, A.I. Fedorenko, A.I. Kopytov, Mine working drivage and support in bump-hazardous zones of iron ore deposits (Science, Novosibirsk, 2008)
4. A.I. Kopytov, M.D. Voitov, S.S. Morozov, Bulletin of KuzSTU, **4,** 32-37 (2014)
5. M.A. Tyulenev, S.O. Markov, M.A. Gasanov, S.A. Zhironkin, Geotech. Geol. Eng., **36:5**, 2789-2797 (2018)
6. *Instructions for the safe mining operations in ore, non-metallic deposits, construction sites of underground structures, liable to rock bump (RD 06-329-99)* (Center for Research and Engineering in Industrial Safety of Federal Mining and Industrial Supervision of Russia, Moscow, 2000)
7. A.I. Kopytov, A.A. Lebedev, B.A. Utrobin, Bulletin of KuzSTU, **5,** 10-15, (2017)
8. M. Cehlár, P. Varga, Z. Jurkasová, M. Pašková, Acta Montanistica Slovaca, **15: SPECIAL ISSUE 2**, 132-138 (2011)

9. A.A. Eremenko, M.V Kurlenya, Journal of Mining Science, **1,** 3-31 (1990)
10. M.V. Kurlenya, V.M. Seryakov, A.A. Eremenko, *Anthropogenic geo-mechanical stress fields* (Science, Novosibirsk, 2005)
11. S.M. Miliy, M.A. Rozhenko, K.A. Prokopenko, G.I. Korotkova, Journal of Mining and Geotechnical Engineering, **2:5,** 4-18 (2019) DOI: 10.26730/2618-7434-2019-2-04-18
12. A.I. Kopytov, G.K. Klyukin, S.S. Morozov, T.E. Tripus, Bulletin of KuzSTU, **6,** 52–53 (2012)

Reseach Old Rubber Tires Utilization to Increase the Life of Support in the Longwall at Quang Ninh Vietnam

Phi Hung Nguyen[1,*], *Cao Khai* Nguyen[1], *Duc Thang* Pham[2], and *Van Kien* Ta[2]

[1]Ha Noi University of Mining and Geology
[2]Quang Ninh University of Industry, Quang Ninh, Vietnam

Abstract. In longwall mining where use frame support and control roof by full caving; rocks could impact the support, damage some its details and reduce support's longevity. Using waste, elastic, and deformable capability materials (like old truck tires) not only reduce impulse which impact the support fatally and increase support's longevity, but also contribute to protect the environment.

1 Introduction

Use the frame support in control roof longwall by full-caving usually occurring the phenomenon of rock impacting into the stone shields during fire caving. The rock's impulse will reduce the life span of the frame support, especially the upper and lower stone shield [1, 2, 3]. How to increase the life span of the stone shield?

Fig. 1. Hydraulic frame support

* Corresponding author: hunguni@gmail.com

 https://doi.org/10.1051/e3sconf/201913401010

During mining operations, there are some waste materials such as rubber tires, conveyor belts... being materials that can be utilized to reduce the stone impulse to impact on the resistance and increase the life of the shield [4].

Fig. 2. Old truck tires

2 Causes of frame support damage

After caving, stone-impulse, one part directly destroys the deformable areas of the front and rear rock shields, and the rest is rapidly propagated through hard-to-deformation areas (where the rock shields contact with hydraulic support) to the hydraulic support, causing the support to be pushed horizontally. The impulse impacting on the shields by the full-caving cycle will cause fatigue breakdown phenomenon [5, 6,7, 8].

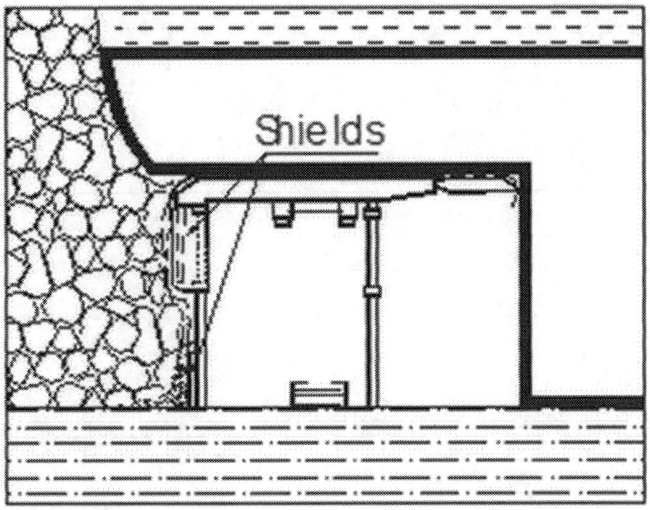

Fig. 3. Process of rock impacting on frame support after caving (Initial status)

 https://doi.org/10.1051/e3sconf/201913401010

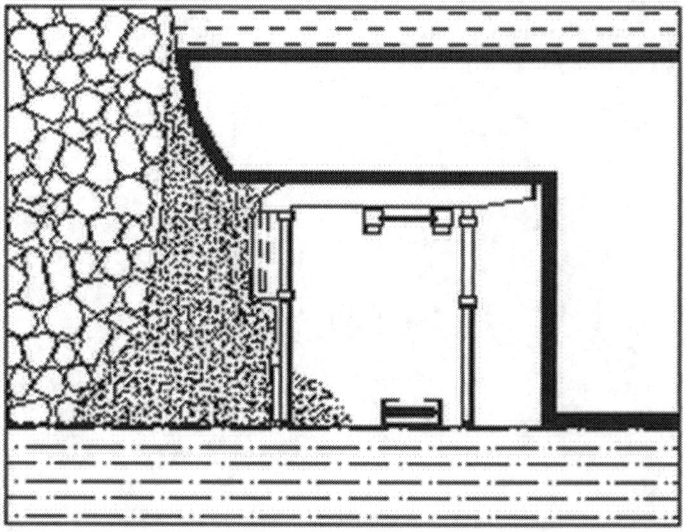

Fig. 4. Process of rock impacting on frame support after caving (Reclaim and preparation for caving)

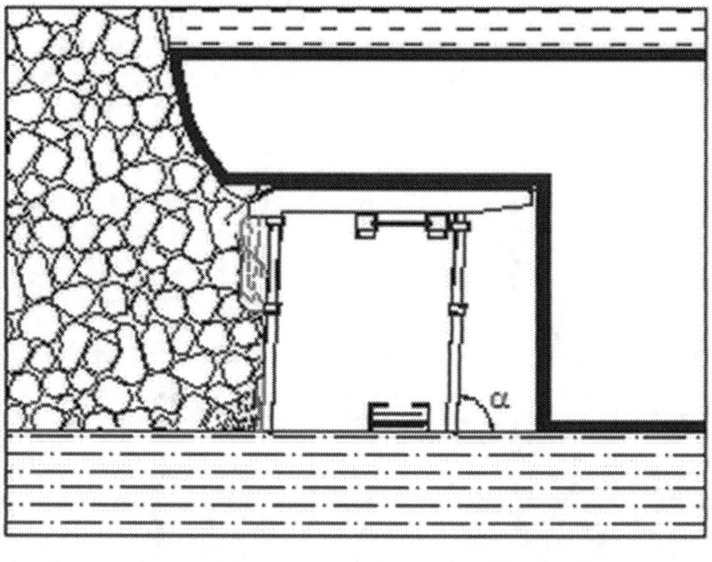

Fig. 5. Process of rock impacting on frame support after caving (Status after caving)

Assumption of impulse problem is in Fig.6.

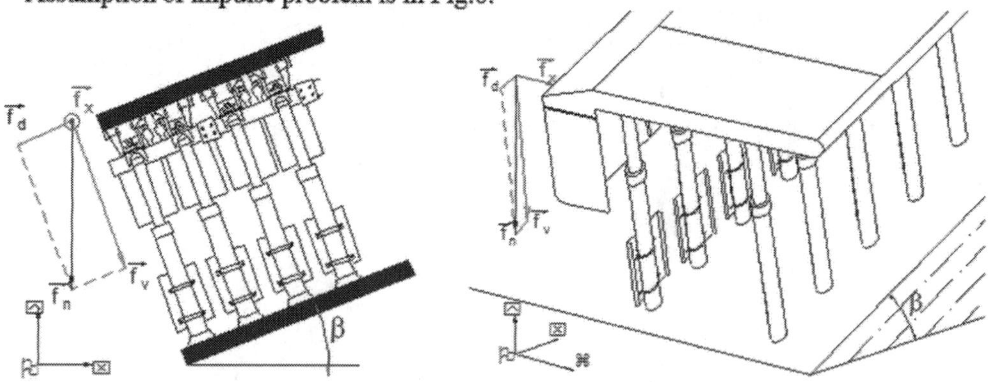

Fig. 6. Analysis of the impact force of the rock on frame support during caving

When rock contacting ground, rock's force as in (1):

$$\vec{F_n} = \vec{F_d} + \vec{F_v} \tag{1}$$

Where: $\vec{F_n}$ - Rock's force during ground contact; $\vec{F_d}$ - Formational slope same direction force; $\vec{F_v}$ - Formational slope perpendicular force.

Rock is pressed by $\vec{F_n}$ then broken and causing $\vec{F_x}$ force impacting on frame support. This force's source is perpendicular to $\vec{F_d}$ and $\vec{F_v}$.

According to the conservation of energy law, the highest force of caving rock on the shield of the frame support is equal to the force acting directly on the base of the floor, as in (2).

$$F_{x\max} = F_n \tag{2}$$

F_n can be calculated by the law of momentum conservation, as in (3):

$$\vec{F_n} = \frac{M \cdot \vec{\Delta v}}{\Delta t} = M \cdot \vec{a} \tag{3}$$

Where: M - Volume of caving rock; $\vec{\Delta v}$ - Velocity reduction; Δt - Impact time; $\vec{a}$ - Acceleration of caving rock from grounding to stopping.

Algebra value of $\vec{a}$ can be calculated as in (4):

$$v_1^2 - v_0^2 = 2a \cdot \Delta s \Rightarrow a = \frac{v_0^2}{\Delta s} \tag{4}$$

Where: v_2 - Velocity when rock stops ($v_2 = 0$ m/s); v_1 - Velocity when rock starts to contact support foundation; Δs - The distance of the rocky road (equal to the average settlement of the foundation).

From (3), (4)and apply the law of conservation of mechanical energy, rock's force as in (5):

$$F_n = \frac{M \cdot v_0^2}{2\Delta s} = \frac{W_d}{\Delta s} = \frac{W_t}{\Delta s} = \frac{M \cdot g \cdot h}{\Delta s} = \frac{\gamma \cdot r \cdot h_z \cdot L \cdot g \cdot h}{\Delta s} \tag{5}$$

Where: W_d - Motional energy of rock when contacting ground; W_t - Potential energy of rock when not caving; γ - Volume weight of rock; r - Width of pull; h_z - Height of collapse and breakdown area; L - Breakage length; g - acceleration of gravity; h - Breakage height.

Height of collapse and breakdown area can be calculated by the Morth formula (1954), as in (6):

$$h_z = \frac{v \cdot L}{2 \cdot (1 - v} \tag{6}$$

Where: ν - Poisson ratio.

So the highest pressure on each shield, as in (7):

$$P_{\max} = F_n \cdot S = \frac{\gamma \cdot r \cdot h_z \cdot L \cdot g \cdot h \cdot S}{\Delta s} \tag{7}$$

Where: S - Shield's area.

3 Measures to increase longevity for frame support

The treatment method is to add an impulse absorption zone (IAZ) with effect [9, 10]:

- Reduce initial impact force F_x;

- Scatter the impulse before impacting the shield.

In essence, the IAZ is the peripheral buffer for the components to be protected.

According to the theory of momentum conservation, the force acting on the shield without absorption as in (8):

$$\vec{F} = \frac{m \cdot \vec{\Delta v}}{\Delta t} \tag{8}$$

Where: $\vec{F}$ - Force acting on frame support; $\vec{\Delta v}$ - Reduction of velocity of load bearing object; Δt - Impact time; m - Volume of load bearing object.

When adding the IAZ, :

- Increase the time (Δt) of the initial thrust impacting the shield causing impulse reduction;
- With the structure of the absorption area, the force is divided into different forces of different senses, so the total impulse (perpendicular to the shield) decreases.

According to the law of energy conservation, rock energy causes more deformation of the IAZ, smaller the energy causing deformation, the frame support movement is.

So choose materials that are capable of large deformation as the absorption. However, this material must be reused after each caving. Only the elastic material can satisfy these two conditions.

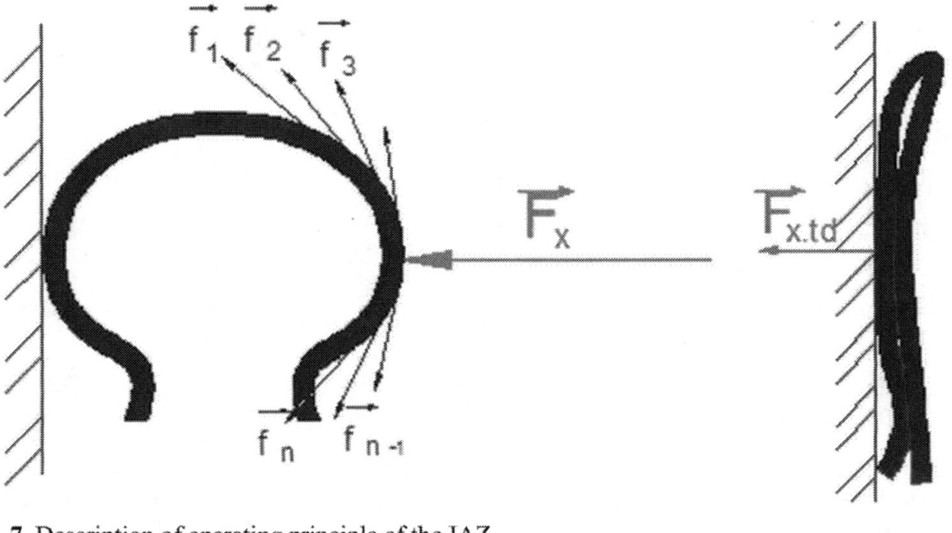

Fig. 7. Description of operating principle of the IAZ

During mine operation as well as in daily life, there are many exhausted elastic materials such as automobile tires, motorcycle tires; conveyors ... These materials are often discarded because they no longer meet the standards of friction, adhesion, but the deformability and elasticity are still very high, so it is entirely possible to use for the IAZ of frame support.

The way to mount this material on the shield of the frame support must ensure:

- Not reduce the bearing capacity of the shield;
- Keep material from falling under rock impact;
- The material has the highest deformability;
- Easy to install, replace.

To meet the above criteria, theyare arranged as in Fig.8 and Fig.9.

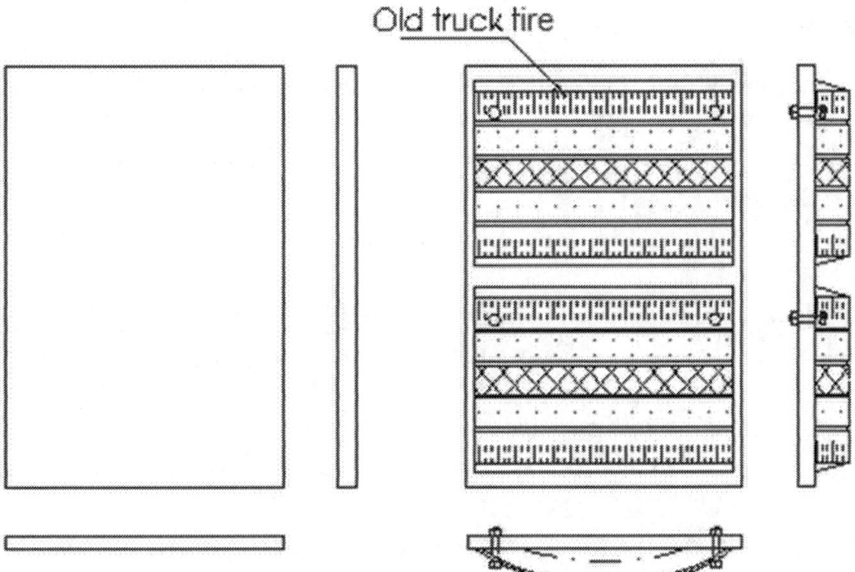

Fig. 8. Installing the IAZ to front shield

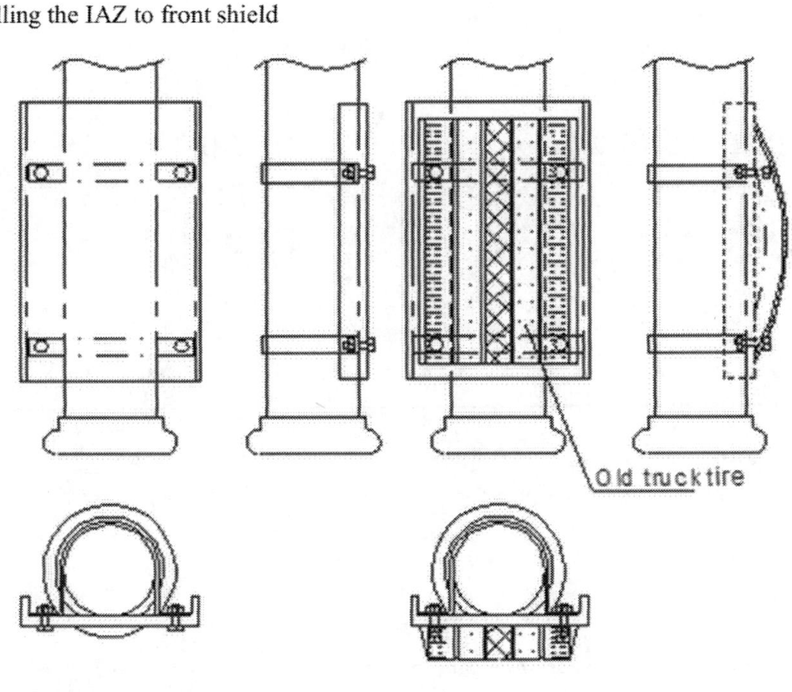

Fig. 9. Installing the IAZ to rear shield

4 Conclusions

This measures are cheap and environmental friendliness. The paper proved that the ability to reduce impulse which impact the support fatally as well as increase support's longevity but didn't calculate how many percent of impact force that IAZ could reduce. Finite Element Method can solve that problem however it's hard for seek parameters of materials. Real test is the method that gives the most accurate results.

References

1. V.V. Kovalenko Mining informational and analytical bulletin Journal, **14,** 267 (2004)
2. Nguyen Phi Hung, Journal of mining and earth sciences, **58**, 80 (2017)
3. Hoang Van Vi. Hanoi.Vinacomin Institute of Mining & Energy Mechanical Engineering (2010)
4. I. Pochinkov, A. Myaskov. Mining informational and analytical bulletin Journal, **5,** 76 (2013)
5. Chen Chong, Yao Enguang, Zhang Zhe. Zhongzhou, Coal, **1**, 4 (2013)
6. Bragin E.P., Vitcalov V.G, Pham Trung Nguyen, Mining informational and analytical bulletin Journal, **6,** 37 (2010)
7. Wang Qingxiong, Ju Jinfeng, Coal Science and Technology, **42**, 125 (2014)
8. Bie Xiaofei, Zhai Xinxian, Zhang Shuai, Coal Science and Technology, **41**, 80 (2014)
9. E. Hoek, Mining Magazine, **282** (1998)
10. L. Martin, Public Health Service, Centers for DiseaseControl and Prevention, National Institute for Occupational Safety and Health, DHHS (NIOSH) Publication, **1** (2000)

The Designing of the New Type of Mine Ventilation Wall

Alexander Nikolaev[1], *Rafail* Gazizullin[2], and *Valery* Nesterov[3]

[1] Perm National Research Polytechnic University, Russia, 614990, Komsomolsky Prospect street, 29
[2] West-Urals Department of Rostechnadzor
(Russian Federal Service for Ecological, Technical and Atomic Supervision), Russia, 614900, Komsomolsky Prospect street, bld, 34 "b"
[3]T.F. Gorbachev Kuzbass State Technical University, 650000 Kemerovo, 28 Vesennya st., Russian Federation

Abstract. Within the underground mining enterprises, in the event of a fire underground, it is necessary to isolate the mining site or mine workings to protect miners from poisoning by flue gases. To do this, the Mine Ventilation Wall (MVW) is used. Its second purpose is to regulate the air distribution underground, to fence sections not designed for ventilating. This survey describes the design of the MVW new type, which will have a number of competitive advantages in regard analogues existing. The results of mathematical modeling of the MVW properties under different conditions are represented.

1 Introduction

Ventilation of underground mining facilities is considered to be the most important technological process, as it is necessary to supply the air into the working area to ensure the possibility of mineral deposits mining. Wherein, the air supply into the mine (ore mine) in the volume required does not guarantee that all mining sites will be supplied with fresh air in the amount necessary to ensure safe working conditions. The reason for this is a large branching of mine workings, as significant part of the air upon moving throughout, is "dissipated" within the space mined. In addition, it is necessary to regulate not only the air distributing amongst underground mine workings to maintain the required operating conditions of miners within mining sites, but also to ensure safety within in the event of a fire in the mine (ore mine), blocking the flue gases spreading [1-4]. The solving of both problems requires the use of devices for rapid and reliable isolation of mine workings, which are taken to be called Mine Ventilation Walls (MVWs).

2 Materials and Methods

Nowadays computer modelling is the main tool for mining operation design [5-6]. It is known that the MVWs have the best properties regarding reliability and quality of insulation [6-9]. This is due to the fact that, inflating, the wall smoothes all the irregularities and roughness of the mine workings, resulting in a maximum area of contact between the walls, soil and roof with the MVW.

However, the air MVWs have got their significant disadvantage. Their construction requires significant amount of air, and hence the energy source for the compressor, as well as significant time costs. As in case of fire and upon MVWs using within the industrial process to control the air distribution, the time factor is considered to be priority. Therefore, the use of air walls, despite their advantages upon isolating the mine workings, is limited.

To solve the problem of rapid and reliable isolation of mine workings, the design of the mine mobile ventilation wall has been developped, as presented in Fig. 1 (the legend: 1 – cloth; 2 – telescopic supports; 3 – rotating masts; 4 – reinforcing frame; 5 – sleds; 6 – locking rings; 7 – pneumohose; 8 – hasps).

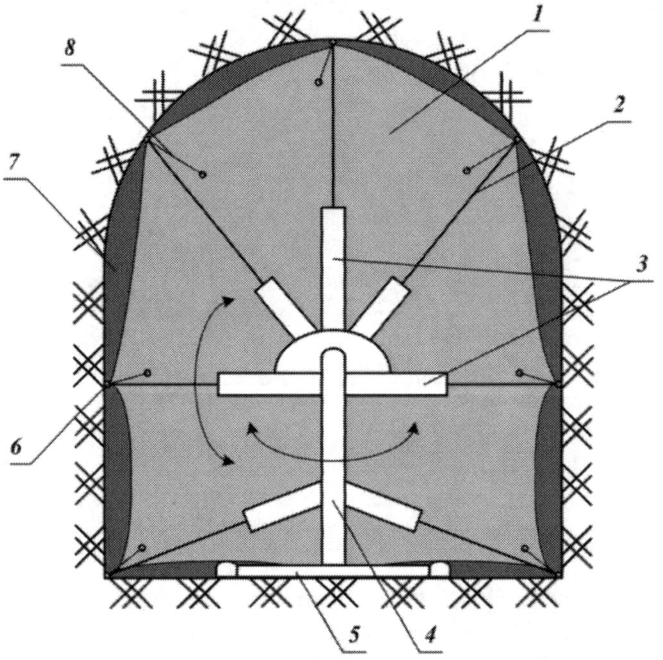

Fig. 1. The Principal Technical Solution Regarding the Design of the Air MVW: 1 – cloth; 2 – telescopic supports; 3 – rotating masts; 4 – reinforcing frame; 5 sleds; 6 – locking rings; 7 – pneumohose; 8 – hasps.

The term for the developping of the new type of MVW is reduced due to the fact that the main part of the production is overlapped by a cloth stretched along its cross section by means of sliding telescopic stops, and the air is pumped into the sleeve only located along the perimetre of the wall (hereinafter – the pneumatic hose). In this case, reliable isolation of the mine workings is provided under a small volume of the air pumped in. This does not require a power supply, since the pumping may be performed via a compressed air cylindre or a low-power compressor (battery-based).

The design of the MVW shall be determined per each individual mining enterprise, depending on the method of mineral deposits mine working. In this case the main idea consists in overlapping the main cross section of mine working with the cloth and pumping of air only into the pneumohose located onto the wall perimetre, as it will persist.

3 Results

The first version of the design of the new type of the MVW is designed for potash mines of the Verkhnekamskoye Deposit of Potassium-Magnesium Salts (VDPMS), which mine workings have got their large cross section and complex configuration. Both of these factors significantly complicate the development of the MVW design, the selection of materials regarding the manufacture of the frame and cloth able to withstand the pressure of the air flow, and as a result, the choice of optimal dimensions and weight of the components is complicated. These problems are solved at the stage of mathematical modelling.

The considered Estimated Cases (ECs) correspond to the following relative overlap of the flow section of mine workings: Ecs No. 1, 2, 3, 4, 5 – with overlap, respectively per 20, 40, 60, 80 and 100 %.

The results of modelling the pressure distribution onto the MVW per all ECs at an air flow rate of 10 m/s are shown in Fig. 2-3.

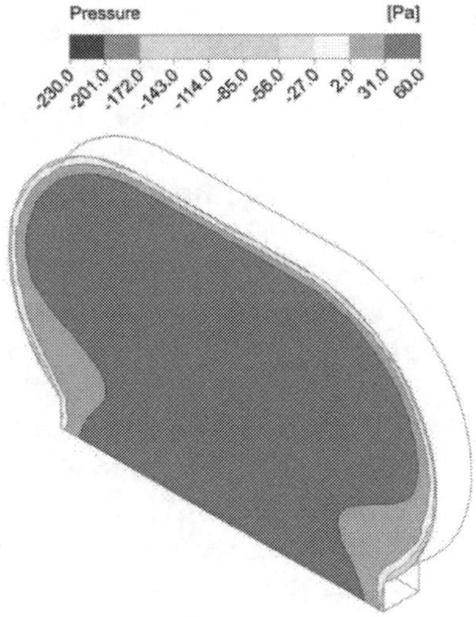

Fig. 2. The Isolines of Excessive Pressure from the Air Incoming Flow for the EC No. 1.

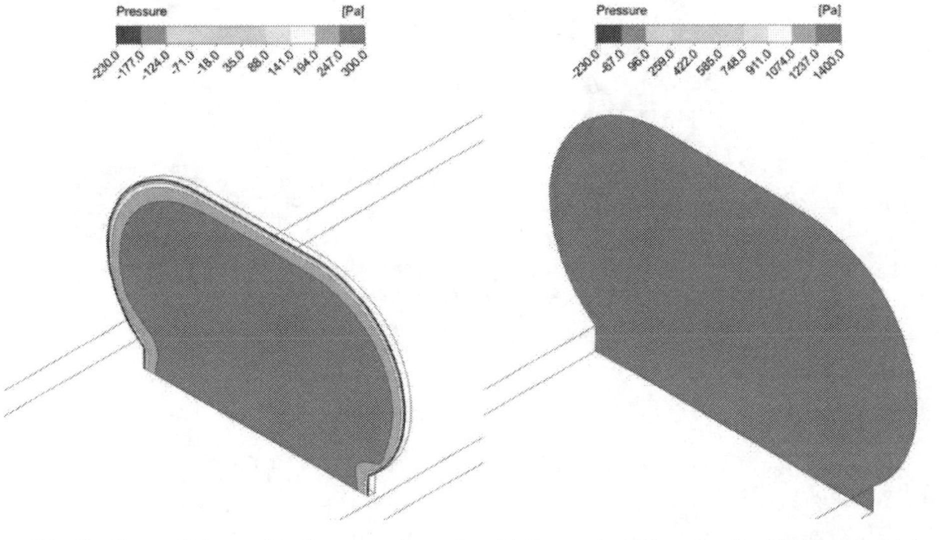

Fig. 3. The Isolines of Excessive Pressure from the Air Incoming Flow for the EC No. 4 and 5.

It follows from Fig. 2-3 that the excessive pressure is distributed unevenly across the MVW. The negative pressure value is established at the edges of the wall due to the uneven flow within the gap through its poorly streamlined shape. The flow, running onto the cloth, is re-transferred and then passes throughout the slit (the classic EC of sharp narrowing of the passage section with its sharp edge). The excessive pressure amount directly onto the wall from the side of the air incoming flow does not exceed 1400 Pa. However, due to the formation of separated flows within the gap and behind the obstacle, where the pressure of the discharge is established, the force acting into the direction of the flow increases significantly. Taking into account the fact that the air flow rate is taken to the maximum rate (according to the Safety Regulations of 10 m/s), the specified pressure value of the air incoming flow may also be considered the maximum and further calculations shall be performed, taking it as the main one.

4 Discussion

The problems need to be solved are related to the pressing of the pneumatic hose along the perimetre of the mine workings and the ability of the MVW to withstand the load of the air flow incoming. To solve these problems, the design of the MVW has been developped, the power frame of which is shown in Fig. 4.

The proposed design is collapsible. At the first stage, the lower part of the MVW is assembled, consisting of the base 1 and the ski sled 2 (Fig. 4). The base 1 is manufactured with the ability to be rotated around its axis. The vertical racks 3 are inserted into the base, the horizontal racks 4 are attached to. The racks consist of two parts: mobile and stationary, so that they may be moved apart regarding the dimensions of the mine workings.

By means of stops 5, the design of the MVW is raised into its vertical position and then they act as a support. To exclude the possibility of the wall tilting, the ski sleds 2 are installed onto the back side are located, which also allow to move the structure from one place into another.

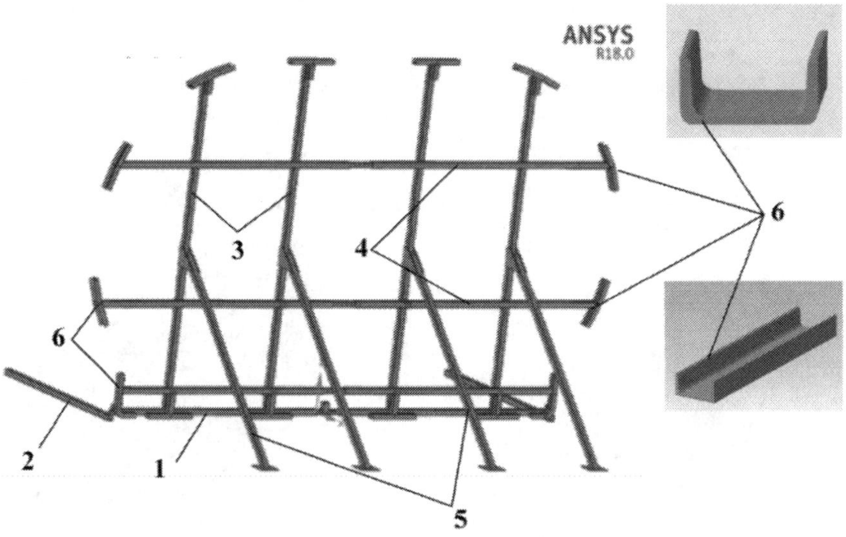

Fig. 4. The Power Frame of the MVW: 1 – base; 2 – ski sleds; 3 – vertical racks; 4 – horizontal racks; 5 – supports; 6 – grooves.

 https://doi.org/10.1051/e3sconf/201913401011

Upon setting the MVW into the grooves 6, the pneumatic hose is inserted (Fig. 4). The grooves are made in such a way that they repeat the contour of the mine workings. The cloth shall be put over the top onto the structure assembled (not shown at Fig. 4) and the air hose is pumped.

The appearance of the experimental (testing) sample of the new type of MVW proposed is shown in Fig. 5.

Fig. 5. The Appearance of the New Type of MVW Power Frame Assembled: 1 – base; 2 – supports; 3 – vertical racks; 4 – horizontal racks; 5 – skiing; 6 – groove.

The design of the experimental (testing) sample was assembled within the manufacturing workshop. As a result, it is established that the term for the MVW adjusting into its working position does not exceed 12 minutes. This result is much better than the analogue walls. The tests are currently planned at one of VDPMS mine workings.

5 Conclusion

As a result of the calculation of the pressure that will be applied to the MVW by air under its flow rate of 10 m/s, mathematical modelling of the clamping forces of the pneumatic hose onto the walls, soil and roof of the mine workings, the new type of the MVW design has been developped. According to the results of calculations, an experimental (testing) sample was manufactured, and was tested within the conditions of the assembly workshop, revealing the advantages of the design proposed regarding the terms of constructing velocity. At the next stage, it will be necessary to test the MVW under the real conditions of the mine to determine the wall's insulating properties.

References

1. Jianwei Cheng, Yan Wu, Haiming Xu, Jin Liu, Yekang Yang, Huangjun Deng, Yi Wang, Tunneling and underground space technology, **45**, 166 (2015)
2. K.G. Wallace, M.J. McPherson, D.J. Brunner, F.N. Kissel, Bur. Mines US Dep. Inter, **1,** 9307 (1990)

3. A. Nikolaev, N. Alymenko, A. Kamenskih, V. Nikolaev, E3S Web Conf., **15**, 7 (2017)
4. S.G. Gendler, S.V. Kovshov. Eurasian Mining, **2017:3**, 3 (2016)
5. F. Abu-Abed, E3S Web Conf., **41**, 01025 (2018)
6. F.N. Abu-Abed, D.V. Martynov, A.V. Ivanova, R.V. Dopira, R.Y. Kordyukov, ARPN Journal of Engineering and Applied Sciences, **11:16,** 9636-9645 (2016)
7. M.A. Trevits, C. McCartney, H.J. Roelots, Printed works of SME Annual Meeting and Exhibit, **9,** 11 (2009)
8. M.Yu. Nazarenko, V.Yu. Bazhin, S.N. Saltykova, F.Yu. Sharikov, Coke and Chemistry, **57,** 10 (2014)
9. A.V. Nikolaev, P.V. Maksimov, R.N. Gazizullin, A.G. Timarov, Bezopasnost' Truda v Promyshlennosti, **4,** 21 (2019)

E3S Web of Conferences **134**, 01012 (2019) https://doi.org/10.1051/e3sconf/201913401012
SDEMR-2019

Computer model of electric drive of 1L100K conveying unit

Denis Zolkin[1,*], and *Vadim* Petrov[1]

[1]National University of Science and Technology, College of Mining, Department of Energy and Energy Efficiency of the Mining Industry, 119049 Moscow, Russian Federation

Abstract. The paper deals with the problems of vibration damping and limiting dynamic loads in the electromechanical system of a belt conveyor by means of an adjustable electric drive. A conveying unit is represented as a three-mass ring system, which corresponds to a conveying unit with cinematically closed tape. The development of the structure of the control system for electric drives of belt conveyors with limited dynamic loads in elastic elements will reduce the dynamic loads on the belt and therefore the wear of the belts. Conditions of damping of oscillations in electromechanical system of belt conveyors are defined.

1 Introduction

The goal of research is determination of conditions of vibration damping in electromechanical system (EMS) of belt conveyors. It may be considered as homogeneous body consisting of elastic bars with distributed mass, which are subject to action of dry and viscous friction. Elastic vibration propagation in such bars is described by second-order partial differential equations.

Analysis of mechanical properties and synthesis electric drive control system is practically impossible at such description of elastic properties of the belt. It is convenient to perform analysis and synthesis when conveyor belt is presented in the form of lumped masses. At that, number of lumped masses is selected from condition of matching of main natural frequencies of vibrations in system with lumped masses and system with distributed masses.

Researches in this area show that at number of masses equal to six, main natural vibration frequencies correspond to system with distributed masses. Analysis at such number of masses is rather difficult because of high order of differential equations. If vibration damping in the belt with achievement of aperiodic nature of transition process will be provided by means of controlled electric drive, then it is sufficient to apply three-mass electromechanical system.

* Corresponding author: zolkin.misis@gmail.com

 https://doi.org/10.1051/e3sconf/201913401012

2 Materials and Methods

Conveying unit may be presented in the form of three-mass calculation model. Kinematic diagrams with three masses may be conditionally divided into linear diagrams and ring diagrams [1-7]. Ring electromechanical systems (hereafter EMS) correspond to conveying unit with kinematically closed belt.

Ring kinematic chain is characterized by third rigidity with c_3 coefficient, which (unlike linear rigidity) realizes elastic connection between outermost masses. Natural vibration frequency of the first mass in this case is defined by expression $\Omega_1 = \sqrt{(c_1 + c_3)/J_1}$.

To visualize and evaluate dynamic loads resulting from vibrations in the belt of conveying unit, it is necessary to perform computer modelling (MATLAB application package, Simulink subsystem), using parameters of block diagram of TEMS with ring kinematic chain.

Belt conveyor 1L100K is considered as simulation object. This type of conveyor is designed for transportation of rock mass with coal lump size not more than 300 mm and solid not more than 150 mm on rectilinear in plan mine workings with inclination angle from minus 16° to plus 18°.

Technical characteristics of 1L100K according to are shown in Table 1.

Table 1. Technical characteristics of 1L100K belt conveyor

Parameters of conveying unit	**1L100K**
Belt speed, m/s	1.6
Maximum output, t/h	475
Intake capacity, m^3/min	9.9
Belt width, mm	1000
Driving drum diameter, mm	800
Quantity of driving drums	1
Quantity of driving motors	1
Reducing gear ratio	40
Maximum conveyor length, m	1000
Conveyor belt length, m	810
Roller diameter, mm	127

Conveying unit utilizes fabric-ply belt of 2ShTK-200-4,5x3,5 type which parameters are shown in Table 2.

Table 2. Parameters of fabric-ply belt of 2ShTK-200-4,5x3,5 type

Parameters of fabric-ply belt	**2ShTK-200-4,5x3,5**
Specific gravity, kg/m^3	18.4
Dynamic module of elasticity, N/m^2	$1.65 \cdot 10^7$
Quantity of spacers, pcs	4

Driver of conveying unit is equipped with asynchronous motor of BP280S4. Parameters of the motor are shown in Table 3.

Table 3. Technical characteristics of asynchronous motor of BP280S4 type

Parameters of asynchronous motor	**BP280S4**
Rated power, kWt	110
Stator nominal power, V	380/660
Stator nominal current, A	200/116
Nominal efficiency factor	93.5
Nominal power factor	0.89
Nominal synchronous rotation frequency, rpm	1500
Nominal slip	0.01
Locked-rotor current ratio	6.8
Initial starting torque ration	2.0
Torque capacity	2.8
Rotor momentum of inertia, kg·m^2	2.93

Technical parameters of driving motor power supply are three-phase circuit 50 Hz, 660 V.

Electric drive is equipped with explosion-proof frequency converter of PChV-K-110 U5 with the following technical characteristics (Table 4).

Table 4. Technical characteristics of explosion-proof frequency converter of PChV-K-110 U5 type

Parameters of frequency converter	**PChV-K-110 U5**
Nominal voltage, V	660
Circuit frequency, Hz	50
Rated power, kWt	110
Current overload ration (60s)	1.5
Frequency regulation range, Hz	0 - 200

Flow-chart of control system of three-mass electromagnetic system (TEMS) of conveying unit with ring kinematical scheme containing elastic mechanical constraints is shown in Fig. 1.

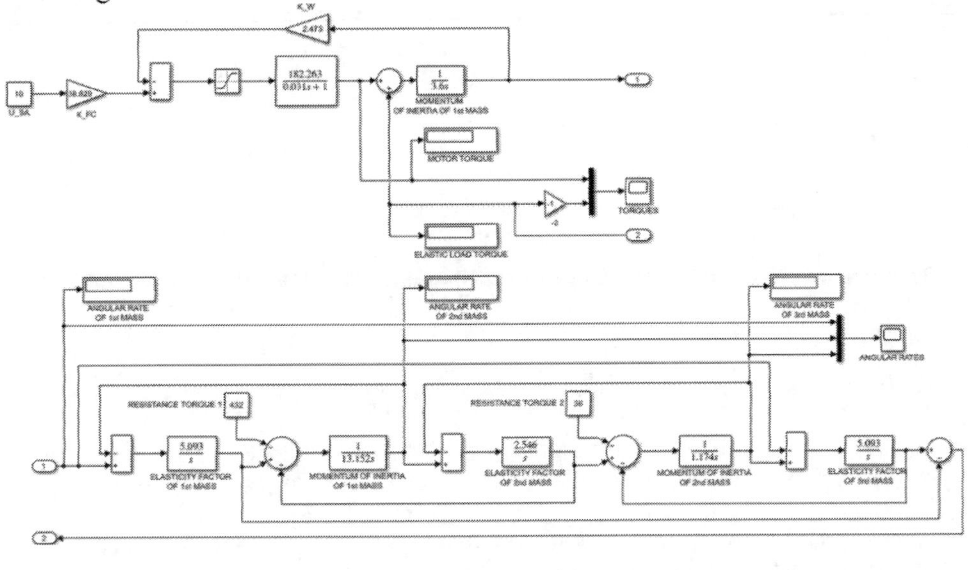

Fig. 1. The flow-chart of the control system of a three-mass electromechanical system of a conveyor system with an ring kinematic scheme containing elastic mechanical connections, implemented in the MATLAB application software package (Simulink subsystem).

 https://doi.org/10.1051/e3sconf/201913401012

3 Results and Discussion

Graphs of transient processes of open-loop control system of 1L100K conveying unit obtained during simulation in *MATLAB* application package (*Simulink* subsystem) are presented in Fig. 2.

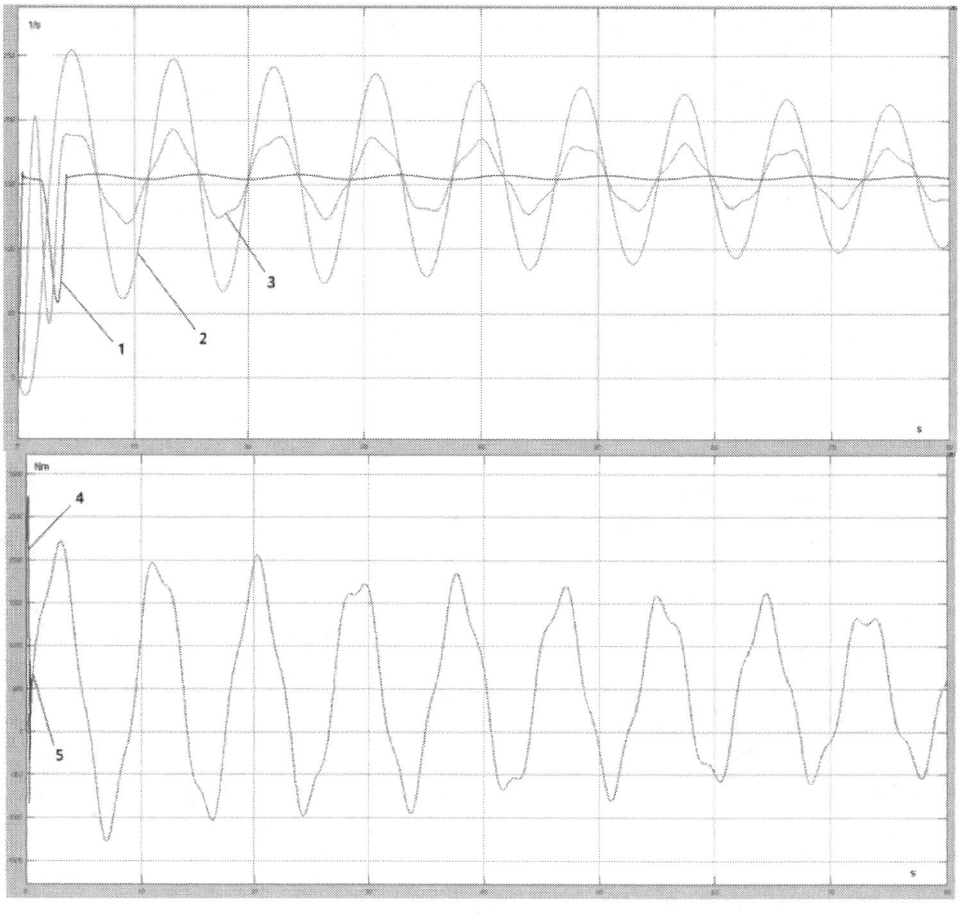

Fig. 2. Transient processes graphs of the open-loop control system of the 1L100K conveyor system, obtained by modeling in the MATLAB application package (Simulink subsystem).

4 Conclusions

1. Mathematical model of elastically connected electromagnetic system is developed. The model takes into account and interrelates coordinates of electric drive and working body of belt conveyor.
2. Interrelation between parameters of electric drive and executive body of mechanism.
3. As a result of solving simultaneous equation of sixth order it was determined that fulfilling of conditions (15) of vibration damping is impossible in the overwhelming

majority of conveying unit operation. So it is necessary create control system providing fulfilling of assigned conditions by means of changes of regulation parameters.
4. Model of conveying unit with ring kinematic chain containing elastic mechanical connections is developed. The model is implemented in MATLAB application package (Simulink subsystem).

References

1. V.N. Fashchilenko, *Development of the structures of the control system for an adjustable electric drive of mine belt conveyors* (Moscow Mining Institute, Moscow, 1986)
2. V.N. Fashchilenko, A.A. Ivanov, *Mathematical modeling of control systems of frequency-controlled electric drive of mine belt conveyors // Improving the design, manufacturing technology and operation of mining equipment and automation equipment* (Moscow Mining Institute, Moscow, 1992)
3. V.S. Kostyuk, L.P. Volkova, V.N. Fashchilenko, *Adjustable electric mine installations* (Moscow Mining Institute, Moscow, 1984)
4. A.V. Lyakhomsky, V.N. Fashchilenko, *Management of electromechanical systems of mining machines* (Publishing House of Moscow State Mining University, Moscow, 2004)
5. N.G. Pereslegin, *Synthesis of electric drive control systems of multi-mass mechanisms with elastic coupling / Automated electric drive* (Atomic Energy Pub., Moscow, 1990)
6. V.I. Klyuchev, U.A. Mateev, E.K. Perepichienko, Electrotechnical Industry. Ser. Electric drive, 7:96, **1-6** (1981)
7. L.A. Plashchanskiy, M.M. Kholmogorov, Mining science and technology, **1**, 20-23 (2016)

Recommendations for determining the thickness of the road pavement of technological roads built from coal-bearing rocks

Victor A. Shalamanov[1,*], *Sergey* N. Shabaev[1], and *Faruz* Alama[2]

[1]T.F. Gorbachev Kuzbass State Technical University, 28 street Vesennyaya, Kemerovo, 650000, Russian Federation
[2]Lebanese International University, 146404 Mazraa, Mouseitbeh, Lebanon

Abstract. With open-pit mining method in Kuzbass is widely used automobile transport, transportation costs up to 70% of the cost of coal production. It is not economically feasible to use asphalt concrete and cement concrete as a material of temporary technological roads, and the use of crushed coal-bearing rocks will make it possible to increase the efficiency of coal mining by open method. Currently, the question of the application of the coal-bearing rocks for building temporary technological roads is poorly studied. The paper proposes a method of determining the thickness of road pavements of technological roads, which can be applied directly by the mining company, their analysis has been carried out, the dependence from the strength factor and the category of blockiness of foundation rocks has been determined.

1 Introduction

Up to 50 % of workers in mining enterprises, with the open-pit mining method are employed in the most expensive and labor-intensive process - transportation of mineral and coal-bearing rocks, transportation costs make up to 70 % of the total costs of mineral extraction. In the implementation of this process is widely used automobile transport, 65 % transportation of rock mass in Russia and up to 90 % abroad.

Asphalt concrete and cement concrete as a material of temporary technological roads cannot be applied, it is not economically appropriate. Application of the fragmented coal-bearing rocks will allow to improve the efficiency of open-pit coal mining and ecological situation in the region. But at present time, the question of using coal-bearing rocks for the building, repair and maintenance of temporary roads has not been sufficiently studied, and therefore the rock is dumped and occupies agricultural land. With that being said determining a road pavements thickness of technological roads built from coal-bearing rocks acquires particular relevance and importance.

* Corresponding author: shalamanov49@mail.ru

2 Materials and methods

In Kuzbass (Western Siberia, Russia) the open pit mining is the most perspective way to extract coal [1-3]. Taking into consideration the annual cost growth of the road-building materials, it is natural that cost of transportation of the extracted rock mass is also growing. The cost of construction, repair and maintenance of temporary technological roads range from 5 to 15 % of the cost of transportation, they can be reduced by using coal-bearing rocks as road-building materials . We found that more than 50% coal-bearing rocks of Kuzbass is physically and mechanically comply with requirements for materials which are necessary for building foundations and coatings of technological roads of Kuzbass open-pit mines.

The main factor, affecting service life of technological road – vehicle speed. With it increase deflections are reduced. They are most noticeable at low speeds of motion (up to 20 km / h), which is typical for heavy vehicles, and at more than 40 km/h deflections stabilizes . This is explained by the fact that the relationship between the loading time and the corresponding value of the modulus of elasticity is non-linear. For fragmented coal-bearing rocks the modulus of elasticity in the transition from instantaneous loading to long-term exposure varies in 1.1-1.5 times.

As a calculated value of the modulus of elasticity of the crushed rock, the value obtained by multiplying the static modulus of elasticity (long-term impact of the load) on the coefficient of transition to the instantaneous loading:

$$E_r = E_{st} \cdot k_r , \qquad (1)$$

where E_r – the calculated modulus of elasticity of the crushed rock, MPa;

E_{st} – static modulus of elasticity of crushed rock, MPa;

k_r – the transition coefficient from the effects of continuous load to instantaneous loading.

When designing technological roads, it is assumed that the velocity of movement of vehicles vary from 10 to 25 km / h, and in the summer period of time for dust removal they are watered with water; in this case, it is advisable to take the calculated elastic modulus value of the crushed rock as the calculated moisture content of the rocks, i.e. .

Features of the technological road pavements calculation, in distinction from the public roads , are as follows:

- loading on the road pavements from rock handlers are significantly higher than loading from motor vehicles moving on public roads;
- calculated values of the strength and deformation characteristics of the coal-bearing rocks differ from the values assumed in the calculation of public road pavements;
- road pavements of technological roads are calculated only by shear stability;
- an indicator of the strength coefficient is introduced when road pavements of technological roads are designed (due to reduced strength coal-bearing rocks during operation).

3 Results

One can determine optimum design value of the coefficient of road pavement strength according to the formula proposed by S. Bogomolov:

$$k_r^{pr} = (0.125 \cdot T + 0.75) \cdot (0.06 \cdot Q + 0.6) \qquad (2)$$

where T – time of operation of the road, years;

Q – the volume of freight on the road, million tons/year.

We propose the following algorithm for calculation the thickness of the road pavement made of coal-bearing rocks:

1. The service life and the load capacity of the technological road are set.
2. The optimum factor of durability of road pavement is calculated by the formula (2).

3. The type of rock handler used for transportation of the rock mass is selected.
4. Depending on the blockiness category of base rocks, their mechanical properties are accepted.
5. The thickness of the road pavement of the technological road is determined.

Performing numerous calculations of road pavements thickness of technological roads of Kuzbass open-pit mines, constructed of coal-bearing rocks in accordance with the proposed algorithm and analyzed the results, we found that it depends from blockiness category of natural base.

The researchers have not disregarded the issue of using coal-bearing rocks as a material for the construction, repair and maintenance of temporary technological roads in the process of mining in the open way [4-17], but recommendations for determining the thickness of road pavements have not been developed. We have proposed a method of calculation, which can be applied directly to the mining enterprise. To do this, use the above algorithm and the nomogram constructed on the basis of numerous calculations (figure).

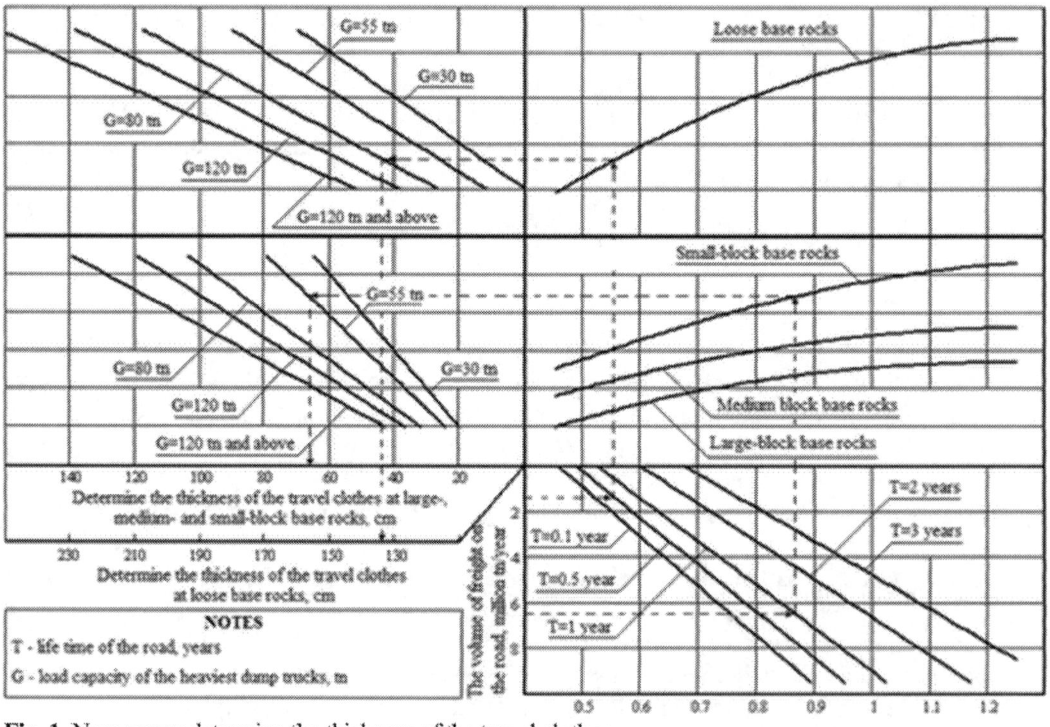

Fig. 1. Nomogram determine the thickness of the travel clothes.

4 Conclusions

Our analysis of the thickness of the road pavements of technological roads, which ensure the normal operation of road transport, with the time of operation and road freight volume, showed that they depend mainly on the blocking category of base rocks and the design strength factor and vary from 0.20 in large-block, up to 2.5 meters in loose.

References

1. T. Gvozdkova, M. Tyulenev, S. Zhironkin, V. A. Trifonov, Yu. M. Osipov, IOP Conf. Ser.: Earth Environ. Sci., **50:1**, 012010 (2017)

2. M. Tyulenev, O. Litvin, S. Zhironkin, M. Gasanov, Acta Montanistica Slovaca, **24:2**, 88-97 (2019)
3. S. Markov, J. Janočko, M. Tyulenev, Y. Litvin, E3S Web of Conferences, **105**, 01021 (2019)
4. V. I. Chernobai, Journal of Coal Science and Engineering, **19:2,** 219-225 (2013)
5. D. V. Moldovan, Journal of Industrial Pollution Control, **33:1,** 1007-1012 (2017)
6. M.A. Tyulenev, S.O. Markov, M.A. Gasanov, S.A. Zhironkin, Geotech. Geol. Eng., **36:5,** 2789-2797 (2018)
7. A.A. Khristoforova, M.D. Sokolova, B. N. Zarovnyaev, A. N. Akishev, Mining Journal, **3**, 47-50 (2016)
8. G. V. Solovyev, K. I. Vatchnadze, Procedia Engineering, **189**, 666-672 (2017)
9. S. L. Jackson, Geoforum, **66**, 94-105 (2015)
10. B. Kinga, M. Goycoolea, A. Newman. European Journal of Operational Research, **257**, 297-309 (2017)
11. V. Czermak, H. G. Huckenholz, L. Rybach, *Physical properties of rocks* (Springer, Berlin, 1982)
12. R. P. Monicard, *Properties of reservoir rocks: core analysis* (Springer, Berlin, 1980)
13. K. Bucher, Kurt, G. Rodney, *Petrogenesis of metamorphic rocks* (Springer, Berlin, 2011)
14. I. W. Farmer, *Engineering behaviour of rocks* (Springer, Berlin, 1983)
15. Lian Ying Zhang, Xian Biao Mao, Ai Hong Lu, Science in China Series E: Technological Sciences, 52:3, 641-646 (2009)
16. A. Kilic, A. Teymen, Bulletin of Engineering Geology and the Environment, **67**, 237 (2008)
17. Zong-Xian Zhang, Journal of Rock Mechanics and Geotechnical Engineering, **8:6**, 886-895 (2016)

The research of operational temperatures of dump trucks tires

Sergey Goryunov[1,*], *Aleksey* Khoreshok[2], *Natalya* Grigoryeva[1], *Elena* Preis[2] and *Oksana* Alitkina[2]

[1]T.F. Gorbachev Kuzbass State Technical University, Prokopievsk Branch, 653039 Prokopievsk, Nogradskayast. 19a, Prokopievsk, Russia
[2]T.F. Gorbachev Kuzbass State Technical University, 650000 Kemerovo, 28 Vesennyayast, Russia

Abstract. Oversized tires play a significant role in the mining process for ensuring continuous and safe operation of open cast mining equipment. The article presents the influence of large tires operating temperatures tolong life operation. The results of studies of the tire heating nature during the operation arealso presented. The regression models of studied multifactor processes of heating tires are obtained by methods of computational experiment and mathematical planning. The influence of the tread height at tire heating is evaluated.

1 Introduction

In Russia an open cast mining has a dominant role in mining industry, as it is safer and more economical [1-2].

High productivity and uninterrupted operation of the rock mass transportation is an important part of mining enterprise functioning. It is ensured by the increase in the technical readiness of the vehicle fleet while maintaining acceptable levels of safe operation. So, the significant role of large tires in ensuring continuous and safe operation of open cast mining equipment becomes apparent. It is also important that, the large tires have the significant role in the structure of the rock mass transporting cost (the cost of large tires varies from 10 to 40 thousand dollars per unit).The consumers of these products are interested in the most efficient management of the tire resource during operation.

The analysis of statistical data, for the typical reasons of large tires failure, showed that about 12% of failures are associated with fatigue and thermal damage caused by excess temperature of large tires during operation for various reasons, and a temperature of 120 ° C is critical for them. If the temperature exceeds the critical temperature, the tire will collapse due to a decrease in the strength of the cord and its connection with rubber, the development of the following defects, such as delamination, bloating of the tread and delamination of the carcass [3]. There were also situations where the cord broke and the tire self-ignited. The significant temperature effect in tire fatigue failure has been confirmed by many researchers. In most performed studies, this fact is explained, so that, when a tire

* Corresponding author: gorjunovsv@kuzstu.ru

is used a high temperature is maintained for a long time in it, there is a sharp decrease in both the strength of the rubber itself and its adhesion to the cord. At the same time, in places of the greatest decrease in adhesion, mechanical friction occurs between the cord and rubber, which additionally causes a sharp increase in temperature and, ultimately, rapid destruction of the tire. In this regard, the fatigue damage of car tires due to the influence of high temperature is often called the thermal damage of tires [4-5].

It should be noted that, during fatigue and temperature failures, we practically cannot extend the life of large tires by restoring the tread, since with these damages, the tires are subject to decommissioning, which leads to significant material losses of the enterprise.

The taking into account and temperature control of large tires at mining enterprises is usually not carried out. As a rule, the mining company is guided only by the operational performance parameter when choosing large TKPH tires (ton-kilometer per hour), which represents the maximum allowable tire load according to the heat generation. The main focus is on factors that influence tire longevity primarily in terms of tread wear, such as tire pressure and average operating speed. All of these factors are controllable and it is possible to influence them. They are monitored at the mining enterprises through the use of navigation and the tire pressure monitoring system "Pressure Pro", or periodic measurements [6-8].

Today, mining enterprises, in connection with plans to increase productivity and reduce fixed costs in the structure of transportation costs, are making changes to the design of mining dump trucks by increasing the height of the side sides of the cargo platform (Fig. 1).

Fig. 1. BelAZ-7513 dump truck with extended side sides of the cargo platform.

Since loading a dump truck is performed with a "cap", this leads to exceeding the rated load capacity of a mining dump truck. Based on the data obtained by weight control at the mining enterprises of Kuzbass, the utilization rate of the carrying capacity of mining dump trucks with extended sides varies from 0.9 to 1.15. Exceeding the permissible load on the tire causes increased heat generation. It is possible to level this out by changing the average operational speed of a mining dump truck.

All mentioned above indicates the relevance of the study of operating temperatures of large-sized tires of mining dump trucks and factors influencing heat generation.

2 Experimental research

The heating of large tires is associated with the following reasons: hysteresis losses, external friction of the tire on the surface of the road surface, as well as the effect of solar radiation on the surface of the tire [9-12].

For the most operating modes, the heat generation is about 85-95 % because of the deformation and internal friction, from 5 to 15%is from external friction. The exceptions are skidding, slipping and applying to the wheel moments exceeding the maximum traction moment significantly. In these cases, the main cause of heat generation in the tire is the external friction on the road surface. However, such modes are implemented quite rarely in the actual operation of dump truck, mainly during the season of sluts and uphillmovement. Solar radiation affects tire heating only in the hottest time of the year. In this case, the temperature of individual sections of the tire illuminated by the sun can increase by 10-15 ° C compared with the surface of the tire in the shade.

Part of the heat generated in the rolling tire array is continuously discharged into the environment. Moreover, if at the beginning of the movement the tire temperature is equal to the ambient temperature and the heat generation is much greater than the heat transfer, then after a certain time the total heat generation and heat transfer are equalized, and the temperature in the tire stabilizes (stationary mode) [13-14].

In the production process of open cast mining enterprises in Kuzbass, mining dump trucks of various capacities from 45 to 220 tons are operated. The dump trucks BelAZ-7555 and BelAZ-7513, with a loading capacity of 55 and 130 tons have the greatest efficiency [15].

Investigation of large-sized tires thermal state of open cast mining dump trucks was carried out in a combination of a computational experiment with methods of mathematical planning, that reduces the number of calculations and facilitates the analysis of the results. Measurement of tire operating temperatures was carried out, using a Raytek-MT6 pyrometer. The main heat generation in the tire occurs in the area of its contact with the supporting surface. In such a zone, the maximum strain rate and the rate of deformation: normal, circumferential and lateralto the tire element act. It has been established that, the largest amount of the heat per unit volume is generated in the middle of the tire's tire layer and at its edges, and the total heat generation in the tire is distributed on its main elements on average as follows: tread (including the undercut) - about 50%, carcass from 12 to 33%, the breaker layer - from 7 to 15%, the sidewalls and the adjoining side from 5 to 25%.

The results of processing experimental studies of pneumatic tires operational temperatures are shown in Fig. 2 and 3. Figure 2 shows the steady-state operating temperatures of the rear axle pneumatic tires, with an average ambient temperature t_{av} = 20-23 °C and equal operating conditions.

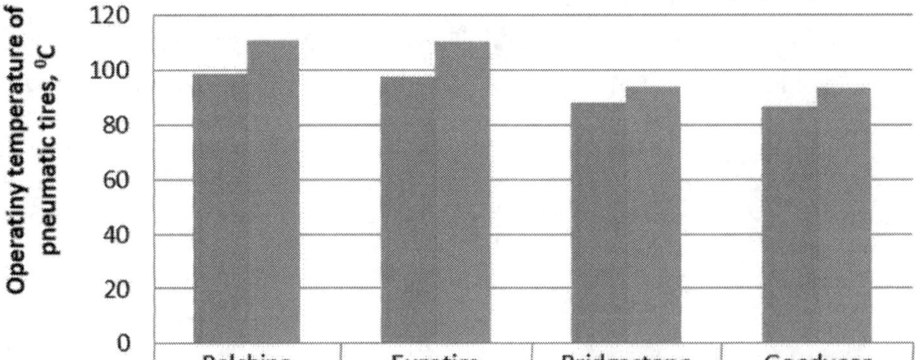

	Belshina	Eurotire	Bridgestone	Goodyear
BelAZ-7555V	98,8	97,7	88,1	86,5
BelAZ-7513	111,1	110,5	94	93,4

Fig. 2. Steady operating temperatures of large tires.

As it can be seen from the graph, the steady-state operating temperatures of pneumatic tires, are significantly different according to the manufacturer. This is primarily due to the design of the tire and its elasticity. It should be noted, that the operating temperatures of pneumatic tires of Bridgestone and Goodyear, Belshina and Eurotire are practically comparable. During the study, it was revealed that the pneumatic tires of the BelAZ-7513 dump truck are additionally heated by the motor-wheel gearbox.

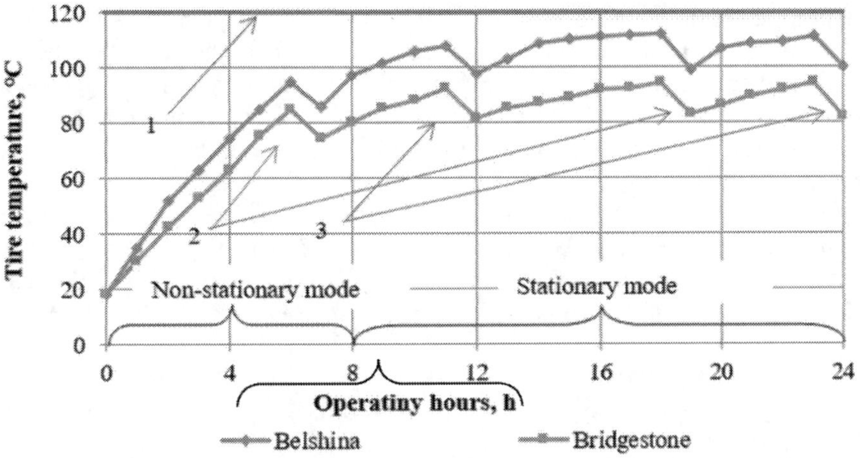

Fig. 3. Dynamics of heating of rear axle large-sized tires of the BelAZ-7513 dump truck (1 - critical temperature, 2 - lunch break, 3 - shift shift).

Based on the analysis of the data obtained, the most intense temperature increase occurs in the first 8operating hours of the truck, that is, unsteady mode. Subsequently, the increase in tire temperature is approximately 1 ... 5 °C/h. During the operation of dump trucks, the tires warm up and cool down in connection with the operating conditions of the dump truck itself.

It has been established that a large tire at an ambient temperature of 20 ... 25 °C is cooled on average 10 hours after shutdown. At the same time, in the first 4 hours of the dump truck sludge, the tire cooling rate is 10 ... 12 °C/h, and later it drops to 5.0 ... 6.7 °C/h. The dynamics of heating the pneumatic tires of the BelAZ-7555V dump truck is similar.

The task of determining the root that causes thermal failures is one of the most important problems to be solved when organizing large-scale tires and dump trucks at the modern level of operation. There are different ways to solve these problem, but in any case, it is necessary to obtain correlation relationships of temperature depending on specific operating conditions, taking into account the wear degree of the tire tread. Without this knowledge, optimization of the modes of movement of dump trucks is impossible.

We define the most determining factors of large tire heat generation [3]:

- the average for the transport cycle radial load on the tire;
- operational speed of a dump truck;
- ambient temperature.

These factors are completely controllable, and we can influence them during the operation of large tires. We'll not consider the internal pressure in a large tire, since it is a completely controllable factor and is currently controlled quite well. Although, special attention must be paid to the work of the rear axle twin wheels. When the pressure difference in the tires of a double wheel is 10%, the temperature of the tire with the highest pressure can exceed 16%.

Experimental studies found that, with a maximum load of 392000 N on a tire of 33.00R51 dimension of the Bel-162 model with an increase in speed from 10 to 15 km / h, the tire temperature increases by 32.9% with a slight increase in air pressure in it by 6%. Exceeding the permissible operating speeds of dump trucks leads to thermal tire failures due to operational reasons. When subject to recommended operating conditions, these types of failures will be determined by production reasons, i.e. insufficient level of workmanship. This confirms the significant influence of the truck speed to the heating of its tires.

The experimental data on the weight per tire on the axis of the BelAZ mining dump truck in operation was determined by weight control, and the average speed of its movement was determined by timing the dump trucks with a fixed speed and comparing it with the GLONASS readings. Information about the air pressure in every tire of mining truck was delivered to the control room of the enterprise through the GLONASS global navigation satellite system.

Technological vehicles at Kuzbasscoal enterprises use pneumatic tires from such manufacturers as: Belshina and Bridgestone. Using the Statistica 6.0 program and the Gauss method with an accuracy of ± 3.9%, the correlation dependences of the maximum steady-state temperature of the rear and front tires were obtained:

BelAZ-7513 -(Belshina)

$$t_{r.ax.t} = 31.7 + 0.6\, t_{av.n.} + 0.101 Q_{r.ax.t.} V_{av.s.} \quad (1)$$

$$t_{f.ax.t} = 26.5 + 0.6\, t_{av.n.} + 0.096\, Q_{f.ax.t.} V_{av.s.} \quad (2)$$

BelAZ-7513 (Bridgestone)

$$t_{r.ax.t} = 30.1 + 0.6\, t_{av.n.} + 0.078\, Q_{r.ax.t.} V_{av.s.} \quad (3)$$

$$t_{f.ax.t} = 25.8 + 0.6\, t_{av.n.} + 0.076\, Q_{f.ax.t.} V_{av.s.} \quad (4)$$

BelAZ-7555V (Belshina)

$$t_{r.ax.t} = 37.2 + 0.6\, t_{av.n.} + 0.097 Q_{r.ax.t.} V_{av.s.} \quad (5)$$

$$t_{f.ax.t} = 22.5 + 0.6\, t_{av.n.} + 0.074\, Q_{f.ax.t.} V_{av.s.} \quad (6)$$

BelAZ- 7555V (Bridgestone)

$$t_{r.ax.t} = 35.4 + 0.6\, t_{av.n.} + 0.081\, Q_{r.ax.t.} V_{av.s.} \quad (7)$$

$$t_{f.ax.t} = 20.3 + 0.6\, t_{av.n.} + 0.064\, Q_{f.ax.t.} V_{av.s.} \quad (8)$$

Where $t_{r.ax.t.}$,$t_{f.ax.t.}$ - tire temperature, respectively, of the front and rear axles of the truck, °C; $t_{av.n.}$ – average ambient temperature, °C; $Q_{r.ax.t.}$, $Q_{f.ax.t.}$ -average operating weight per tire of the front and rear axles of the truck, respectively, t; $V_{av.s.}$ - average operating speed of the dump truck, km/h.

A comparison of the results of experimental studies with the results of calculations based on the obtained regression dependences showed that the error does not exceed 6.5% for BelAZ-7513 dump trucks, and for BelAZ-7555V dump trucks they obtained comparable results and the error is about 5.2%. Consequently, the obtained regression dependences can be used to estimate the maximum steady-state temperatures for heating pneumatic tires during operation and to determine average operating speeds.

To assess the influence of the tread height of a large tire on its heating in operation, experimental studies were conducted, the results of which are shown in Fig.4.

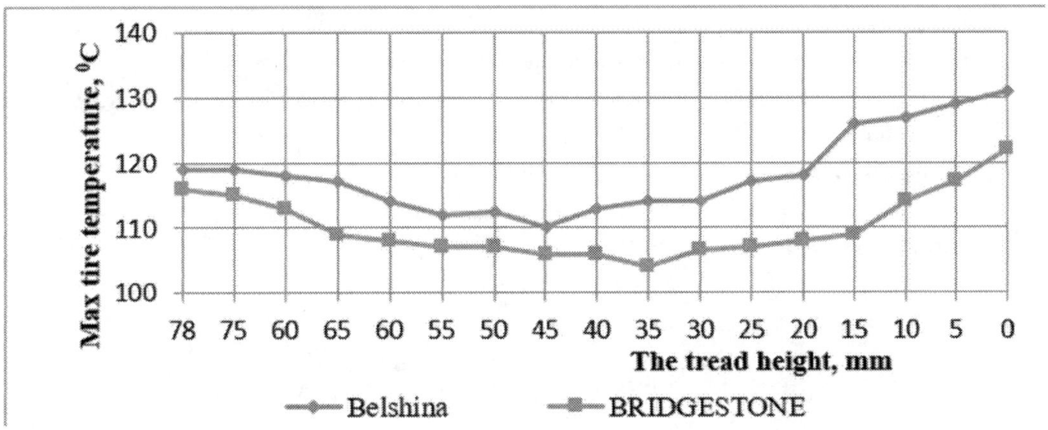

Fig. 4. Effect of tread height on the maximum temperature of tires of the BelAZ-7513 dump truck.

It has been established that during operation of tires of the open cast mining truck, the temperature of heating of the tires changes depending on the amount of wear. In the process of tire wear, their heating temperature decreases to a certain value first, and then increases. Worn out (bald) tires are subject to higher heat. Moreover, Bridgestone tires undergo less heat compared to similar tires of the Bel-162 model of Belshina OJSC. Obviously, this is due to lower hysteresis losses in Bridgestone tires.

3 Conclusions

In general, the studies conducted allow us to assess the main factors that affect the operating temperatures of large tires, and to manage them to realize the most complete tire life. In the course of the study, the influence of the tread height on the heat generation in the tire was established, thatmust be taken into account during the operation. The use of research results in the organization of production processes for mining will reduce the cost of rock mass transporting.

References

1. M. Tyulenev, S. Markov, M. Cehlar, S. Zhironkin, M. Gasanov, Acta Montanistica Slovaca, **23:4**, 368-377 (2018)
2. A.I. Kopytov, Ugol', **3**, 72 (2017)
3. S.V. Goryunov, V.M. Sharipov V.S. GornyiZhurnal, **11**, 127-130 (2015)
4. I. Lapczyk, J. Hurtado, S. Govindarajan, *A parallel rheological framework for modeling elastomers and polymers* (American Chemical Society. Cicinatti, 2012)
5. N. Saintier, G. Cailletaud, and R. Piques, International Journal of Fatigue, **28:1**, 61-72 (2006)
6. S. Tan, *Production planning and operational control; new algorithms with application to the mining industry* (Pennsylvania State University, Logan, 1992)
7. Y. Kuo, *Highway earthwork and pavement production rates for construction time estimation* (University of Texas, Huston, 2004)
8. A. Thomas, Journal of Polymer Science, **31:123**, 467-480 (1958)

9. D.F. Moore, *The friction of pneumatic tires* (Springer, Berlin, 1975)
10. F.J. Kovac, *Tire technology* (Penguin, London, 1973)
11. W.E. Meyer, J.D. Walter, *Frictional Interaction of Tire and Pavement: A Symposium* (ASTM International, Glasgow, 1983)
12. R. Ogden, Engineering Analysis, **1:2**, 119 (1984)
13. S. Alarie, S. M. Gamache, International Journal of Surface Mining, Reclamation and Environment, **16:1**, 59-76 (2002)
14. N. Trivisonno, *Thermal analysis of a rolling tire* (SAE Technical Paper, Chicago, 1970)
15. A. B. Efremenkov, A. A. Khoreshok, S. A. Zhironkin, A. V. Myaskov, IOP Conference Series: Earth and Environmental Science, **50:1**, 012009 (2017)

Open pit mine truck robotization at the stage of rock transportation

Vyacheslav Zhdanov[1], *Andrey* Kosolapov[1], and *Flyura* Ibragimova[2]

[1]Kuzbass State Technical University named after T.F. Gorbachev, Russia, 650000, Кемерово, Vesennyaya street, 28
[2] Kazakh Humanitarian Juridical Innovative University, Kazakhstan, Semey, Abaya street, 94

Abstract. The prerequisites for robotization of such a component stage of mining as mined rock transportation are considered in the article. Digitization of dump truck positioning, maneuvering and 3D map tracking forms the basis of approaching robotization. The equipment required for accident prevention in an open pit is discussed. A simplified algorithm of automated vehicle operation is offered.

1 Introduction

Currently, technological modernization of open pit coal mining is a key area of development for the industries of the basic complex of the Kuzbass economy (Russia's main coal cluster) [1-2]. The process of robotization and digitalization is now widespread in the mining industry [3-5]. A new challenge is the creation of unmanned equipment with high specific productivity for open cast mining.

2 The prerequisites for mine truck robotization

2.1 Accident prevention when transporting mined rock

Nowadays a lot of automobile manufacturers are staking on digitization and new services in principal counting on future where automobiles will have an ability to communicate with their operators, all their control and tracking systems will be electrified, and they will become autonomous.

It refers not only to the automobiles constructed for operation on public roads. The introduction of unmanned technologies for rock transportation amounting 30-50% of coal mining cost will solve several system tasks in mining industry, which is on a global scale based on open pit mining. Firstly, it is cost saving; secondly, it is furthering safety of mining including potential exception of truck operator traumatizing at the stage of loading and unloading, and rock transportation as well. The gravity of any accident consequences caused by dump trucks, which total mass exceeds 500 tons, is known to be very high. Even if there are no human casualties, the expenses on evacuation and the repair of damaged machinery (Fig. 1) are very high as well.

Fig. 1. A dump truck bogged down when unloading.

Thus, one of Australian companies is one of the pioneers of using unmanned dump trucks at their open pits [6]. The company continues developing technological processes to protect its employees from the harm and dangers of mining in the first place.

Similar approach is shown by Sweden company Volvo, one of the world leaders in the sphere of labor safety. The peculiarity of technological approach is the usage of a dump truck in Boliden mine, located on the north of Sweden, 100 km from Arvidsjaur. It is the first autonomous truck in the world [7] operating underground in the conditions of almost total darkness. The trial route is 7 km to the lowest point of the mine, which depth reaches 1320 m (Fig. 3).

Fig. 2. Autonomous Volvo FMX used as a robot for transportation inside a mine.

Volvo FMX dump truck operating in a mine is equipped with a set of sensors. They are used not only in this vehicle, but also in other Volvo models constructed for mass market. The sensors constantly identify the conditions of truck operation environment and prevent collisions. On-board system collects data and chooses the speed of the truck, taking into account traffic conditions and fuel saving. Volvo's attitude to unmanned driving is extremely cautious, and the unmanned truck testing in real conditions has become Volvo's first pilot project of such a level.

Similar trials are conducted by Belorussian company BelAZ.

One the one hand, a human has more experience in driving. They are creative and resourceful at avoiding accidents on a road. One the other hand, it is a human factor, which causes the vast majority of car accidents. The robot, in its turn, is capable of carrying out the actions prescribed by the algorithm precisely, and it is still to be taught to cope with nonstandard situations.

2.2 Sensors and 3D maps

The operation of mine trucks has some distinctive features that differ it from public traffic. Firstly, the traffic intensity is sharply different; secondly, when rock transportation, trucks execute rather simple maneuvers such as one lane traffic with no need to overtake or pass road crossing; thirdly, there are practically no unforeseen obstacles on pit roads, which are non-public technological lanes. It rather simplifies the development of traffic management algorithm.

The robot-automobile under development is to be equipped with already well-known and common in the sphere of autonomous driving sensors (a GPS device, odometers, lasers, lidars and front, rear and side cameras) to detect obstacles, which are to be avoided, and to define a traffic lane, an on-board computer for data processing and decision making, wires providing acceleration and deceleration and actuating steering to operate a vehicle and communication devices (Fig. 3) [8-10].

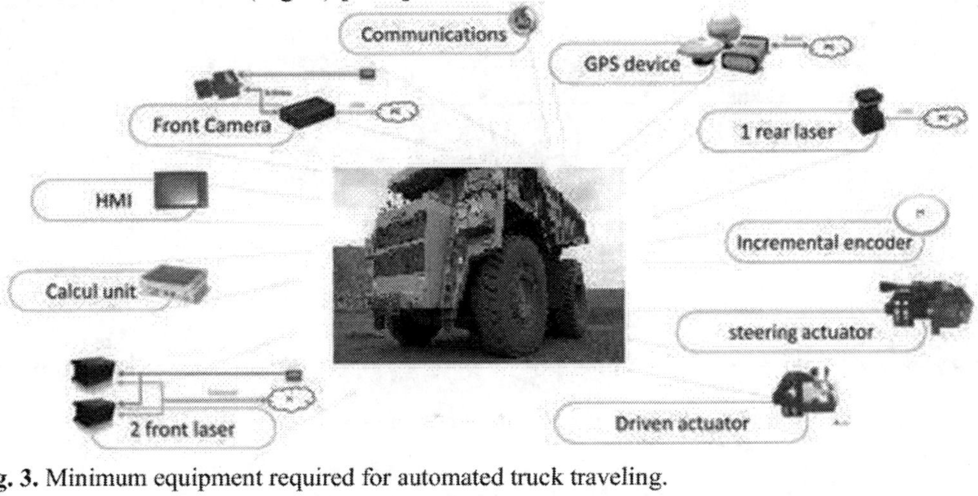

Fig. 3. Minimum equipment required for automated truck traveling.

One of essential elements for autonomous driving is a digital model of road network, which trucks operate in. The design of the model requires the whole wealth of knowledge gained in mine surveying, geodesy, cartography, and satellite positioning. The accuracy of existing systems is to be raised to a centimeter range. However, there are practically no signal suppression zones in an open pit, which is another advantage of using robots for transportation in open pit conditions. The necessity of designing the digital map of an operational zone, which, unlike standard maps, has to be three-dimensional, might become a significant problem.

3 The mine truck robotization algorithm

The assigned task to automate the management of mine truck traffic at the stage of rock transportation with the development of communication and artificial intelligence technologies, machines, and systems of unmanned engine control, transmission, deceleration and steering (by wire) will be solved by creating big data bases (big data), emergence of new professions necessary for new information flow servicing, and transportation modeling.

Mining companies will get particular benefits including financial ones as there won't be any need to deliver operators to the site, organize their feeding, recreation, and etc. Operators will move into the category of office workers working shifts while trucks will operate round-the-clock with short stops for refueling. Operators won't be in danger of having an accident, won't get tired, as it happens to drivers, or be affected by harmful burnt gases, or etc.

As a result, an optimal route will be planned according to the destination of a particular truck at present moment of time and to environment condition data. Having located itself, the robot will choose a strategy to reach the destination and the tactics of driving on the route and will control self-driving real time [10-15]. All necessary actions of truck control systems will correlate with the following simplified flowchart (Fig. 4)

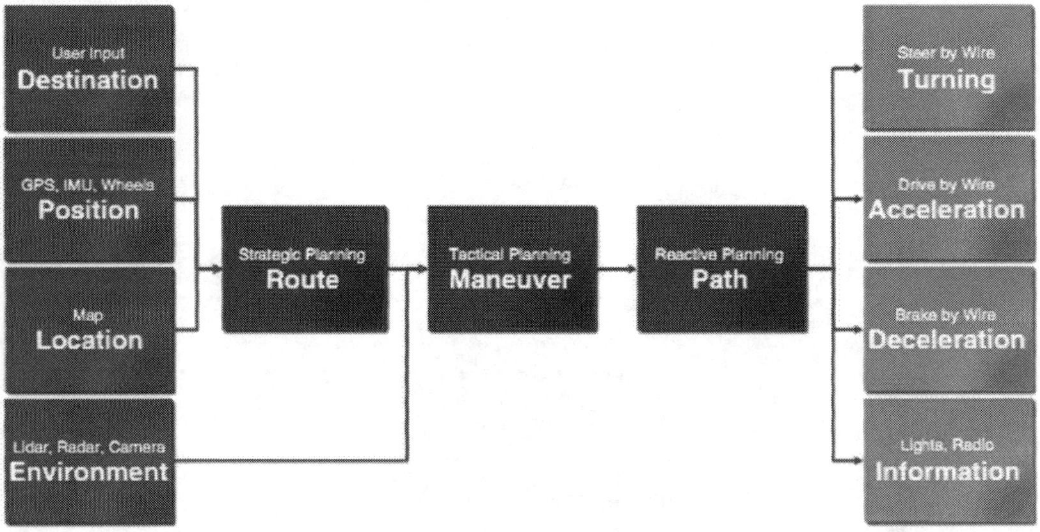

Fig. 4. The flowchart of an automated vehicle (borrowed from [10]).

4 Conclusion

Robotization will influence the design of dump trucks. For example, the unmanned truck won't require a driver's cabin, an I-shift and pedals, a ladder and railing or a large-tonnage cabin guard which protects a driver from rocks falling from above as well. It will make the truck lighter and, thus, more efficient.

References

1. M. Tyulenev, S. Markov, M. Cehlar, S. Zhironkin, M. Gasanov, Acta Montanistica Slovaca, **23:4**, 368-377 (2018)
2. M.A. Tyulenev, S.O. Markov, M.A. Gasanov, S.A. Zhironkin, Geotech. Geol. Eng., **36:5**, 2789-2797 (2018)
3. F. Abu-Abed, N. Borisov, E3S Web Conf., **21**, 01019 (2018)
4. Zhironkin, O. Aleshina, V. Gorev, Y. Gunyakov, O. Zhironkina, E3S Web of Conferences, **105**, 04001 (2019)
5. F. Abu-Abed, A. Khabarov, E3S Web Conf., **41**, 01028 (2018)
6. *Smart mining. Operating with pioneering spirit.* URL: https://www.riotinto.com/ourcommitment/smart-mining-24273.aspx
7. J. Milne, *Connected fleets and the smart mine* (Global mining, Houston, 2017)
8. E. Pollard, F. Nashashibi, P. Resende *ABV – A Low Speed Automation Project to Study the Technical Feasibility of Fully Automated Driving* (ERCIM News, New York, 2013)
9. H. Li, F. Nashashibi, *Robust real-time lane detection based on lane mark segment features and general a priori knowledge* (IEEE ICRB, London, 2011)
10. P. Resende, *Low Speed Automation: technical feasibility of the driving sharing in urban area* (IEEE ITSC, Amsterdam, 2013)
11. *High Definition Lidar. Velodyne corporate website.* URL: http://velodynelidar.com/lidar/lidar.aspx
12. S. Beiker, *Implementation of an Automated Mobility-on-Demand System, Autonomous Driving* (Springer, Berlin, 2015)
13. S. Beiker, *Deployment Scenarios for Vehicles with Higher-Order Automation, Autonomous Driving* (Springer, Berlin, 2015)
14. *Taxonomy and Definitions for Terms Related to On-Road Motor Vehicle Automated Driving Systems* (SAE International, Chicago, 2014)
15. A. Kosolapov, S. Krysin, E3S Web Conf., **21**, 03012 (2017)

Substantiation of Optimal Design Parameters for New System Angle Headframes in State-of-the Art Mines

Elena Kassikhina[1,*], *Vladimir* Pershin[1], and *Yunliang* Tan[2]

[1]T.F. Gorbachev Kuzbass State Technical University, 650000 28 Str., Kemerovo, Russia
[2] Shandong University of Science and Technology, Qingdao, China

Abstract: The article formulates a new concept for design of the steel angle headframes of the new system for state-of-the art mines. One of the important structural elements of the steel headframe is a rig, installed above mouth of the vertical shaft and designed for equipment and passage of mine conveyances in shaft. A traditional frame rig of rectangular section is the most cumbersome and heavy-weight part of the headframe due its load-bearing. The proposed by the author headframe of multifunctional purpose, that includes a non-load bearing rig of circular cross-section, will allow reducing load on mine mouth and forgoing the sub-frame that greatly simplifies the task of finding an optimal design solution for the structure. There is an analysis of the most significant factors that significantly affect formation of the main design parameters of the sheave wheels and the rig of the examined headframe. The proposed idea of the independently operating structural units, one of which is a circular rig, significantly reduces a number of controlled variables, with which a designer can vary designing a headframe.

1 Introduction

Headframe takes special place in the complex of mining facilities. First of all, headframe is an engineering structure designed to maintain pulleys of winder or the winder itself at a certain point. Secondly, in headframe the dumping of minerals lifted to the surface is carried out, so it needs to ensure the smooth operation of conveyances. For this purpose headframes are furnished with special guides: conductors and dumping curves. Thirdly, headframe structure acts as a continuation of a shaft; therefore, it is fully involved in the process of ventilation of underground workings and should be well sealed. Design of the headframe is classified as: angle headframe (A-shaped, four-post, tent) and tower-type [1-12]. Design of the steel angle headframe. Options of the constructive embodiment of steel angle headframes, their advantages and disadvantages in dealing with basic tasks.

2 Materials and Methods

Substantiation of optimal design parameters for angle headframes of the new system to meet given tasks:

[*] Corresponding author: kalena-07@mail.ru

- formalize search for an optimal technical solution in lieu of alternate design in view of loads caused by: lifting, guides, compression and depression, provided that lining is sealed,
- articulate a new concept of steel angle headframe design of the new technological level.

The assigned tasks are most fully solved in the steel angle four-post headframes of the rig system, which is confirmed by long practice of their design and operation both in our country and [1-12].

In headframes of this system (Fig. 1) rig *1* of the rectangular cross-section is a load-bearing part of headframe and together with jib *2* receives load from hoist and transfers it through base frame *3* on mine mouth *4*.

Angle headframe of the rig system installed in almost all mines of Kuzbass, indicating their undoubted advantages, but there are a number of disadvantages.

Design solutions of some units require revision, as they contribute to development of corrosion. Presence of a large number of braces in the rig frame contributes to accumulation of coal dust, slag and lube at the nodes of connection and, as a consequence, appearance of significant corrosion in stagnant places

Support rig unit on base frame requires special attention:

– base frame is located below zero level in the zone of high humidity, that leads to development of corrosion of its elements and rig supporting elements;

– base frame should be mounted into mine mouth, what complicates assembling process of the rig;

– base frame due to its size, restricts operating conditions and performance of reconstruction and shaft deepening.

Also, it is necessary to improve situation with the rig lining. The matter is that its rectangular configuration dictates application of standard wall panels as an external protection. It creates a large number of joints between them and leads to increased air permeability barriers, and under the influence of dynamic loads - also to the damage of joints, that significantly reduces the service life of the rig.

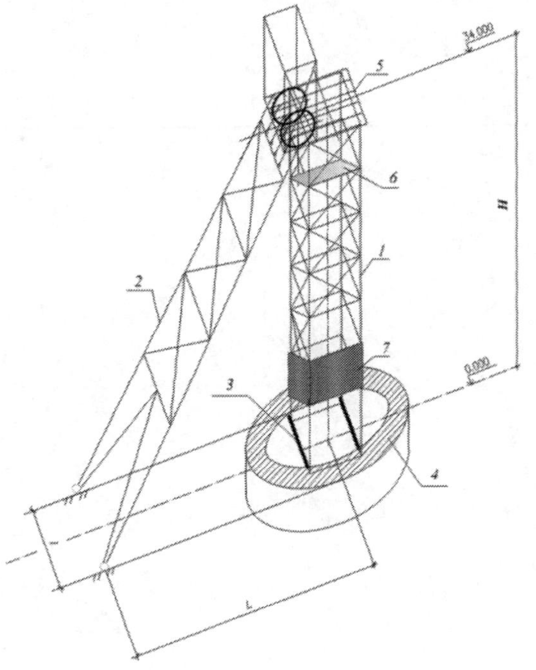

Fig. 1. Schematic of Steel Four-Post Headframe of Rig System: *1* – rig; *2* – jib; *3* – support frame; *4* – mine mouth; *5* – pulley pad; *6* – shock absorbers pad; *7* – lining.

In addition, rectangular rig configuration is connected with a possibility of conveyance dumping in two directions only: in the lifting plane and from the lifting plane.

Attempts to eliminate these disadvantages have been taken by M.M. Zerbinin in headframe of cylindrical system used in the mines of Donbass during the postwar years [12]. The main drawback of this headframe is support of cylinder on the mine mouth, which should take significant bending moments from lifting, which greatly complicates the design of headframe and mine mouth.

The Department of Underground Facilities and Mines Construction at KuzSTU developed a constructive solution for the headframe of multifunctional purpose [15, 16, 11]. The developed design made it possible to eliminate disadvantages of cylindrical headframe in parts of the upper and lower support units and preserve all its advantages.

The headframe (Fig. 3) consists of sheave wheels (it includes frame jib *1* of variable box cross-section, central tubular post *2*, pulley pads *3* and *4*, the annular spacer *5*) and rig of circular cross-section *6*, supported on mine mouth *7* via walling curb *8*.

The presented design of multifunctional headframe solves the design problems as follows:

1) loads from lift works are taken by independent block – sheave wheels *1-5*, and not transmitted to mine mouth *7*;

2) loads from guide conveyances *9* are taken by frameless round rig *6*, and transmitted to mine mouth via walling curb *8*;

3) compression and depression loads are taken by ring elements *10* (see Fig. 3) of rig.

Fig. 2. Schematic of steel headframe of multifunctional purpose for operational time: *1* – jib; *2* – central tubular post; *3*, *4* – pulley pads; *5* – ring base; *6* – rig; *7* – mine mouth; *8* – walling curb; *9* – conveyance; *10* – ring element.

3 Results

The offered technical solution is protected by patents of the Russian Federation [10-12] and has the following advantages versus cylindrical headframe and rig type headframe:

– frameless round rig [11] is separated from the sheave wheels, does not perceive and transmit load from pulleys to mine mouth, that greatly simplifies support unit on mine mouth;

– absence of base frame simplifies rigging up process of rig, makes operating conditions easier;

– absence of braces on frame jib and rig provides easy access for inspection and protection of elements and structure units, as well as it eliminates accumulation of coal dust, slag and lube at the nodes of connection;

– frameless rig *6* (see Fig. 3) consists of ring elements *10*, which have a much smaller number of joints and more reliable sealing versus to frame rigs *1* (see Fig. 1) with traditional fencing panels;

– ring elements have high resistance to bending and torsion, are more resistant to action of dynamic and wind loads;

- round cross-section on cut-away view of the frameless rig meets the layout requirements with some reserve in areas and volumes, which ensures the maximum possible equipment location;

- dumping of conveyances is possible in any direction, as openings in circular rig has no restrictions.

The main disadvantage of the headframes of multifunctional purpose is the lack of school for their design. If the designing of the rig headframes is based on the verified methods applying the vast domestic experience, then there is no technology for designing of the proposed headframe system.

Design of a headframe of any system starts with the determination of the headframe height *H*.

Determining the rig headframe height *H*, it is assumed that the height of the rig and the height of the headframe are identical concepts, and are measured as the difference between the marks of the axis of the upper head-block and mine mouth [12].

In case of placement of the head block on the same level:

$$H = h_1+h_2+h_3+0.75R_{sc\kappa}, \tag{1}$$

where h_1 – dumping place height, m; h_2 – conveyance height, i.e. distance from the dumping place to upper wire rope clip of suspension gear, m; h_3 – overwind height, m; $R_{sc\kappa}$ – head block radius, m.

Based on this, possible removal of winding machine from shaft ***l*** is accepted, that often causes difficulties in the already existing layout of the mine surface.

In angle headframes of multifunctional purpose the headframe height *H* and the rig height H_{cm} are different parameters. The rig height is calculated as per the formula as follows:

$$H_{cm} = h_a - 2\text{ m} \tag{2}$$

where h_a – height of the sinking sheave wheels pad as per technological requirements, m;

The headframe height *H* means the sheave wheels height *1-5*(see Fig. 2). In this case, removal of winding machine from shaft *l* is assigned by a customer, based on specific conditions of the mine surface layout, and the headframe height *H* is determined by computation chart (Fig. 4).

Analysis of the acting Federal regulations and rules in terms of industrial safety, as well as a priori sources [2, 3, 4] has shown that the allowed value area (AVA) $H \geq H_i^{min}$ (Fig. 3) of the headframe of multifunctional purpose is determined based on the following conditions:

a) maximum angle of inclination of the rope string to the horizon $\beta_1 = 50°$;
b) minimum angle of inclination of the rope string to the horizon $\beta_2 = 30°$;
c) maximum permissible length of the rope string $L_\beta^{max} \leq 65$ m;
d) compliance with the maximum deviation angles $\alpha_{dev} \leq 1°30'$;

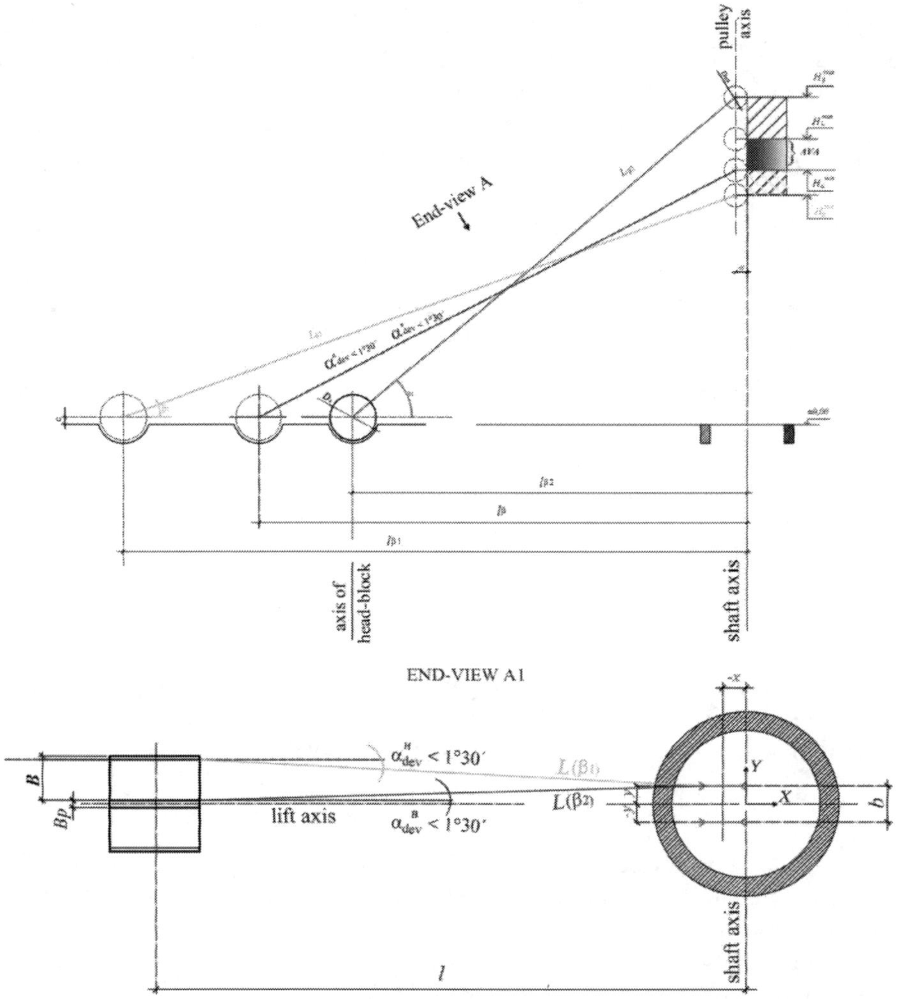

Fig. 3. Scheme to determine the allowed value area (AVA) of the sheave wheels height H depending on the critical deviation angles α^H_{dev}, α^B_{dev} on critical angles of the rope string inclination to the horizon β_1, β_2,° on critical rope string length L_β

The obtained values of H ($H_{\alpha i}^{max}$, $H_{\beta i}^{min}$, $H_{\beta i}^{max}$, H_L^{max}) are calculated as dependences on distance l, which in turn has *boundary conditions* [10-12]:

- maximum possible closing of the winding machine base and jib base as per the YUZHGIPROSHAKHT formula:

$$l'_{min}=0{,}45H + D_b+0{,}5D_{scк}+6; \quad (3)$$

- maximum possible closing of winding machine and headframe to eliminate vibration from headframe to winding machine

$$l''_{min} = 0{,}6H + D_b + 3{,}5 \text{ м} ; \quad (4)$$

- maximum possible removal of winding machine from headframe, guaranteeing absence of vibration of the inclined part of the hoisting line without installing additional support pulleys

$$l_{max} = 1{,}4(H - c) + 0{,}5D_{scк} \quad (5)$$

Taking into account these conditions, the boundary conditions are computed for H ($Hl'min$, $Hl''min$, $Hlmax$) on l.

Meanwhile it is required to consider the boundary conditions for H as per technological requirements:

$$Ht \geq h_a^{min} + 3 \text{ м} \quad (6)$$

where h_a^{min} = 19m – minimum height of the sinking sheave wheels pad as per technological requirements, m.

Based on the conducted factor analysis and as a result of numerical experiments the allowed value area (AVA) of the headframe heights H (see Fig. 4) has been determined.

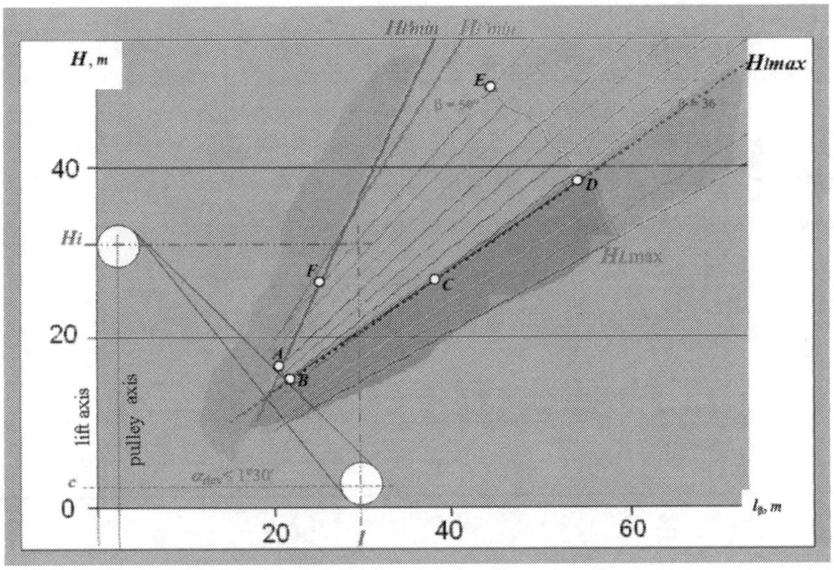

Fig. 4. Computation chart for the AVA graphical determination of the headframes heights H at the wheel width B = 3m, b = 1.45m, D_{sck} = 5m, D_b = 5m. The coordinates of the AVA boundary points: A(21, 15; 7), B(22, 1;15), C (33.0; 22.82), D(55.0; 38.5), E(45.0; 49.65), F (25.45; 26.50).

The obtained computation chart allows determining the headframe height ***H***, depending on the winding machine type and layout ***l*** relatively to the shaft, based on the specific conditions of the mine surface layout.

4 Conclusion

The proposed idea of independently operating structural units, one of which is a round rig, significantly reduces the number of controlled variables with which a designer can vary while the headframe designing. This greatly simplifies automation of the design process in search of an optimal solution, not a typical one, but an individual one.

This approach allows us to formalize the search for optimal technical solutions in lieu of the optional design and formulate a new design concept for steel angle headframes for state-of-art.

References

1. D. W. Butler, A. C. Schneyderberg, Mining Congress Journal, **7:1**, 15 (1982)
2. F. A. Edwards, Society for Mining, Metallurgy, and Exploration, **2,** 125 (1992)
3. J. I. Rojas-Sola, J. M. Montalvo-Gil, M. Castro-Garcia, **181**, 118 (2013)
4. H. L. Hartman, S. G. Britton, J. M. Mutmansky, Society for Mining, Metallurgy, and Exploration, **3,** 342 (1998)
5. I. Ross, *Goldcorp going electric with Chapleau gold mine* (Northern Ontario Business, Ontario, 2016)
6. J. I. Rojas-Sola, I. Palomares-Muñoz, Dyna, **6**, 602, (2015)
7. F. Wei, I. F Wang, R. *3D Parametric Design for Steel Headframe of Coal Mine Based on Solidworks* (Key Engineering Materials, London, 2010)
8. L. Kelly, Histor. End., **34**, 31 (2014)
9. E. U. Reuther, Erster Band, **12,** 37 (2010)
10. E. G. Kassikhina, Gornyi zhurnal, **8,** 56 (2017)
11. E. G. Kassikhina, Qiao Weiguo, V. V. Pershin, N. O. Butrim, *Selection of a Rational Form for the Steel Winding Tower as a Preventive Measure to Increase its Industrial Safety* (Atlantis Press, Paris, 2014)
12. E. G. Kassikhina, V. V. Pershin, J. F. Glazkov, E3S Web Conf., 41, 01025 (2017)

 https://doi.org/10.1051/e3sconf/201913401017

Increasing safety at rock preparation by blasting in open pits

Igor Katanov[1,*], *Sergey* Kondratyev[2], and *Andrey* Sysoyev[1]

[1]T. F. Gorbachev Kuzbass State Technical University, 650000 Kemerovo, 28 Vesennya st., Russian Federation
[2]JSC NMZ Iskra, Novosibirsk, Russia

Abstract. The topicality of the present work lies in justification of the means of increasing safety during mass explosions in opencasts. The solution to this problem is possible due to the orientation of downholes in the array, taking into account the angle of incidence and extension of the strata, high-quality stemming and initiation of downhole charges by delay detonators with a minimal deviation from nominal values of their firing time. The most promising way to increase the safety and efficiency of blasting is an integrated approach, including drilling downholes, normal to the strata, and using the design of the downhole charge, rationally distributing the explosion energy in the space of the rock mass due to the use of stemming of a variable density, low-density gap between the parts of the explosive charge and the bottom a downhole compensator based on foam gel and the use of hybrid electronic detonator Iskra-T.

1 Introduction

Kuzbass has very diverse and complex mining and geological conditions for the development of coal opencasts. The seams of coal lie in the form of suites with different angles of incidence. Overburden consists of sandstones, siltstones and mudstones. The structure of rock masses sometimes even within the limits of a single blasting block varies from small-block (with sizes of natural units 200-400 mm) to large-block (with sizes of units 3000-5000 mm and more). In some cases, rock crushing, especially large-block structures, using the existing blasting technology is not enough to ensure productive operation of mining and transport equipment. One of the primary tasks of open cast mining is to improve the quality of rock preparation for excavation by excavator with the creation of safe conditions during blasting.

2 Materials and Methods

An important tendency in improving blasting is to increase the efficiency of the explosion by means of rational redistribution of explosive energy. By changing the spatial arrangement

*Corresponding author: noa-0025@yandex.ru

of downhole charges in a rock mass, it is possible to rationally distribute the energy of the explosion in the array.

F. A. Baum presented the mechanism of rock destruction by explosion as the combined effect from detonation products, shock waves and rarefaction waves [1]. In time and in the nature of physical phenomena, an explosion is characterized by several successive stages. At the initial stage of the process, the main role is played by expanding products of detonation, which at the first moment are under pressure of about 10^5 kg / cm^2. At this stage of explosion development in the rock array, there occurs plastic flow of the rock, emergence of radial cracks and its crushing. At the charge-rock contact, the compressibility law is applied, which determines the initial parameters and certain patterns of shock wave propagation. The second stage of the process begins after the shock waves exit to the free surface. It is associated with the propagation of the rarefaction wave reflected from this surface and its interaction with the tail of the traveling compression wave. As a result of this process, significant tensile stresses arise in the rock and, consequently, new cracks appear. The intersection of these cracks and the natural micro- and macrocracks present in the rock with the system of radial cracks arising in it at the previous stage of the explosion leads to the formation of a spatial network of cracks, the nature of which to a certain extent determines the possible size and number of rock pieces, crushed during the explosion. The third, final stage of the process is reduced mainly to the effect on the rock of already expanded products of detonation. By the time of its full expansion, detonation products still contain about 50% of the explosion energy of its total amount. Under the influence of this energy, various types of rock destruction occur during the penetration of gases into cracks, separation of individual pieces from the array, etc., displacement of the destroyed mass of rock and scattering of individual pieces of the array.

In the works of G. I. Pokrovsky [2], who considered the principle of self-similarity of the shock wave propagation during an explosion and main energy dependences, physically deeply substantiated representations of a qualitative picture of the effect of an explosion on a medium, established the parameters of a shock wave arising in a medium during the explosion of a spherical explosive charge.

A. N. Hanukaev [3] adheres to the theory according to which the process of destruction occurs under the action of elastic waves. Basing on the theory of rock destruction by the reflected wave, a proposition is put forward that the process of rock destruction largely depends on their acoustic rigidity.

G.P. Paramonov [4] shows that the destruction mechanism of a fractured array depends on what the cracks are filled with. If the cracks are filled with solid material, then the destruction occurs mainly due to stress waves. When cracks are filled with air, the stress wave destroys only the part, in which the charge is located, further destruction occurs due to the products of detonation.

Adhering to the classical ideas about the mechanism of destruction of rocks with a layered structure, explosive preparation of rocks for excavation by the method of downhole charges includes drilling parallel rows of vertical or deviated downholes, loading them with explosive followed by stemming, assembling blast pattern and short-delay blasting.

Downholes are drilled at a calculated distance according to a rectangular or checkerboard pattern. Moreover, in the array downholes are located so that the longitudinal axis of each well on the plane of the ledge slope is designed as a perpendicular to the generators of the upper and lower edges of the ledge. In this position, the axis of the well intersects the bed plane at a certain angle. In most cases, the most favorable conditions for crushing the array occur when the fracture load from the blast of an explosive charge in the rock layer is strictly perpendicular to the bedding or when the bedding extends parallel to the charge [5, 6].

At present, the design of explosions does not take into account the relationship of shear stresses in the direction of layering and the direction of fracturing of the array. If the angle of incidence of the strata is between 35 and 65 degrees with the downhole axis, then the stress wave attenuates in a separate part and the quality of the array fragmentation deteriorates. Therefore, for uniform crushing of the array with an unfavorable location of the charge relative to the strata, an increased specific consumption of explosive is required. Charges with the same specific explosive consumption on one block without taking into account the structure of the array crush it with unpredictable results.

3 Results and Discussion

In order to study the possibility of uniform crushing of the rock within the entire ledge, before designing drilling and blasting operations at one of the casts, the array was surveyed using the OKO-2 georadar with a screened collapsible antenna unit AB-9. Analysis of radarograms and photos of the slope of the ledge showed that the angle of incidence of the strata surface was 10-22 degrees, and the azimuth of their fall 120-140 degrees. At a certain distance from the inspection site, the surfaces of the strata flatten out, and after another 85-100 m the angle of incidence of the strata changes to 146 degrees.

Quality analysis of the explosive preparation of the rock for excavation at Kuzbass coal opencasts allowed us to obtain a pattern of change in the size of the piece in the broken rock disintegration depending on the spatial location of the downholes in the layered array (Fig. 1).

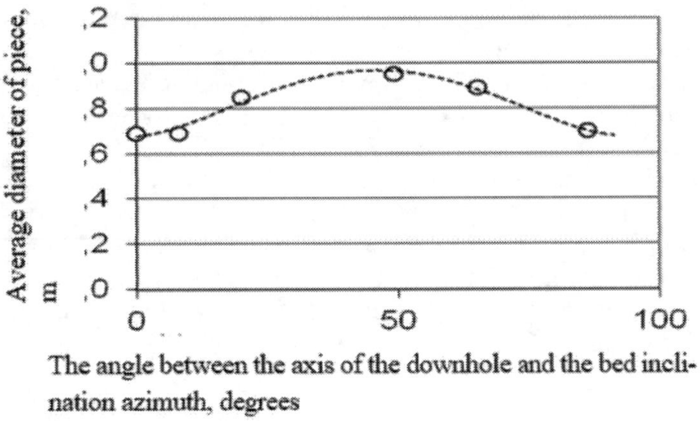

Fig. 1. Change in piece size in broken rock disintegration depending on the angle of the downhole charge with the bedding of the array.

Analysis of the graph shows that a good quality of overburden preparation for excavation was obtained when drilling downholes relative to the bedding plane at an angle of 85-90 degrees and if the drilling direction coincides with the bedding plane.

Since [7] allow the drilling rig to be installed at any angle with respect to the slope of the ledge, downholes can be drilled at an angle close to normal with respect to the bedding azimuth.

All other things being equal, the material and stemming quality play a significant role in the design of the downhole charge. Stemming of downholes is designed to increase the momentum of detonation products due to the duration of their impact on the array. In Kuzbass coal opencasts, as a rule, drill cuttings are used as a stemmer. It is recommended that the magnitude of the stemming from drill cutting be proportional to the diameter of the downholes. Drill cuttings and other high-density materials used as stemming during mass explosions have high inertial characteristics, so they fly out of the downhole without properly locking the detonation products.

The greatest effect is obtained by stemming, which ensures the locking of detonation products due to clamping of the downhole channel. These properties are possessed by a variable density stemming consisting of a combination of a low-density porous material and a layer of drill cuttings. The property of low-density porous stemming to enhance the explosive effect of an explosion on an array in combination with the possibility of dust suppression.

The mechanism of this stemming construction is as follows. After the downhole is loaded with explosive, the space above the charge is filled with foam gel in the sleeve, the top of the sleeve is filled with a drill cuttings to the level of intense fracturing of the rock array from the previous explosion.

The effect of clamping the downhole channel in the upper part is explained by the fact that the propagation velocity of the stress wave has a finite value, and, depending on the blockiness of the array, amounts to 800–3000 m / s in the rock, 330 m / s in air, and 1500 m / s in water, and in the low-density composition only 30–50 m / s.

In addition to low-density stemming, the design of the downhole charge may consist of a low-density gap dispersing the downhole charge. This is carried out using a sleeve with its filling with foam gel [8].

Blasting of downhole charges is usually carried out by short-delayed blasting with non-electric initiation systems with shock tubes.

The main technological drawback of non-electric initiation systems is the low accuracy of the moment of charges initiation, which does not always allow achieving the desired results of a mass explosion.

When using short-delay blasting with non-electric initiation systems the main contribution to the deviation from nominal delay time is made by downhole detonators with a long delay time (1000-2000 ms). "Novosibirsk Mechanical Plant "Iskra" developed and is launching a new generation of initiation means - hybrid electronic detonators ISKRA-T [9].

Unlike pyrotechnic initiation systems in hybrid electronic detonators ISKRA-T, instead of the delay composition, where the deviation of the delay time reaches ± 7% of the nominal value, an electronic micromodule is used which provides an accuracy of ± 1 ms with any specified delay time in the range from 0 up to 2000 ms and with a programming step of a delay time of 1 ms (Fig. 2).

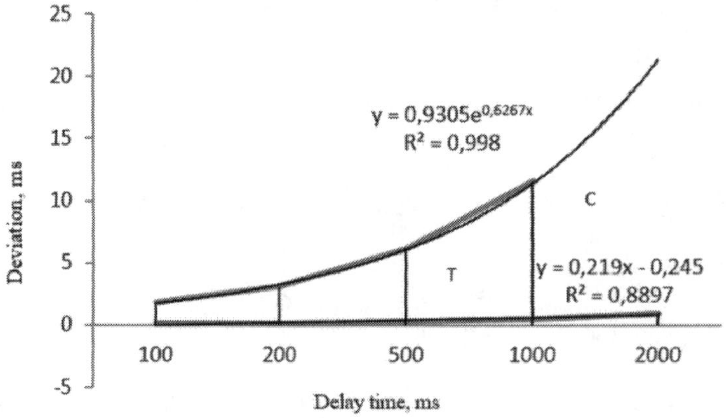

Fig. 2 Delay time deviation depending on the delay time.

4 Conclusion

As the experimental results show, the use of ISKRA-T downhole detonator caps with electronic delay is more preferable from the point of view of seismic effects of industrial explosions.

Thus, an integrated approach to the design and conduct of mass explosions with the above measures will improve the safety of blasting in coal opencasts.

References

1. Baum, M.A. Berezhets, Digest: Blasting, **49:6**, 109-121 (1962)
2. G. I. Pokrovsky, *Blast* (Nedra, Moscow, 1973)
3. A. N. Hanukaev, *Energy of stress waves during the destruction of rocks by explosion* (Gosgortechizdat, Moscow,1962)
4. G. P. Paramonov, Notes from Mining Institute, **204**, 294-296 (2013)
5. I. B. Katanov, Patent Bul., **34** (2016)
6. S.O. Markov, M.A. Tyulenev, R.Yu. Aliyarov, Journal of Mining and Geotechnical Engineering, **1:4**, 69-78 (2019). DOI: 10.26730/2618-7434-2019-1-69-78
7. *Federal norms and rules in the field of industrial safety "Safety rules for open pit mining of coal deposits"* (Federal Service for Ecological, Technological and Nuclear Supervision Moscow, 2017)
8. I. B. Katanov, Patent Bul., **25** (2018)
9. S. A. Kondratyev, S. A. Pozdnyakov, A. S. Ivanov, K.A. Vandakurov, Blasting, **123:80**, 136-144 (2019)

Criteria for the difficulty of working out coal-bearing zones of quarry fields

Alexei Selyukov[1,*], and *Nuray* Demirel[2]

[1]T.F. Gorbachev Kuzbass State Technical University, 650000 Kemerovo, 28 Vesennya st., Russian Federation
[2] Middle East Technical University, Üniversiteler Mah., Dumlupınar Bulvarı, No: 1, 06800 Çankaya, Ankara, Turkey

Abstract. Coal-bearing zones of open pit fields are represented by formations of coal seams from inclined to steep bedding with unstable thickness, both in dip and strike with different dip angles even within the same strata, with varying rock inter-layers and strength. In addition, most coal seams have a complex structure, including rock layers inside. In most cases, there is an uneven distribution of coal seams, and, consequently, coal reserves over the area of quarry fields. In this regard, for the analysis of quarry fields, there is a need to select criteria for assessing the complexity of coal-bearing zones for the purpose of systematizing them and choosing directions for constructing technological schemes for excavation and loading operations and mining methods.

1 Introduction

The development of technical equipment and the scale of open pit coal mining draws increased attention to the Kuzbass open pits, as the Kemerovo Region (Western Siberia, Russia) has limited agricultural resources [1-3]. The deep-seated longitudinal single and double–sided system for developing inclined and steep coal deposits that is currently widely used contributes to the progressive disruption of the earth's surface, both by mining operations and external dumps [4-6], while the specific land intensity reaches 55 ha per 1 million tons of coal. The disadvantages also include the long range of transportation of overburden rocks to the dump, both by rail and by road. All this increases the cost of coal mining and reduces its competitiveness in the market [7-8].

The negative impact on the environment can be minimized by changing the mining order of the existing open pits of Kuzbass coal basin, while using mining methods with internal partial or complete dumping. If this is not foreseen at present, then all sections in the coming decades will be limited by their own dumps and their further development will be problematic.

Constraining factors for the development of new technology are: the lack of options for the formation of initial mine workings and new varieties of development systems with internal dumping in the conditions of longitudinal deep, and recommendations for the formation of technological stages and transition schemes; substantiation of mining

* Corresponding author: sav.ormpi@kuzstu.ru

technology in the transition period and during the completion of quarry fields and, ultimately, insufficient knowledge of the transformation of deep longitudinal into transverse mining method in the mode of an active section.

Therefore, on the existing open pits of the Kuzbass coal basin, there has been a tendency to use systems for the development of inclined and steep deposits with the placement of overburden in the worked out space. At the same time, the studies carried out on the above mining method are based on the use of high-cost technological systems, however, at the same time, the open pit mining conditions allow the use of direct dumping technology [9-11].

2 Materials and Methods

Currently, the mining–and–geometrical analysis of quarry fields mainly uses the indicator "stripping ratio". However, this indicator does not characterize the difficulty of the quarry fields mining.

In this regard, the following criteria are proposed for assessing the technological complexity of the quarry fields mining.

1. The density distribution of coal saturation of the quarry field. The physical meaning of this criterion is to assign coal reserves in blocks or in the entire quarry field to the total volume of P_{yi} blocks or P_y quarry fields.

$$P_{yi} = \frac{V^b_{yi} \cdot \rho^b_{yi}}{V^b_{yi} + V^b_{ni}} \quad (1)$$

$$P_y = \frac{V^{kp}_y \cdot \rho^{kp}_y}{V^{kp}_y + V^{kp}_n} \quad (2)$$

where V^b_{yi}, V^b_{ni} – is the volume of coal and rocks, respectively, in the *i–th* block, m^3; ρ^b_{yi} – the average density of coal in the *i–th* block, t / m^3;

V^{kp}_y, V^{kp}_n V^{kp}_y, V^{kp}_n – the volume of coal and rocks, respectively, within the boundaries of the quarry field, m^3; – the average density of coal within the boundaries of the quarry field, t / m^3.

Dividing the numerator and denominator by the result of multiplication and by V^b_{yi}, ρ^b_{yi}, we obtain:

$$P_{yi} = \frac{1}{\frac{1}{\rho^b_{yi}} + K^{cp}_{Bi}} \quad (3)$$

$$P_y = \frac{1}{\frac{1}{\rho^{kp}_y} + K^{cp}_b} \quad (4)$$

where K^{cp}_{Bi} and K^{cp}_B are the average stripping ratio within the boundaries of the *i–th* block and the career field, respectively.

This criterion can be used when choosing a technological strategy for developing a quarry field and a development system.

2. The share of the coal-bearing zone in the total quarry field. The essence of the criterion is to assign the volume of the coal-bearing zone to the total volume of the quarry field, i.e.:

$$\delta_{cb} = \frac{V_{cb}^{kp}}{V_{y}^{kp} + V_{n}^{kp}} \tag{5}$$

where V_{cb}^{kp} is the volume of the coal-bearing zone within the boundaries of the quarry field, m^3.

This criterion is used in assessing the overall efficiency of mining a quarry field using various technological options.

3. Brand composition of coals in the layers of the coal-bearing zone and their share in the total reserves of the quarry field. The criterion is supposed to be used when planning mining operations and evaluating the effectiveness of technological options for mining quarries.

4. The angle of incidence of the formation. This indicator affects the choice of technological schemes for excavation and loading operations and the structure of the excavation and transportation complex.

5. The thickness of the rock interlayers. The criterion determines the possibility of selective extraction of coal seams of the strata, characterized by the degree of proximity.

6. The degree of proximity of coal seams. According to this criterion, all the layers of the strata are divided into close and dispersed. Close coal seams of inclined and steep dipping include seams between which the thickness of the rock interlayer does not allow preparation for their excavation by working cutting trenches from the hanging side. The dispersed seams are characterized by the possibility of their separate excavation with the cutting of a trench from the side of the hanging side. Close coal seams of a flat bedding in the transport zone include seams whose distance between them does not provide the formation of a transport platform for a dead end turn and the movement of trucks.

This criterion is decisive in deciding on the technological option of mining coal seams and the selection of mining and loading and transport equipment.

7. The proportion of close and dispersed coal seams in their total number in the coal–bearing zone of the quarry field. This indicator is used in a comparative evaluation of the efficiency of technological schemes of excavation and loading operations by complexes of excavation and transportation equipment and the overall efficiency of mining coal–bearing zones of open pit fields.

3 Analysis of coal-bearing and coal-less zones of the quarries of Kuzbass

Open pit mine "Kedrovskiy." For the "Main field" section the share of the coal–bearing zone in the quarry field is 0.48. The density of coal saturation is 0.129 t/m^3. The mutual position of the seams within the quarry field is: close – 27% and dispersed – 15% of the volume of the quarry field. For "Latyshevsky field": the density of coal saturation – 0.11 t/m3; share of the coal–bearing zone – 0.59; close seams – 33%; dispersed – 10%. For "Khorosheborsky field," the density of coal saturation is 0.136 t/m^3; Fraction of coal–bearing zone – 0.51; close seams – 30%; dispersed – 10%. As a whole, the weighted averages are: density of coal saturation – 0.125 t/m^3; fraction of coal–bearing zone – 0.53; close seams – 30%; dispersed – 8.3%.

Open pit mine "Bachatsky." "Block North." Density of coal saturation – 0.136 t/m3; Fraction of coal–bearing zone – 0.59; At that, the close seams – 13%; dispersed – 9%. It should be noted that when calculating for 2–3 weighted average geological sections, seams with thickness of 1m and less were taken into account. "Central Block." Of the quarry field blocks analyzed, this site is the largest. Thus, the fraction of the coal–bearing zone is 0.62; Density of coal saturation – 0.151 t/m^3. At the same time the share of close seams – 32%, and dispersed – 4%. "Southern block" density of coal saturation – 0.128 t/m^3; Mutual position of the close seams – 29%; dispersed – 10%. Weighted average characteristics of parameters of the carbon–bearing zone. Density of coal saturation – 0.138 t/m^3; fraction of coal–bearing zone – 0.57; close seams – 24.6%, dispersed – 7.6%.

Open pit mine "Krasnobrodsky." "Krasnobrodsky field" of the section. For this field, the density of coal saturation within the boundaries of the quarry field is – 0.133 t/m^3; Share of coal-bearing zone in quarry field – 0.53; close seams – 38%; dispersed – 9.3%. "Nowergeyevsky Field." For this field the density of coal saturation – 0.121 t/m^3; fraction of carbon–bearing zone – 0.48; close seams – 26%; dispersed – 8.9%. The average values for the two fields of the section are: density of coal saturation – 0.127 t/m^3 of the coal-bearing zone – 0.5; dispersed seams – 9.1%; close – 32%.

Open pit mine "Osinikovsky." When considering the field of this section, the following weighted averages are established: density of coal saturation within the field boundaries – 0.125 t/m^3; Share of the coal-bearing zone in the quarry field – 0.59; close seams – 8%; dispersed – 22%.

Open pit mine "Vakhrushevsky." Parameters characterizing the coal-bearing zone: density of coal-bearing – 0.116 t/m^3; zone's share within the quarry field – 0.46; close seams – 32%; dispersed – 13%.

Open pit mine "Kaltanskiy". Density of coal saturation within the field boundaries – 0.119 t/m^3; fraction of coal-bearing zone – 0.45; close seams – 18.5%; dispersed – 21%.

4 Conclusion

The effectiveness of transverse mining methods depends to a large extent on the share of internal dumping in the total opening volume in the boundary contours of the open pit, the method of internal dumping, the parameters of the quarry field and the development options determining the volume of internal dumping. The volumes occupied by mining taking into account the developed space do not exceed 10%–30% of the total volumes of the quarry field.

The control over mining operations can be achieved by concentrating them in a coal-bearing zone. The duration of the technological transition depends on the depth of the seams bedding, the thickness of the strata, the productivity of the technical complex. The annual production capacity and for the cuts of Kuzbass is 8–11 years (dipping-and-longwork mining method) and up to 2–6 months (shuttle-and-layer mining method). Technologically, the transition to internal dumping is possible when the level of reserves covers the duration of the transition with compliance with the criteria of the algorithm for selecting the location of workings.

References

1. M. Tyulenev, S. Markov, M. Cehlar, S. Zhironkin, M. Gasanov, Acta Montanistica Slovaca, **23:4**, 368-377 (2018)
2. T. Gvozdkova, M. Tyulenev, S. Zhironkin, V. A. Trifonov, Yu. M. Osipov, IOP Conf. Ser.: Earth Environ. Sci., **50:1**, 012010 (2017)
3. M.I. Agienko, E.P. Bondareva, G.V. Chistyakova, O.V. Zhironkina, O.I. Kalinina, IOP Conference Series: Earth and Environmental Science, 50:1, 012022 (2017)
4. M. Prokudina, O. Zhironkina, O. Kalinina, M. Gasanov, F. Agafonov, E3S Web of Conferences, **21**, 04003 (2017)

5. S. Markov, J. Janočko, M. Tyulenev, Y. Litvin, E3S Web of Conferences, **105**, 01021 (2019)
6. A.B. Palamarchuk, A.V. Stukan, T.N. Gvozdkova, Journal of Mining and Geotechnical Engineering, **1(4)**, 50-68 (2019). DOI: 10.26730/2618-7434-2019-1-50-68
7. S.A. Zhironkin, Ugol', **4,** 29-31 (2001)
8. M., Cehlár, K. Teplická, A. Seňová, 11th International Multidisciplinary Scientific Geoconference and EXPO - Modern Management of Mine Producing, Geology and Environmental Protection, SGEM 2011, **1,** 913-920 (2011)
9. M. Tyulenev, O. Litvin, S. Zhironkin, M. Gasanov, Acta Montanistica Slovaca, **24:2,** 88-97 (2019)
10. M.A. Tyulenev, S.A. Zhironkin, E.A. Garina, Int. J. Min. Miner. Eng., **7:4**, 363-370 (2016)
11. M. Tyulenev, S. Zhironkin, E. Tyuleneva, S. Anyona, M. Hellmer, Coal International, **265:3**, 30-34 (2017)

The research of environmental-and-economic risks of the coal mining enterprise impact on water resources

Vladimir Mikhailov[1,*], *Natalia* Kudrevatykh[1], and *Tatiana* Tyuleneva[1]

[1]T.F. Gorbachev Kuzbass State Technical University, 650000 Kemerovo, 28 Vesennya st., Russian Federation

Abstract. Coal mining enterprise is a high-tech complex with a significant negative impact on water resources in the form of wastewater discharge. To improving the environmental safety of enterprises having a negative impact on the aquatic ecosystem, it is necessary to detail its features in order to increase the efficiency of environmental management. The article analyzes known approaches to the definition and features of formation of systems of assessing the negative impact on water resources. On the basis of synthesis of theoretical developments of domestic and foreign researchers' study and analysis the actual data of an industrial enterprise authors researched its negative impact on water resources to form a program of effective water protection measures. This researching has practical importance for large industrial enterprises that have a negative impact on water resources for effective environmental management decision-making.

1 Introduction

Coal mining is one of the most dangerous sectors of real sector of economy that is associated with a wide range of negative effects, one of which is pollution of water resources concerning not only waste-water discharges, but also coarse air pollutants, and "flushing" waste production and consumption, including ones characterized by chemical and biological activity [1-4]. Tightening national environmental legislation and international requirements [5, 6] motivate coal mining enterprises to use innovative modern technologies, which are called BAT (best available technologies). Despite of intensification this activity, coal mining enterprises are a potential source of high environmental and economic risks, and this fact requires increasing the efficiency and diversification of management methods [7, 8]. In such circumstances, it is necessary to identify the most significant environmental and economic aspects of negative impact on water resources and their assessment on the basis of developed system of environmental and economic indicators. For large industrial enterprises carrying out a complex of basic, auxiliary and service technological processes, the problem of wastewater discharge is urgent, and determines the feasibility of analyzing this type of negative impact with development of measures to ensure environmental safety.

* Corresponding author: mvg.eohp@kuzstu.ru

2 Materials and Methods

The object of research is a coal mining enterprise as a source of negative impact on the environment in the form of wastewater discharges.

The subject of the research is the system of basic ecological and economic indicators of an enterprise characterizing pollution of water sources, including ecological and economic risks.

Researching is based on domestic and foreign literature about problems of ecological and economic analysis of enterprises that discharge wastewater, with using adequate evaluation indicators. Special attention is paid to researching the dynamics of pollutants` discharge as sources of environmental and economic risks. For the purpose of practical implementation assessment methodology, statistical processing of environmental data of the coal mining enterprise on the basis of official reporting forms.

3 Results and Discussion

There are many scientific points of view to the formation of a system of indicators characterizing the negative impact on water resources, which depend on the industry enterprise specifics and other factors [9, 10].

Researching this enterprise was carried out with the help of a number of environmental and economic indicators, where the central place is occupied by economic damage from negative impact on water resources (ED_{WR}) – economic losses of enterprises caused by negative environmental impact on water sources. At the same time, it is advisable to conduct an independent calculation for each pollutant of water resources with the determination of the most significant component [11-13]. Another important indicator is charge for excess negative impact on water resources (C_{WR}), which compensates for the economic damage caused. To increase the assessment efficiency, the following indicators are calculated: the charge for excess negative impact on water resources, its share in the total amount of charges for negative impact on water resources, and coefficient of compensation for economic damage caused to water resources ($CCED_{WR}$), reflecting the effectiveness of the existing mechanism of environmental payments - formula (1):

$$CCED_{WR} = C_{WR} / ED_{WR} . \tag{1}$$

Additional relative environmental and economic indicators - water damage capacity (WDC) of coal production - formula (2) and water charge capacity (WCC) of coal production - formula (3), characterizing corresponding environmental and economic impacts per 1 ton of coal production:

$$WDC = WCC / PO, \tag{2}$$

where PO - production output (coal production), tones,

$$WCC = C_{WR} / PO . \tag{3}$$

In table 1 results of calculation main ecological and economic indicators of the coal mining enterprise are presented [14-16].

Table 1. Results of calculation of the main ecological and economic indicators of the enterprise JSC "SUEK-Kuzbass" S.M. Kirov Mine.

Indicator	2013	2014	2015	2016	2017
Economic damage from negative impact on water resources, million rubles	143.89	60.49	4.72	6,30	6.39
Total charge for negative impact on water resources, thousand rubles	993.5	1465.8	316.0	312.0	42.9
Charge for excess negative impact on water resources, thousand rubles	949.3	1432.1	266.0	0	0
Charge for negative impact on water resources` share, %	95.55	97.70	84.18	0	0
Coefficient of compensation for economic damage caused to water resources, %	0.69	2.42	6.69	4.95	0.67
Production output (coal production), thousand tones	3747	4198	4668	5720	7177
Water damage capacity of coal production, rubles / tone	38.40	14.41	1.01	1.10	0.89
Water charge capacity of coal production, rubles / tone	0.27	0.35	0.07	0.05	0.01

From table. 1 it can be seen that the value of $CCED_{WR}$ for the period significantly decreased from 143.89 million rubles in 2013 to 6.39 million rubles in 2017. The analysis of C_{WR} shows a sharp decline in the indicator-from 1465.8 thousand rubles in 2014 to 42.9 thousand rubles in 2017. Charge for excess negative impact on water resources and its share in the total amount of charges change "nonlinearly".

To assess the compensation function of charge for negative impact on water resources, $CCED_{WR}$ has great importance, and its value was maximum in 2015 (6.69 %), and then it decreased to 0.67 % in 2017.

Improving adequacy of the assessment can be implemented with indicators WDC and WCC coal production. WDC has a maximum value in 2013 - 38.40 rubles/tone, and then a sharp decline begins to 0.89 rubles/tone, which may be characterized as a positive trend. In terms of WCC a similar situation is observed - a decrease of value from 0.35 rubles/tone in 2014 to 0.01 rubles/tone in 2017.

The analysis of ecological and economic indicators of negative impact of water resources requires a more detailed researching pollutants as a source of environmental and economic risks.

In Fig. 1 the dynamics of pollutants` concentration with the largest mass in wastewaters of JSC "SUEK-Kuzbass" S. M. Kirov Mine by quarters taking into account "representatives" of each hazard class (nitrite nitrogen-II, nitrate nitrogen-III, sulphates-IV, suspended substances - without class of danger).

From Fig. 1 it can be seen that quarterly values of pollutant concentrations differ significantly. In particular, the maximum value of nitrite nitrogen is observed in the 2nd quarter of 2014 (0.0509 t) and its minimum one is done in the 4th quarter of 2017 (0.0042 t). Also, a large range of values takes place for sulfates, respectively it is 83.34 t and 12.78 t. This situation is observed in the implementation of environmental policy of the enterprise associated with the technological process of coal mining.

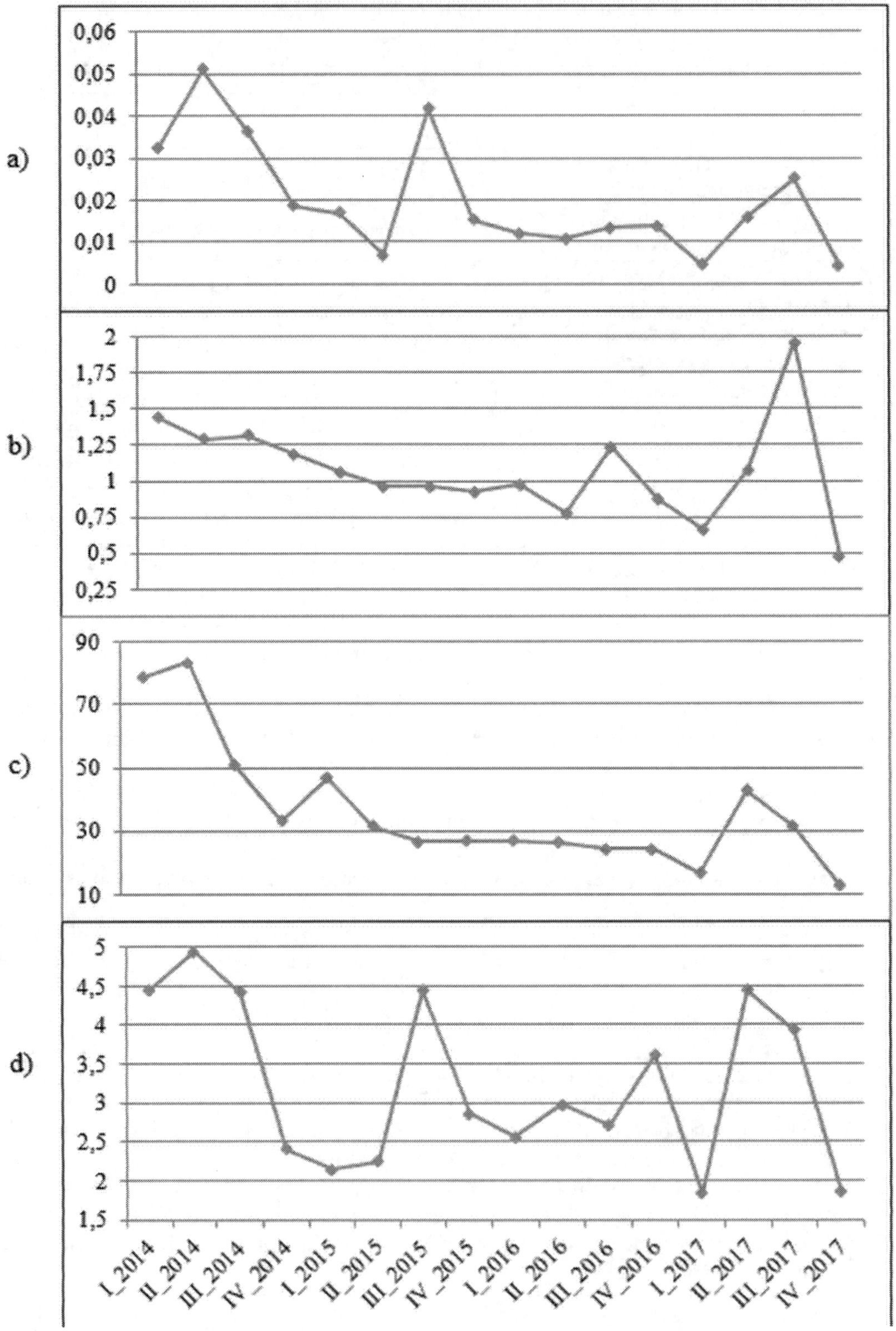

Fig. 1. Dynamics of pollutants` concentration with the largest mass in wastewaters of JSC "SUEK-Kuzbass" S. M. Kirov Mine by quarters: a) - nitrite nitrogen, b) - nitrate nitrogen, c) - sulfates, d) - suspended solids.

A wide range of pollutant mass values requires an analysis of potential environmental and economic risks and it can be performed statistically [17-21] on the basis of checking the data sample for homogeneity. For this purpose it is necessary to estimate maximum, minimum and average values of the characteristic (x_{max} , x_{min} и x_{cp}), scope of variation (R), variance (σ^2), the standard deviation (σ) and variation coefficient (V).

Table 2. Main statistical indicators of environmental and economic risks assessment for selected pollutants contained in wastewater of JSC "SUEK-Kuzbass" S. M. Kirov Mine

Pollutant	x_{max}, t	x_{min}, t	R, t	x_{cp}, t	σ^2	σ, t	V
Nitrite nitrogen	0.0509	0.0042	0.0467	0.0199	0.0002	0.0132	0.66
Nitrate nitrogen	1.9541	0.4812	1.4729	1.0722	0.1105	0.3324	0.31
Sulfates	83.34	12.78	70.56	36.39	376.88	19.41	0.53
Suspended solids	4.9465	1.8271	3.1194	3.2368	1.0632	1.0311	0.32

From table. 2 it can be seen that values of variation coefficient for these pollutants are very high, and characterize a significant probability of environmental and economic risks. For example, obtained value V for nitrite nitrogen (0.66) exceeds standard one (0.25) by more than 2 times, and it determines the importance of researching causes of this situation. In general obtained V values show the need for a detailed analysis of significant range of these pollutants` quarterly values.

4 Conclusion

The researching allows us to draw the following conclusions:

- known approaches to the formation of a system of indicators characterizing the negative impact on water resources are analyzed;
- absolute and relative ecological and economic indicators for assessment negative impact of coal mining enterprise on water resources are proposed;
- calculation and interpretation of ecological and economic indicators of negative impact of coal mining enterprise on water resources for the purpose of preparation of management decisions aimed at ensuring environmental safety are carried out.;
- graphs of the dynamics pollutants of each hazard class` quarterly values according to the actual data of the enterprise are constructed;
- statistical analysis of changes in polluting water resources selected substances` quarterly values to assess potential environmental and economic risks and make environmentally safe management decisions is carried out.

References

1. M. Tyulenev, Y. Lesin, E. Tyuleneva, E. Murko, E3S Web of Conferences, **15**, 02003 (2017)
2. N.M. Kachurin, V. I. Efimov, S.A. Vorobev, Gornyi Zhurnal, **9**, 138-142 (2014)
3. M. Tyulenev, S. Markov, E. Makridin, Y. Lesin, V. Gogolin, E3S Web of Conferences, **105**, 02022 (2019)
4. M. Cehlár, P. Varga, Z. Jurkasová, M. Pašková, Acta Montanistica Slovaca, **15: SPECIAL ISSUE 2,** 132-138 (2011)

5. M. Cehlár, L. Mihok, Theoretical and Empirical Researches in Urban Management, **8:4**, 60-72 (2013)
6. V.M. Tumin, A.G. Koryakov, E.P. Nikiforova, World Applied Sciences Journal, **25:6**, 645-649 (2013)
7. V.N. Burkov, I.V. Burkova, Game Theory and Application, **17**, 17-36 (2015)
8. G.A Ugolnitskii, A.B. Usov, Automation and Remote Control. **70:5,** 897-906 (2009)
9. E.A. Tretyakova, World Applied Sciences Journal, **25:5**, 756-759 (2013)
10. O.V. Kosolapov, M.N. Ignatyeva, A.A. Litvinova, Economy of Region, **1,** 158-166 (2013)
11. G.M. Mkrtchyan, T.O. Tagaeva, Yu.O. Tsvelodub, World of Economics and Management, **17:1**, 57-69 (2017) (In Russ.)
12. A.S. Tulupov, Gornyi Zhurnal, **8**, 61-65 (2017)
13. B.N. Porfiryev, A.S. Tulupov, Studies on Russian Economic Development, **28:6**, 600-607 (2017)
14. V.G. Mikhailov, T.V. Kiseleva, V.A. Karasev, G. S. Mikhailov, V.A. Skukin, IOP Conf. Ser.: Earth Environ. Sci., **66:1**, 012008 (2017)
15. V. Mikhailov, T. Kiseleva, S. Bugrova, A. Muromtseva, Ya. Mikhailova, E3S Web of Conf., **21**, 02004 (2017)
16. V.G. Mikhailov, S.M. Bugrova, Yu.S. Yakunina, A.K. Muromtseva, Ya.S. Mikhailova, Ugol, **9**, 106-111 (2019)
17. V.P. Avdeev, V. N. Burkov, T.V. Kiseleva, Automation and Remote Control, **62:10**, 1645-1650 (2001).
18. V.P. Avdeev, V.N. Burkov, T.V. Kiseleva, Avtomatika i Telemekhanika, **10**, 118-125 (2001)
19. I.A. Ageev, V.N. Burkov, V.I. Zinchenko, T.V. Kiseleva, Automation and Remote Control, **66:6**, 995-1002 (2005)
20. A. Pimonov, E. Raevskaya, T. Sarapulova, E3S Web of Conf., **15**, 01021 (2017)
21. M. Cehlár, J. Janočko, Z. Šimková, T. Pavlik, M. Tyulenev, S. Zhironkin, M. Gasanov, Resources, **8:1**, 21 (2019)

Improving the quality of professional drivers' selection in haulage companies transporting minerals

Yuri Semenov[1,*], *Olga* Semenova[1], and *Daniela* Marasová[2]

[1]T.F. Gorbachev Kuzbass State Technical University, 650000 Kemerovo, 28 Vesennyaya St., Russian Federation
[2]Technical University of Košice, Faculty of Mining, Ecology, Process Control and Geotechnologies, Letná 9, Košice, Slovakia

Abstract. The given article analyses the connection between drivers' psychophysiological parameters and road safety. Psychological methods were used to determine the psychophysiological features affecting the drivers' reliability. The parameters having the greatest impact on road safety include attention, memory, emotions, temperament, thinking, and stress resistance. To assess the professional qualities of drivers working on quarry transport in complex mining and road conditions it is highly recommended to use The Diagnostic&Training Software Package Traffic Psychology. It includes the modules Attention, Emotions, Temperament, Memory, Assessment of Working Modes Influence upon the Operator's Activity and others. The authors present the research results and give instructions to use during the interview for selecting and recruiting drivers for Kuzbass coal enterprises and haulage companies transporting minerals.
Keywords: the professional selection of drivers of technological transport for coal enterprises, drivers' testing, sensorimotor reactions, drivers' psychophysiological parameters

1 Introduction

Heavy trucks use for rock mass transportation on technological roads places special demands on professional training of drivers and their psychophysiological parameters. The peculiarity of road accidents involving quarry transport is the high costs of restoring the operability of both the transport unit and the elements of the technological infrastructure [1-2]. Such accidents often result in staff grave injuries and even deaths. The injured, the haulage companies and the authorities have to cover considerable costs of treatment and recovery while these means could have been used differently.

However, at the present stage of social development, quarry motor transport has no alternative due to its autonomy, mobility, and high-performance rate in practically any topographical, geological, and climatic conditions.

* Corresponding author: semenov63@mail.ru

To reduce the number of accidents and the severity sets of actions are being developed that aim at quarry transport design perfection, an increase in active and passive safety. Great attention is paid to traffic management enhancement and improvement of traffic conditions in quarries. [3].

Drivers' selection and recruitment when scientifically grounded increase the operational reliability of the system Driver – Vehicle – Road – Environment. The selection and staffing problem is extremely acute in those companies where transport vehicles operate in complex conditions [3-5]. Quarry transport operation demands much of the drivers and necessitates that they enjoy good health and physical condition, high psychophysiological reliability, and quality vocational training.

One of the most complicated stages in identifying the staff suitability for operating quarry vehicles is to determine and assess drivers' psychophysiological parameters [6-11]. The methods applied during the procedure include making up questionnaires and surveying by these, using worksheets with test tables and heaps of specially designed equipment, however, the greater part of these methods are labour- and time-consuming.

Computer aids introduced into the process of selection and recruitment allow diminishing labour and time input and simplify the analysis of the results.

2 The Research Method

At the Motor Transport Department of Kuzbass State Technical University a new diagnostic & training software package Traffic Psychology was developed (Fig. 1). The software product includes a test programme and a database and allows not only determining quarry transport drivers' psychophysiological parameters but also analyzing the collected data.

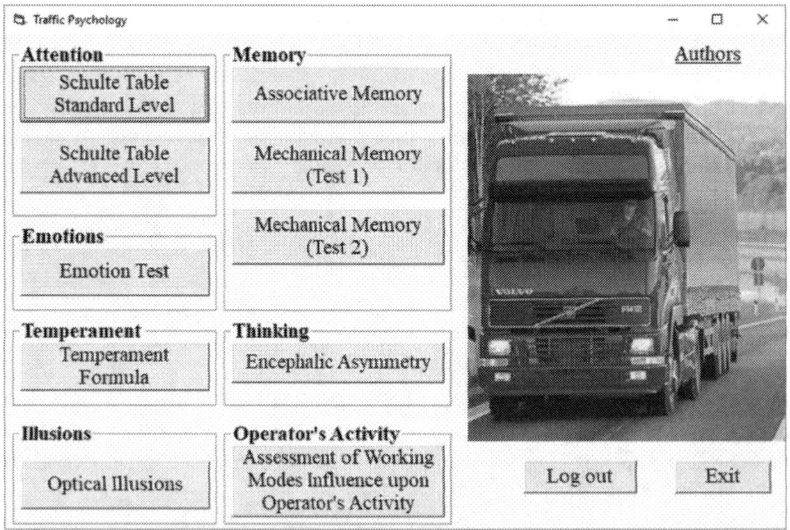

Fig. 1. Diagnostic&Training Software Package Traffic Psychology.

The programme consists of ten specialized modules aimed at assessing drivers' attention, memory, emotions, temperament, thinking, and stress resistance. The module Attention uses automatically generated Schulte Tables. To imitate extreme conditions the table complexity was increased by adding distracting factors/noises (Fig. 2). The Bass-Darky Questionnaire is the basis for the module Emotions. The module Temperament allows assessing drivers' character traits based on the Eysenck Personality Questionnaire. The module Memory evaluates associative and mechanical memory. The evaluation of the first is based on the memorization and reproduction of logically related information, while the second evaluates the driver's ability to memorize and reproduce both numerical and textual information. The module Assessment of Working Modes Influence upon Operator's Activity allows assessing the driver's adaptability to working environment changes.

Research and evaluation of psychophysiological parameters using the given software package were carried out in the group of professional quarry transport drivers aged 24 to 52. Their parameters were determined and recorded from 9 a.m. to 11 a.m. since the highest working capacity is observed in the group during this period.

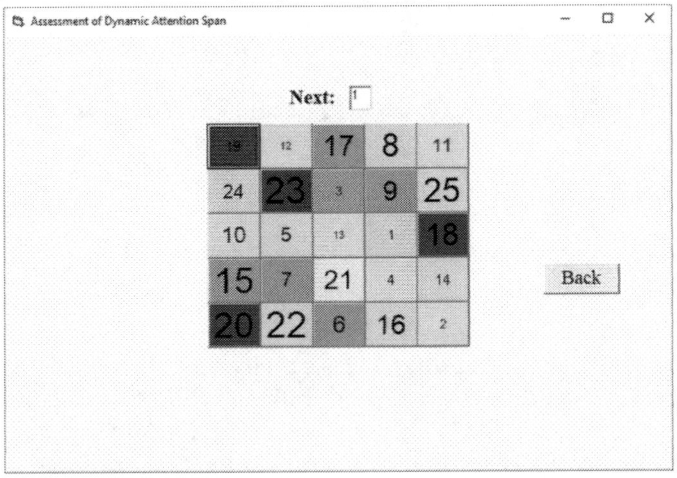

Fig. 2. Schulte Table Advanced Level.

3 Results and Discussion

The results of the research allow us to conclude that about 25% of the subjects demonstrate low attention index, especially in extreme conditions, which reveals insufficient concentration, lack of attention stability and slow switching (Fig. 3-4).

The drivers' age also affects attention parameters. The trend towards shortening attention span and lengthening the span for information processing was determined, which is especially characteristic of extreme working modes.

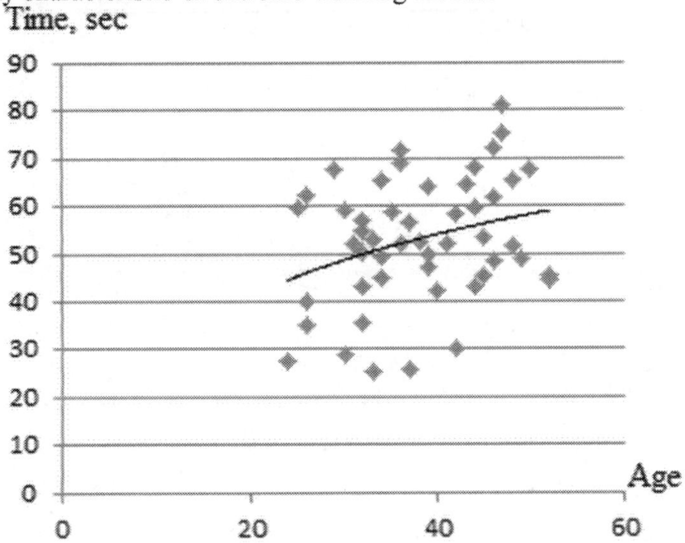

Fig. 3. Dependence of testing time on age (optimal working regime).

 https://doi.org/10.1051/e3sconf/201913401020

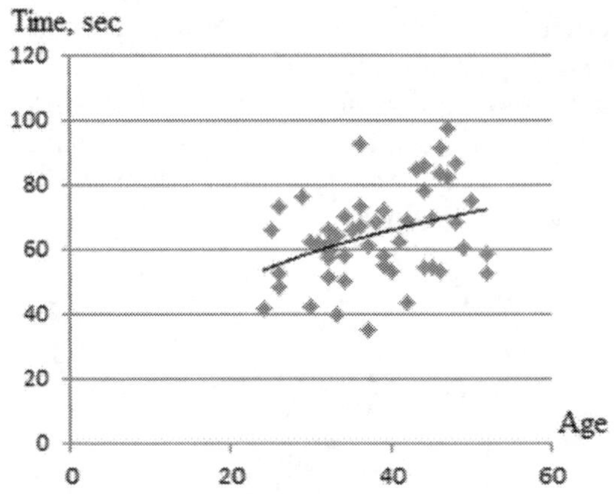

Fig. 4. Dependence of testing time on age (extreme working regime).

Research and evaluation of associative memory allowed concluding that it impairs after the age of 40 (Fig. 5). Above all, physiological changes in the human body and the deterioration of health, caused by professional activity, are responsible for this. However, mechanical memory parameters hardly change with age (Fig. 6). According to the results of the memory research, about 25% of the subjects have a low memory index. So these memory malfunctions are likely to produce a harmful effect on traffic safety.

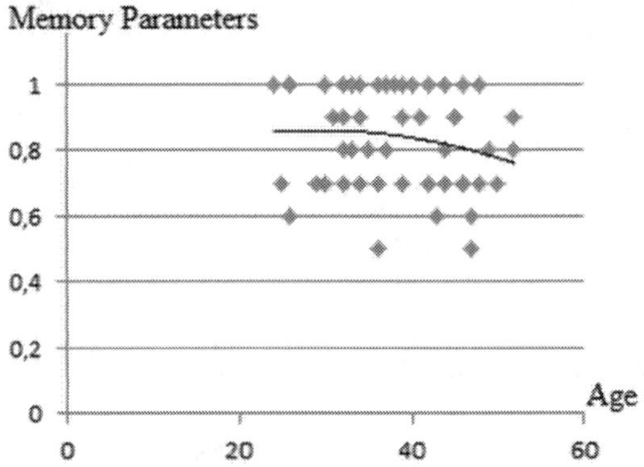

Fig. 5. Dependence of memory parameters on drivers' age (associative memory).

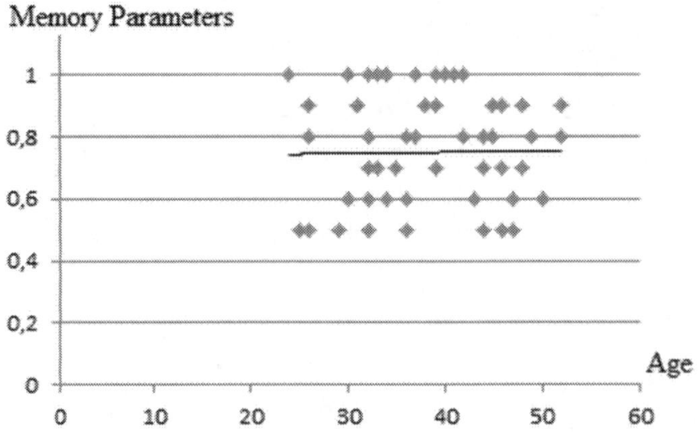

Fig. 6. Dependence of memory parameters on drivers' age (mechanical memory).

The working modes influence upon operational activity was evaluated in optimal and extreme regimes. The optimal working regime implies a normal working environment, where the operator performs his functions automatically at a comfortable speed. In such conditions, the driver maintains a high working capacity for a long time due to a low level of psychological stress.

Testing a driver in the extreme working regime simulates driving a vehicle in complex road conditions and evaluates how accurate he is when driving if there is the poor state of the road pavement, road complex geometry, a lot of noises, etc. Analysis of the results displayed a faster decrease in working capacity than in the optimal regime.

The results of the research into how the working mode influences the operator's activity enabled concluding that about 25% of the subjects had the indices that do not meet the requirements for the position, and this concerns both the optimal and the extreme operational modes. With age, the tracking index of the subjects increases, which in turn negatively affects the perception and processing of information (Fig. 7).

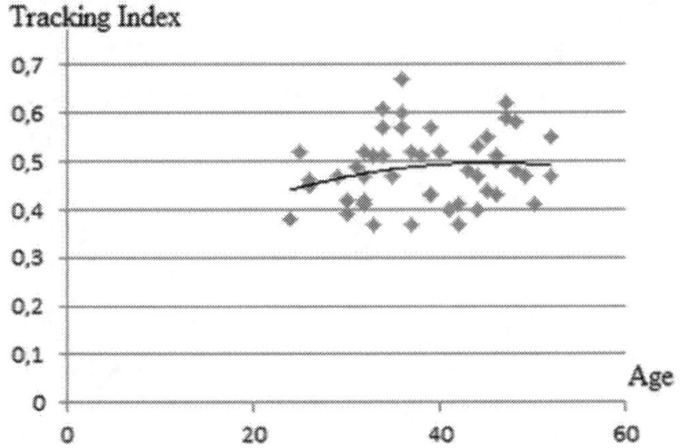

Fig. 7. Dependence of tracking index on age.

4 Conclusion

Professional selection and recruitment of drivers, a better quality of their training can considerably reduce the number of accidents on technological roads of Kuzbass coal enterprises and the severity of the accidents as well.

The Diagnostic&Training Software Package Traffic Psychology allows testing the drivers at the recruitment stage. The programme modules enable evaluating psychophysiological parameters of the potential employees to hire for minerals transportation. Analysis of the applicant's attention, memory, emotions, temperament, thinking, stress resistance will produce a conclusion about the possibility of employment.

After the five-year use of the software the following conclusions were drawn: approximately 25% of the drivers do not qualify for the position because of their psychophysiological properties; certain properties, such as attention, sensorimotor reactions, associative memory tend to impair gradually; however, the drivers' mechanical memory does not seem to undergo significant changes with time.

References

1. A. Parikh, A.M. Jain, International Journal of Engineering and Technical Research, **2:3,** 86-89 (2014)
2. H. Frumkin, *Urban sprawl and public health: Designing, planning, and building for healthy communities* (Island Press, New York, 2004)
3. T. Carter, Fitness for Work, **22**, 564 (2013)
4. F. Gérardin, Atmospheric Environment, **127,** 46-54 (2016)
5. B.K. Borland, B. G. Funke, M. Chacko, Traffic management, **30**, 112-117 (2014)
6. S. Linda, T. Rothengatter, Applied social psychology, **15:8**, 234-241 (2017)
7. M. Ciceri, M. Rita, G. Franco, R.S. Feldman, Psicologia generale (DTRS, Paris, 2013)
8. D. Shinar, Traffic Psychology and Behaviour, 1:2, 137-160 (1998)
9. M. Power, T. Dalgleish, *Cognition and emotion: From order to disorder* (Psychology press, London, 2015)
10. E.A. Morris, J.A. Hirsch, Travel Behaviour and Society, **21**, 98-104 (2015)
11. T.C. Lansdown, A.N. Stephens, H.G. Walker, Accident Analysis & Prevention, 74, 360-367 (2015)

Application potential analysis of heat accumulator for mining mechanisms

Radim Rybár[1*], and *Martin* Beer[1]

[1]Technical University of Košice, Faculty of Mining, Ecology, Process Control and Geotechnologies, Institute of Earth Resources, Letná 9, Košice, Slovakia

Abstract. The paper deals with the application potential analysis of heat accumulators for mining machines working in cold climatic conditions. The proposed concept works with the idea of accumulating waste heat generated by the operation of an internal combustion engine, with use of exhaust heat, heat from the radiator and heat from the drive of the hydraulic system. An important innovative element is the use of the heterogeneous structure of phase change material - sodium acetate as a storage substance and a metal foam matrix. The use of sodium acetate enable to store heat for a long period of time without heat loss and the metal foam matrix significantly increases the thermal conductivity. The presented paper deals with the analysis of the functionality of the heat accumulator prototype and with the description of the key phase of the working cycle - heat accumulation.

1 Introduction

World energy consumption, demand for raw materials and negative environmental impacts have seen a significant increase in tens of percent over recent decades [1]. Thanks to this fact, it is necessary to carry out geological exploration work and, then, mining itself in an increasingly wide scale and in places that pose a challenge not only for the operators but also for the machinery itself. Such places undoubtedly include Arctic regions with cold climatic conditions. The operation of machines in such conditions represents a significant increase in costs, due to increased fuel consumption or the need to preheat the machine part after prolonged downtimes. One possibility is the utilization of waste heat and its subsequent accumulation.

Useful heat resulting from the operation of mining machinery with combustion engine (stationary or mobile) is unused energy potential in the form of exhaust gas, coolants or air, which are heat sources with varying energy levels. This potential can be suitably converted to further uses and bring immediate savings in fuel consumption by improving the operating parameters of the machines, respectively extending the lifecycle in terms of reducing wear of individual parts, and thus extended service intervals [2,3]. Since mining industry is energy intensive sector, a lot of approaches that can reduced energy demands are studied including use of renewable energy sources or energy storage [5-7]. A small range of

* Corresponding author: radim.rybar@tuke.sk

industrial applications and technology is dealing with useful heat, i.e. heat recuperation with direct use, thermoelectric generators [8] or heat storages with delayed use [9].

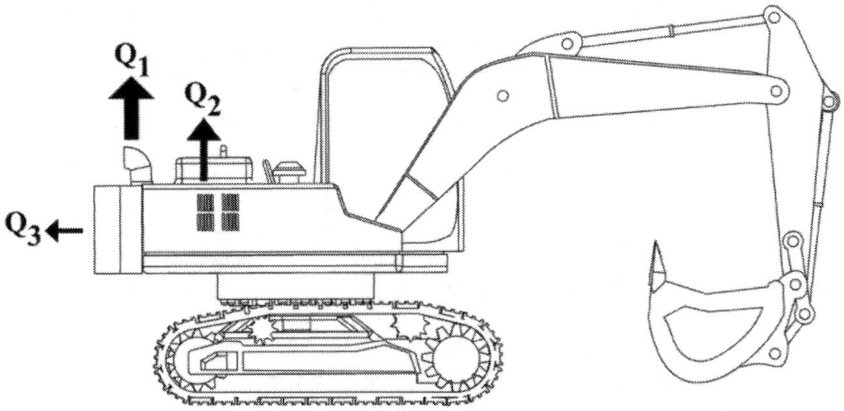

Fig. 1. Example of the most important sources of useful heat from excavator, as typical mining machines, Q_1 - heat from exhaust gas, Q_2 - coolants, Q_3 - hydraulic circuit.

Identifying sources of useful heat of mining machines is mainly focused into the engine compartment, where are located heat exchangers of operation fluid and gases, outlet of exhaust gas and drivetrain of hydraulic circuit (see Fig. 1). In terms of energy flow is dominant exhaust gas outlet from combustion engine that provides constant heat flux with sufficient thermal capacity. To a lesser degree is possible to use heat fluxes from the cooling space of operation fluids and gases, respectively from a hydraulic pump in hydraulic circuit (see Fig. 2). The last mentioned sources represent sources with low energy potential which should be advantageous for designs with different concepts of heat accumulators. The heat energy contained in the exhaust gas and cooling systems generally represents 60-70% of the total energy content of the fuel [10].

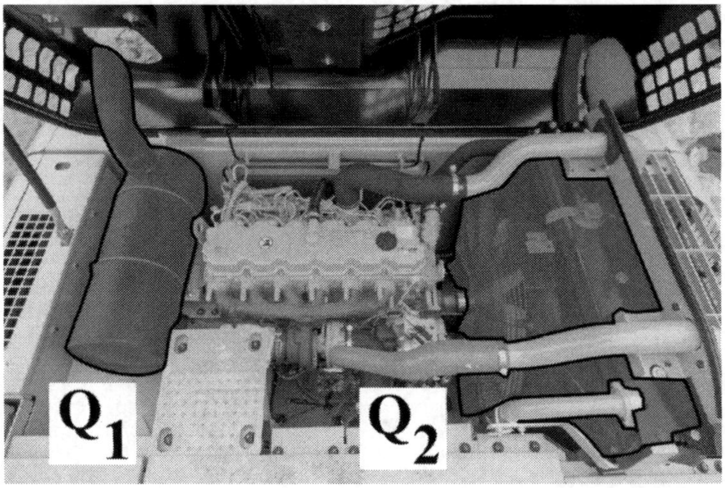

Fig. 2. Graphical illustration of waste heat source in engine area.

Heat accumulation and storage unit with appropriate design and thermal capacity located in the corresponding position of the mining machine can improve economy of operation and the life of the construction parts particularly in difficult climatic conditions. A good example is the possibility of elimination of cold engine starts where the heat-storage unit allows heating of the engine prior to the start of operation with heat stored from the previous operation. The inclusion of the heat-storage unit allows obtaining a balanced operation of engine (or machine) in terms of reduction of thermal shocks and stress of working fluids, which represents negative side of operation. However, such constructed heat-storage unit is bounded to assumption that it will be able to overcome longer periods with stored heat without progressive reduction of the amount of accumulated heat. This allows some materials using phase change, which also may be in the state of supercooling. Concept of heat-storage unit proposed by authors that works with phase change materials (hereinafter PCM) in combination with metal foams is a suitable technical solution that can be applied at the above mentioned operating conditions of machines, which are not limited to the mining machines, but span a wide range of machinery producing significant amounts of energy in the form of useful heat. The incorporation of the proposed heat-storage units into the mining machines is also suitable due to their significant operation weight and thus ratio of weight of heat-storage unit to total weight of machine is a negligible.

From physical point of view is accumulation of heat possible in several ways: accumulation of sensible heat, latent heat, sorption heat and chemical heat [11]. In recent years is getting into attention use of heat accumulation operating on the principle of phase change of PCM where latent heat is stored. As reported in [12] latent heat is one of the most effective ways to store thermal energy. Materials suitable for heat storage should achieve the greatest values of accumulation of heat energy in a minimum volume of material. Additional requirements for PCM include: cyclic stability, flame resistance, low volume changes associated with phase changes, non-corrosive properties and low tendency to supercooling.

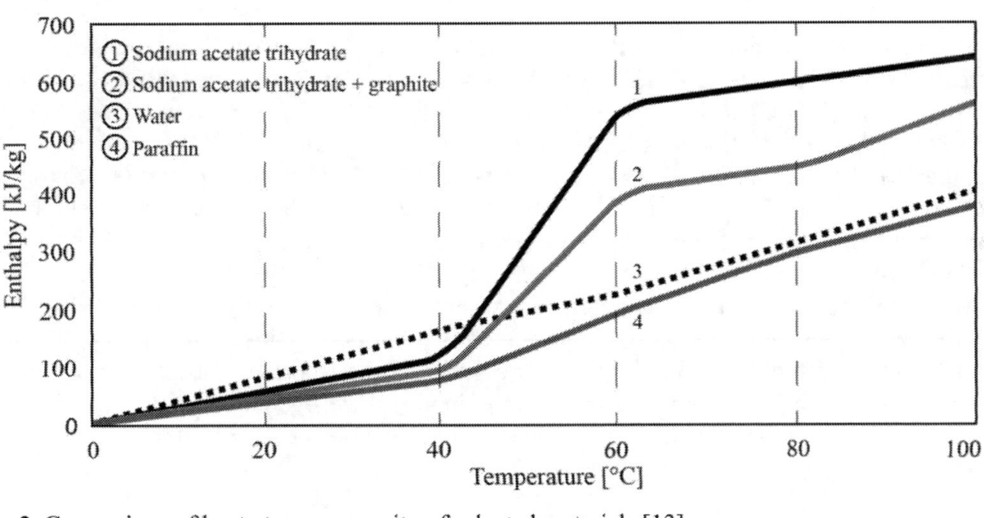

Fig. 3. Comparison of heat-storage capacity of selected materials [13].

In terms of the ratio between the storage capacity of the materials and their weights are organic (waxes) and inorganic (hydrated salts) PCM almost comparable. In comparison of the amount of accumulated heat to the volume of accumulation material there is a clear significant difference in density between the accumulation of organic (waxes) and inorganic (hydrated salt) PCM.

The operating cycle of such conceived heat-storage units can be divided into three phases, the first phase involves the accumulation (charging) of the heat-storage unit, in the second phase there is a stabilization of heat-storage unit and in the third phase is the release of stored energy. The temperature of accumulation material (PCM) is increasing in the charging phase. In first part of this phase is accumulated sensible heat without phase change of PCM, for sodium acetate trihydrate is this part limited with temperature of 58°C. Above this temperature is accumulated latent heat, this accumulation is ended at the moment when phase change from solid to liquid phase is completed in whole volume of heat-storage unit, i.e. when PCM receive all latent heat in relation to its volume (e.g. for the sodium acetate trihydrate it is 265 kJ.kg-1). If the heat-storage unit received more energy the temperature of PCM is increasing and only sensible heat is accumulated. First phase is followed by supercooling phase, which stabilizing heat-storage unit by preventing unwanted nucleation and thus premature release of the stored energy [13]. When PCM is in supercooling phase, heat-storage unit is in state of liquid phase below to point of phase change (58°C for sodium acetate trihydrate). The course of the temperature for the phases of accumulation and the supercooling are shown in Fig. 3.

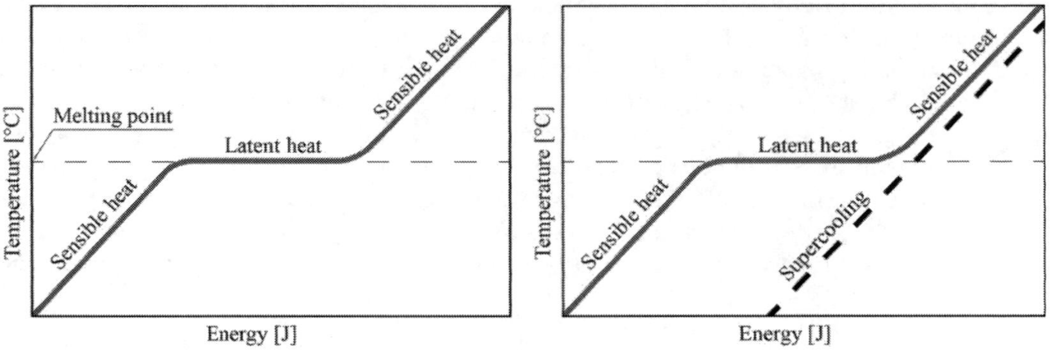

Fig. 4. Temperature curves of PCM for accumulation phase (left) and supercooling phase (right) [13].

An important characteristic of PCM materials in terms of thermal properties is their low thermal conductivity [14]. This feature presents some conceptual limitation of creating heat-storage unit based on PCM, which is reflected in the length of time for melting and solidifying of PCM. Problems related to the heat distribution in PCM are presented, for example in [15] wherein author experimentally tested possibility of improving heat transfer in PCM (paraffin wax) with use of the open cell structure of the copper foams. By adding metal foams in paraffin, thermal conductivity of the created structure increased from 16 to 18 times compare to pure paraffin. Another way of increasing the thermal conductivity of PCM is application of additive, which produces heterogeneous mixture with PCM [16-18].

A specific concept of the storage unit, which is presented in this paper, is a combination of PCM material - sodium acetate trihydrate and spatial structure consisting of blocks of metal foam with an open pore structure that in this heterogeneous spatial arrangement creates heat exchanger. This unique combination is designed with intent to eliminate characteristically low thermal conductivity of PCM materials.

2 Methodology

The manufactured and assembled prototype of heat-storage unit has a cylindrical shape with base in form of annulus. Used metal foam has an outer diameter of cylinder of 50 mm and inner of 12 mm. Height of the metal foam cylinder, and thus the whole heat-storage unit, is 103 mm. The metal foam is characterized by an open structure of pores with pore density of 10 PPI (pores per inch) and total porosity of 96%, the basic material of metal foam is copper. The second component of the heat-storage unit, thus PCM, is sodium acetate trihydrate ($C_2H_3NaO_2.3H_2O$) with melting temperature of 58°C and latent heat capacity 265 kJ.kg^{-1}. Volume of used PCM that fills pores of metal foam was 184.3 cm^3, respectively with using volumetric mass density 1.45 g.cm^{-3}, total mass of used PCM was 267.235 g. In order to improve operational and technical characteristics of presented prototype was used sodium acetate trihydrate in the composition of 59.86% sodium acetate anhydrous and 40.14% water. Heat-storage unit consists of a copper pipe to which was soldered copper metal foam annulus. This object was enclosed with acrylic panels, respectively acrylic tube with air layer left above the metal foam, which compensates volume changes of PCM caused by phase changing and temperature increase. Manufactured prototype of heat-storage unit is shown in Fig. 5.

Fig. 5. Prototype of heat-storage unit based on PCM.

Table 1. Key parameters of experimental heat-storage unit

Parameter	Values	Unit
Shape of heat-accumulation unit	Annulus cylinder	/
Dimensions of heat-accumulation unit	Φ50(12)x103	mm
Type of PCM	Sodium acetate trihydrate	/
Composition of PCM	59.86% sodium acetate anhydrous, 40.14% water	/
Volume of PCM filling	184.3	cm^3
Latent heat of PCM	265	$kJ.kg^{-1}$
Temperature of phase change	58	°C

The total possible amount of accumulated energy of heat-storage unit is determined by calculation according to Eq. 1 to Eq. 3:

$$Q_A = Q_{solid} + m \cdot Q_L + Q_{liquid} \quad (1)$$

$$Q_{solid} = m \cdot c_{solid} \cdot \Delta t_{solid} \quad (2)$$

$$Q_{\text{liquid}} = m \cdot c_{liquid} \cdot \Delta t_{liquid} \qquad (3)$$

where, Q_A - total amount of accumulated energy, Q_{solid} - sensible heat of solid phase, Q_L - latent heat of PCM, Q_{liquid} - sensible heat of liquid phase, m - weight of PCM. In Eq. 1 and Eq. 2 is c_{solid}, respectively c_{liquid} specific heat capacity of solid, respectively liquid phase and Δt_{solid}, respectively Δt_{liquid} is difference between temperatures of solid, respectively liquid phase given the temperature of phase change of sodium acetate trihydrate.

The principle of physical experiment was based on flow of heat transfer medium (water) with temperature of 80°C through inner copper pipe of heat-storage unit what simulates source of useful heat from combustion engine. Heat flux from heat transfer medium flows through structure of metal foam where was heat transferred to sodium acetate trihydrate what change its temperature and phase with simultaneous accumulation of sensible and latent heat. The task of thus conceived experiment was to test the capability of heat transfer from the heat source through structure of metal foam to the PCM - sodium acetate trihydrate.

The presented experiment was performed at the Centre of Renewable Energy Sources of the Faculty BERG on experimental apparatus designed and manufactured by authors in which was installed heat-storage unit described in the previous section. Design of experimental apparatus was based on analysis of the input conditions in order to induce and control non-stationary processes from view of changing temperatures and phase changing of PCM. The experimental apparatus allowed supplied a constant heat flux from heat transfer medium which circulated in hydraulic circuit that is depicted in Fig. 6, where 1 - heat-storage unit, 2 - flowmeter, 3 - flow control, 4 - flow heater, 5 - expansion vessel, 6 - hydraulic pump, TC1 to TC4 - thermocouples probes.

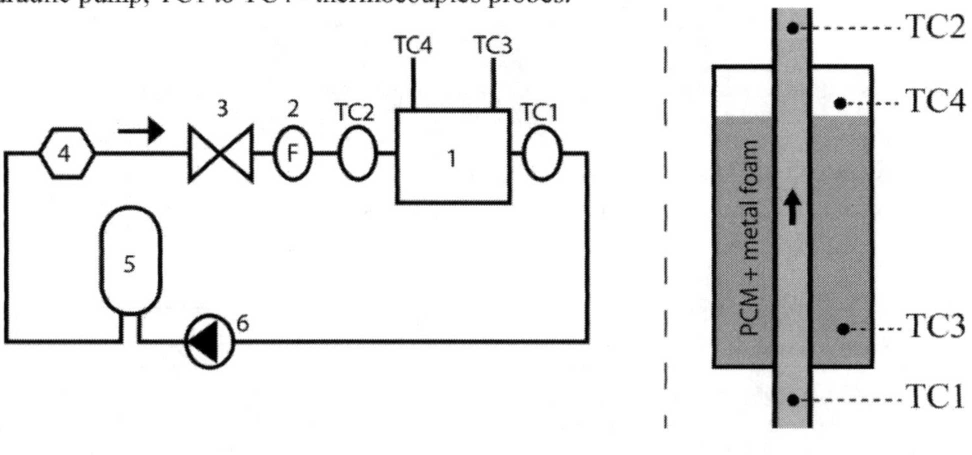

Fig. 6. Schematic diagram of hydraulic circuit (left) and heat-storage unit with depicted position of thermocouple probes (right).

In terms of verifying the functionality of the combination of the metal foam structure and sodium acetate trihydrate as a thermal-storage unit was necessary to measure the temperature of the PCM, i.e sodium acetate trihydrate as well as the temperature of the metal foam using thermocouple probes that are labeled as TC3 and TC4. Thermocouple probes labeled as TC1 and TC2 measured temperature of heat transfer medium before and after heat-storage unit, as it is depicted in Fig. 6. Experimental apparatus was supplemented with measuring devices, which consist from thermocouple probes KIMO TTKE-363 (type K, range from -40°C to +400°C), data acquisition system KIMO AMI 300 with temperature range from -100°C to +250°C with total uncertainty ± 0.4%. Step of data recording was 10 s. Flow rate of heat transfer medium was monitored with analogous rotameter with range 0.3 to 1.5 l/min and uncertainty ±10% F.S.

Status of phase changing of sodium acetate trihydrate was also monitored and recorded in the form of time-lapse video with using of camera located in the immediate vicinity of the heat-storage unit. The frequency of image capture was determined with respect to the rate of phase change to interval of 5 minutes. Camera lens was equipped with polarization filter, which eliminates the glare from outer acrylic tube of heat-storage unit. On the body of heat-storage unit was placed measuring tape allowing accurate tracking of phase change of sodium acetate trihydrate.

3 Results and Discussion

The presented analysis of proposed heat-storage unit is focused on part of verification of its functionality, which is in this case characterized by the first phase of working cycle, where heat-storage unit receives and accumulates heat energy from selected source. This process involves phase change of PCM, which has at the beginning solids phase and with raising temperature is changing to liquid phase. Melting point of sodium acetate trihydrate, which was used in presented work, is 58°C.

In first phase of working cycle are sensible and latent heat accumulated. Their amount was for proposed heat-storage unit calculated using Eq. 1 to Eq. 3. Results of the calculation that reflects the real dimensions and characteristics of sodium acetate trihydrate is for amount of accumulated sensible heat Q_{solid} equal 28,64 kJ. This heat is delivered within temperature difference Δt = 58 - 20°C, since the ambient temperature at the start of experiment was 20°C and sensible heat is accumulated to the temperature of 58 °C. Amount of accumulated latent heat $m \, . \, Q_L$ is with used weight of sodium acetate trihydrate equal to 70.8 kJ. Because in the heat-storage unit was used sodium acetate trihydrate in the composition of 59.86% sodium acetate anhydrous and 40.14% water, it was necessary according to [19] rise temperature of PCM to 65°C to ensure phase change from solid to liquid in whole volume of heat-storage unit. Amount of sensible heat accumulated in this process that occurs in temperature difference Δt = 65 - 58°C is $Q_{liquidus}$ = 5.7 kJ. The resulting delivered heat thus represents a value of 105.1 kJ from which useful heat that is stored for later use is 70.8 kJ (described for a proposed heat-storage unit).

Fig. 7 shows the temperature curve of the heat transfer medium simulating the useful heat of the combustion engine (TC1) and the temperature of PCM - sodium acetate trihydrate (TC3). Due to the used construction of heat-storage unit in terms of heat transfer from the inner copper tube, through the structure of the metal foam to the PCM, the temperature of the heat transfer fluid was selected to 80°C. The temperature was experimentally determined at trial operation of experimental apparatus and provides sufficient heat flux and therefore temperature increase of the PCM. Visible anomaly of thermocouple probes TC1 in initial part of the curve represents the error in controlling of flow heater, but as can be seen, was quickly fixed.

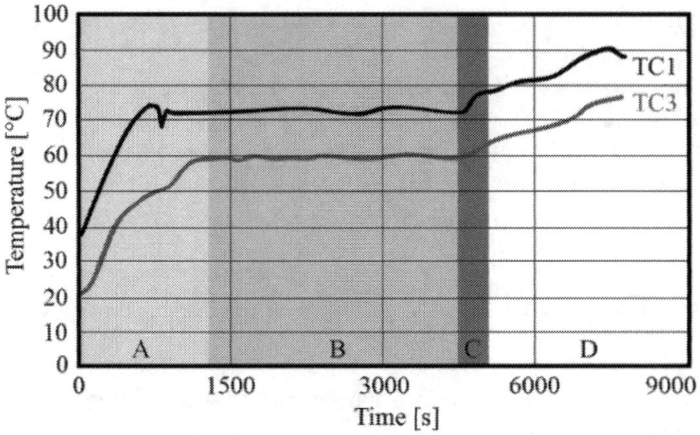

Fig. 7. Temperature curves of PCM and heat transfer medium at inlet to heat-storage unit.

From the shape of temperature curve for TC3 thermocouple probe, which measured temperature of sodium acetate trihydrate can be clearly recognize phases of accumulation. First part when sensible heat is delivered (labeled as A) is characterized with steep rise of temperature to the point at 58°C. The second part is the accumulation of latent heat (labeled as B), which takes place at a standard temperature of 58°C, horizontal course without major fluctuations proves the success of the concept in terms of its function which is storing latent heat of used PCM. After this part, which in the shown experiment ended at time t=+4500s, the temperature of the heat-storage unit increased to 65 °C ensuring complete phase change from solid to liquid state (labeled as C). The last part shown in Fig. 7 represents the experimental introduction of the extreme operating conditions in order to detect design and construction errors of the heat-storage prototype based on PCM.

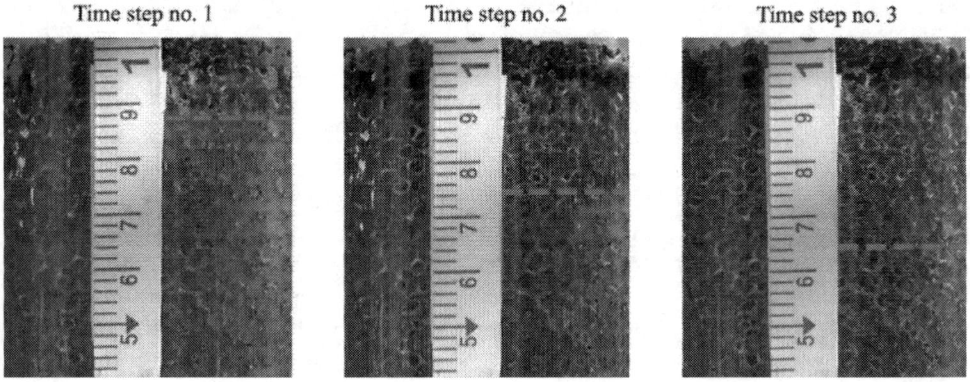

Fig. 8. Time-lapse image of phase change in heat-storage unit (red lines points to level between solid and liquid phase)

In addition to recording of the temperature from thermocouple probes TC1, TC2, TC3 and TC4, the process of heat accumulation of the proposed heat-storage unit was monitored by capturing the image at regular intervals without changing the unit position to the axis of the lens. In this way obtained time-lapse recording is possible to observe gradual phase change of PCM in the direction of the vertical axis of the heat-storage unit. Process of phase change started at the point with highest temperature, in this case at the point of entry of heat transfer medium into the inner copper pipe. Fig. 8 shows the phase change process in second part of accumulation, when sensible heat is already accumulated and the heat-storage unit begins to accumulate latent heat and thus changing its phase from solid to liquid. The rate of phase change of sodium acetate trihydrate was in vertical axis 3 $cm.h^{-1}$.

However, it is necessary emphasized that the inner configuration of described heat-storage unit was not designed for quick heat accumulation (i.e. fast phase change of PCM), but for testing the functionality of the connection of PCM and metal foam structure.

Next steps in the development process leading to the final product are directed to the experiments related to the supercooling phase of working cycle. In term of design, it is necessary to ensure inner configuration of heat-storage unit in way that there will not occur uncontrolled nucleation of PCM and thus release the accumulated heat. The final step of the proposal, which is dependent on the successful supercooling and stabilization of PCM, is ensuring fully controllable nucleation (when heat is released) in appropriate way and at any time period.

4 Conclusions

Accumulation of useful heat of mining machines or combustion engines and its re-use is a relatively new concept that allows eliminating some of the negative operating conditions related primarily to unfavorable climatic conditions (cold engine start, temperature variations of operation mediums, elimination of thermal shocks, etc.). The presented concept of heat accumulation with use of heat-storage units based on PCM allows this energy store and uses it with time delays when it is needed. The uniqueness of the proposed concept lies in the simultaneous use of metal foam structure and PCM - sodium acetate trihydrate, which ensures sufficient thermal conductivity, which has low value when the sodium acetate trihydrate is use alone, and was limiting factor for further expansion of PCM storage. The functionality of this concept was verified by physical experiment that simulates the accumulation phase of heat-storage unit with use a heat source in the form of flowing heat transfer medium giving an analogy to conventional mining machines or combustion engines. Experiment was carried out in the Centre of Renewable Energy Sources Faculty BERG on experimental apparatus designed and manufactured by authors. Published results have shown the functionality of the concept in the first phase of accumulation of useful heat, where authors were able to successfully accumulate latent heat to PCM, which demonstrates temperature curve of PCM with flat part of curve at level of 58°C, which is related to the phase change of PCM, and therefore, accumulation of heat.

References

1. H. Pavolová, T. Bakalár, D. Kudelas, P. Puškárová, Journal of cleaner production, **209**, 1119-1125 (2019)
2. S. Elevli, B. Elevli, Acta Montanistica Slovaca, **15**, 95-101 (2010)
3. M. Mohammadi, P. Rai, S. Gupta, Acta Montanistica Slovaca, **21**, 1-8 (2016)
4. P. Křivík, P. Bača, M. Calábek, K. Micka, P. Král, Journal of Power Sources, **154**, 518-522 (2006) doi: 10.1016/j.jpowsour.2005.10.029
5. J. Zimáková, S. Vaculík, P. Bača, D. Fryda, Acta Montanistica Slovaca, **20**:3, 165-171 (2015)
6. Z. Dostál, M. Ďulík, Acta Montanistica Slovaca, **17**:4, 310-314 (2012)
7. R. Rybár, D. Kudelas, M. Beer, Acta Montanistica Slovaca, **20:**3, 172-180 (2015)
8. T.Y. Kim, A.A. Negash, G. Cho, Energy Conversion and Management, **124**, 280–286 (2016) http://dx.doi.org/10.1016/j.enconman.2016.07.013
9. J. Shon, H. Kim, K. Lee, Applied Energy, **113**, 680–689 (2014) http://dx.doi.org/10.1016/j.apenergy.2013.07.049

10. F.R. Spellman, *Water & Wastewater Infrastructure: Energy Efficiency and Sustainability* (CRC Press, Boca Rator, 2013)
11. G. Bajnóczy, Gagyi E. Pálffy, E. Prépostffy, A. Zold, Periodica Polytecinica SerChemEng, **43**, 137 - 147 (1999)
12. H. Mehling, L.F. Cabeza, *Heat and cold storage with PCM: An up to date introduction in to basics and applications* (Springer, Berlin, 2008)
13. L. Huang, E. Günther, CH. Doetsch, H. Mehling, Thermochimica Acta, **509**, 93-99 (2010) http://dx.doi.org/10.1016/j.tca.2010.06.006
14. R. Srikanth, C. Balaji, International Journal of Thermal Sciences, **111**, 188–203 (2017) http://dx.doi.org/10.1016/j.ijthermalsci.2016.08.018
15. P. Himanshu, *Thermal Energy Storage in Copper Foams filled with Paraffin Wax* (Mechanical& Industrial Engineering University of Toronto, Toronto, 2011)
16. D.H. Choi, J. Lee, H. Hong, Y.T. Kang, International Journal of Refrigeration, **42**, 112–120 (2014) http://dx.doi.org/10.1016/j.ijrefrig.2014.02.004
17. A. Ansone, M. Dzikevics, A. Zandeckis, Energy Procedia, **95**, 153–158 (2016) http://dx.doi.org/10.1016/j.egypro.2016.09.037
18. V. Kumaresan, P. Chandrasekaran, N. Maitreyee, A.K. Maini, R. Velraj, International Journal of Refrigeration, **36**, 1641–1647 (2013)
19. E. Nohejl, *Sezónní akumulátor tepla s minimálními ztrátami.* URL: http://oze.tzb-info.cz/akumulace-tepla/11626-sezonni-akumulator-tepla-s-minimalnimi-ztratami

Allocation of the geo-dynamically hazardous zones during intensive mining of flat-lying coal seams in the mines of SUEK-Kuzbass JSC

Evgeny Yutiaev[1], *Anatoly* Meshkov[1], *Anton* Popov[2], and *Arcady* Shabarov[2]

[1]SUEK-Kuzbass JSC, 652507, Kemerovo region, Leninsk-Kuznetsk, Vasilyev str., 1
[2]Saint-Petersburg Mining University, 199106, Saint-Petersburg, Line 21st, V.I., building 2

Abstract. Hazardous and ineffective mines were closed, and a number of underground and strip mines of a modern technological level were built as a result of the restructuring of the coal industry in Russia. However, safety in the coal mines continues to be one of the urgent issues nowadays. In underground coal mining, the main types of risks are processes occurring in the rock mass as a result of mining operations. It is necessary to realize timeous and the most accurate prediction of harmful events, which occur during a development of coal fields, to increase the efficiency and reduce the cost of coal mining. The most optimal and, at the same time, comprehensive approach to predicting possible natural hazardous events in the process of modern coal mining are considered in this article. The results of implementation of integration of different methods of geodynamic and geomechanics researches are presented in this paper.

1 General terms

Until quite recently, it was believed that mainly technological reasons were the cause of the vast majority of accidents and their extent of damage. However, the practice of construction, operation and conservation of industrial facilities and productions associated with the development and usage of subsoil and land surface has shown that the effectiveness and safety of their functioning can be ensured only with the preliminary identification and assessment of geodynamic processes occurring in geological environment. It can also be achieved by assessment of natural and technology related risks, including environmental review and monitoring of geodynamic processes.

2 Allocation of geo-dynamically hazardous zones based on the integration of geodynamic and geological and structural methods

Prediction and assessment of hazardous geodynamic processes are carried out on the basis of the geodynamic zonation. Geodynamic zonation is a method of studying the block structure, stress and gas-hydrodynamic states of a rock mass; it is a method of identifying, evaluating and monitoring currently active raptured tectonic displacements – active faults, geodynamic zones and associated risk zones, in order to increase the geodynamic and environmental safety of industrial and civilian facilities during the development of mineral resources and land surfaces [1,2].

Modern tectonics considers four main types of major geodynamic regimes. Segregation of these regimes is founded on the basic concepts of plate tectonics, and this segregation is associated with variations in plate motions. There are regimes of prevalence of compressive horizontal stresses, tensile horizontal stresses, shear stresses, and, finally, prevalence of compressive or tensile vertical stresses [3-4].

Aforementioned geodynamic regimes, presented within the fields as a whole and the mines particularly, determine the tendency of the territories under development to the occurrence of emergency situations of a different nature.

The analysis of geodynamic safety includes the process of identifying geodynamic hazards, assessing the frequency (or probability) of their occurrence and assessing the effectiveness (adequacy) of the applied preventive measures to manage (minimize) geodynamic risks.

Firstly, assessment of the geodynamic state of the rock mass blocks is carried out within the mining allotment of a coal field, and then the selection of a list of hazardous events takes place. It is necessary to conduct risk analysis of geodynamic hazards for these events.

Next, identification of the types of geodynamic hazards is executed. This measure is designed to establish the presence or absence of a geodynamic hazard in the mine. A description of all types of geodynamic hazards in the considered mine is presented. A list of possible types of geodynamic hazards that may occur during the mining is given. Cause and effect relationships for emergencies are established according to the characteristic of the source of the occurrence of the geodynamic hazards (stressed rocks, gas pressure, seismic waves, shaking deformation, water pressure, etc.).

A hazardous geodynamic event is an event of natural or man-made origin, or a result of natural, man-made and anthropogenic-produced geodynamic processes; these processes can cause an emergency in anthropogenic and natural systems [4-5].

A characteristic of geological and mining conditions (their combinations) are given for each type of the geodynamic hazard.

The main methodological principle of geodynamic zonation is based on a system approach – integration, continuity and coordination of working scale from the general to the particular. Zoning according to the risk of emergencies at the regional, district and local levels is uniform, regardless of the working scale.

Base, peak, residual, difference and goniobasis surfaces of various orders are obtained from topographic maps of different scales, and then they are compared to the data of the whole system of available geological, geophysical, geochemical and geodynamic data. Currently, morphometric analysis of these compared surfaces allows revealing tectonic faults and structures of various orders and their motions, both in platform and folded areas.

However, it was found that within the mine fields during the exploration and development of coal deposits, as well as during observations to solve various problems, in particular, when determining the basic values of the process of deformation of land surface as a result of underground mine workings and predicting the zones of collapses and screes occurring in mines, emanations of a linear character of localization can be detected in areas (zones) where according to geological and exploration data there are no tectonic faults. Moreover, the nature of emanations is similar to abnormalities above tectonic faults [6-7].

These geodynamic zones are associated with processes, which occur in the overburden as a result за their current deformations. Stress fields are responsible for changes in magnetic, electromagnetic, atmochemical, physical and mechanical and other properties of geological environment.

The upper levels of the soil mass to a depth of 10-15 m are experiencing reciprocating motions along three spatial axes.

The amplitude of vertical movements of the land surface varies between 10-30 mm, depending on the density of soils. The relative horizontal displacement of two fairly close points of the land surface reaches 70 mm [8]. In addition to the above, the greatest movements are observed within geological discreteness – faults, folds, fractured zones, fractures.

Features of rock bedding: fractured zones, buried residual topography, discordant occurrence of rocks, – or industrial processes associated with operation of mining facilities are considered to be the reason for the appearance of active geodynamic zones within the mine that are not related to fault.

The structure of deep-seated rocks is defined by the presence of buried residual topography, which is not expressed on topographic maps, but is revealed as a result of subtraction of the base surface from the hypsometric, moreover, their structure can be described by the presence of areas of discordant occurrence of rocks [8]. These abnormalities of the structure of the coal-bearing rock mass are effectively determined using methods of geological and structural simulation.

An example of the results of morphostructural analysis of the territory is presented in the form of a map of faults of rank V, zones of influence of active faults of rank III-IV, determined as a result of morphostructural analysis on a topographic base of a scale of 1:2000, and lineaments (Fig. 1).

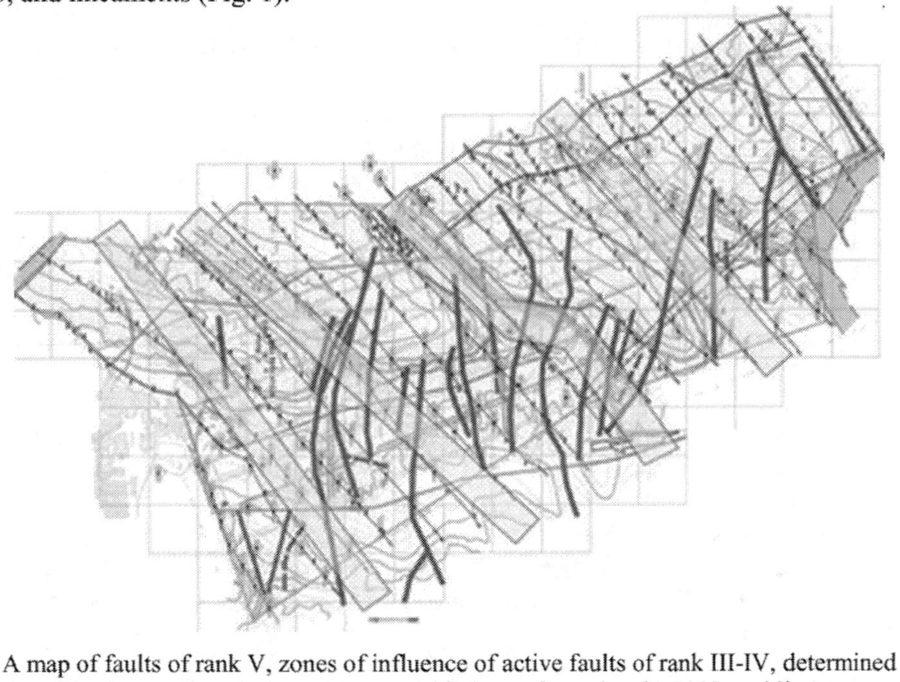

Fig. 1. A map of faults of rank V, zones of influence of active faults of rank III-IV, determined as a result of morphostructural analysis on a topographic base of a scale of 1:2000, and lineaments (minefield n.a. V.D. Yalevskogo).

Faults are shown in red. Yellow color is designated to zones of active faults of rank III-IV and lineaments.

Intense tectonically stressed zones (TSZ) are associated not only with single faults, but also with systems of faults. The most dangerous structures, which form intense TSZ, were revealed. These primarily include systems of echelon-like faults, fault convergences, partition at sharp angles (10° - 15°), and wedge-shaped structures.

Local TSZ are also formed in the places of pinches and swells of the seams and any significant changes in their hypsometry.

Therefore, tectonically stressed zones (TSZ) are mainly confined to the faults of block rock mass; their potential danger is determined by neotectonic activity, morphology, and structure of the fault displacer, as well as the direction of rock deformation under the displacement of the side of the fold. There are two main types of TSZ associated with faults, differing in geological features and natural stress state.

Zones of type I are adjacent to those sections of faults, in which displacer is in the form of tightly closed fracture surrounded by an area with small number of fault fractures; it is described by a high level of tectonic stresses, low porosity and humidity, increased values of strength characteristics and Young modules of rock mass, as well as by a high level of electromagnetic emission.

Zones of type II are separated from the displacer by a disturbed and relieved area; it is described by significantly lower stresses and changes in the state and properties of rocks than in zones of type I [9-10].

The geodynamic zonation technique is not used to identify separate, independent, and randomly distributed lineaments and faults; the feature of this technique is the identification of regularly ordered and ranked systems of faults, which are characteristic of the selected region (district) and reflect allied and hierarchical distribution of stress fields and display in basement and sedimentary cover deformations.

3 Three-dimensional structural and geological simulation of minefields

The results of geodynamic zonation were compared to the simulation results. Structures identified during zonation process were discovered and confirmed. An analysis was carried out to indicate forms of manifestation of lineament structures of rank III-V in the morphology of coal seams and to indicate changes in thickness of seams. These data provided the basis for further identification of risk zones, which are planned to be exposed during the development process of the mine.

Initially, a number of operations, which are necessary for the thorough preparation of the initial data at the core of the model, were carried out during simulation process of the minefield.

All of the geological information available today: geological profiles, geological maps, tectonic frameworks, coring data, data of geophysical studies, sketches of mines, – was used to develop geological and structural model of the field.

A common database of exploration wells drilled within the scope of the mining allotment of the minefield was formed as a part of the model development process. A unique code, that is uniform for the entire area, was assigned to each lithological variety identified during drilling.

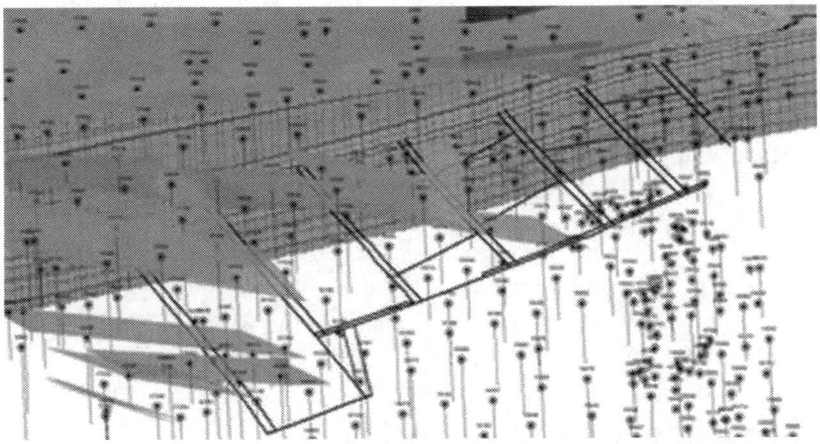

Fig. 2. The block model of the minefield with faults and mine workings of seams 50 and 52.

This was followed by the development and generation of a mesh of the marker horizon, where the block size for the further model was determined. The accuracy of the model depends on the block size – the smaller the block, the more accurate the model will be. The selection of the optimal size of the building block is the main difficulty at this stage of the simulation process due to the fact that the calculation can take several days for small sized blocks.

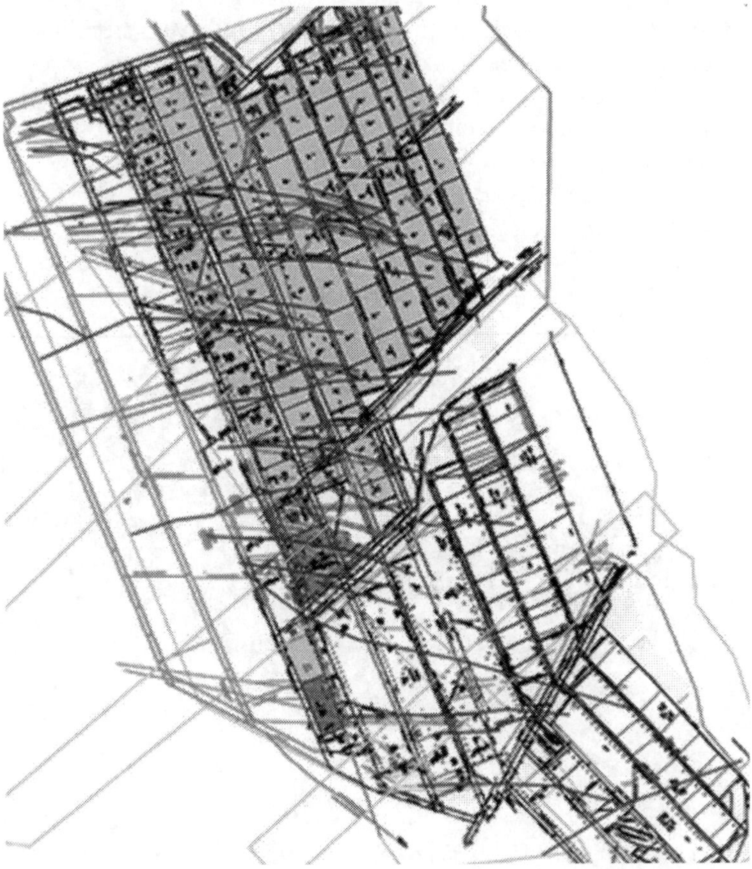

Fig. 3. Conformation of faults identified during zonation and simulation processes.

The results of geodynamic zonation on the surface were coordinated with discovered faults in the roof and the bottom of coal seams during the development process of the three-

https://doi.org/10.1051/e3sconf/201913401022

dimensional model of the fault displacer within the minefield. It is worth noting that all the faults indicated in the modern topography and confirmed by the geodynamic zonation are exposed in the coal-bearing stratum (figure 2). In figure 2, the red color indicates the faults identified during zonation process, the green lines correspond to the zones of influence of active faults, the red ones indicate faults that were confirmed during the development and exploration of the field; the azimuths and slope angles of the faults and fault zones are almost completely maintained.

The result of the development process of the three-dimensional model of the minefield is the correspondence between the results of the geodynamic zonation, the fact of the coal-seam uncovering and the position of the fault displacer in the block model of the minefield. The obtained results are presented in figure 3.

4 Geodynamic model of the minefield

Geometric layout of the numerical model, which includes the existing topographic surface of the field region, working coal seams No. 50 and No. 52, two large-scale faults and tectonic faults of a lower rank, is presented in the figure 4.

The existing topography of the region was taken into account during the development process of the numerical model. The area of the mine "n.a. V.D. Yalevskogo" has a landscape close to plain-like, and does not have any significant difference in elevation. However, taking into consideration the relatively shallow depth of working coal-seams, even a slight difference in elevation (up to 100 meters) can have a significant impact on the natural stress state of the rock mass. Therefore, idealization of the land surface and its approximation to the plane can substantively simplify the numerical model, but, at the same time, it can lead to a distortion of the simulation results.

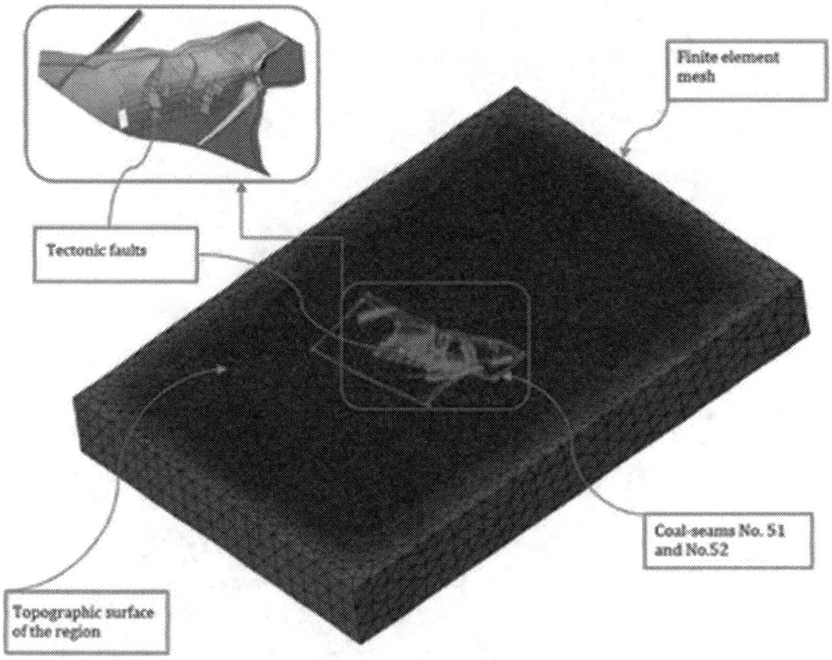

Fig. 4. Geometric layout of the numerical finite-element model of the mine "n.a. V.D. Yalevskogo".

Isochromatic curves of the initial stress field in the rock mass, which were obtained as calculation results, for coal-seams No.52 and No.50 are presented in the figure 5.

Tectonic faults, which were indicated during geodynamic zonation of the minefield, have a significant impact on the redistribution of the maximum principle stress conditions in the rock mass.

The natural stress field of the mine is classified as a gravity type. Nevertheless, the isochromatic curves in the figure 5 show the areas, which have stress spikes and relieved zones.

Quantitative analysis of the stress state of the mine n.a. V.D. Yalevskogo suggests that the average value of the principle compression stresses, which are outside the zones of influence of the tectonic faults, does not exceed 3.0-3.5 MPa in the areas planned for stoping.

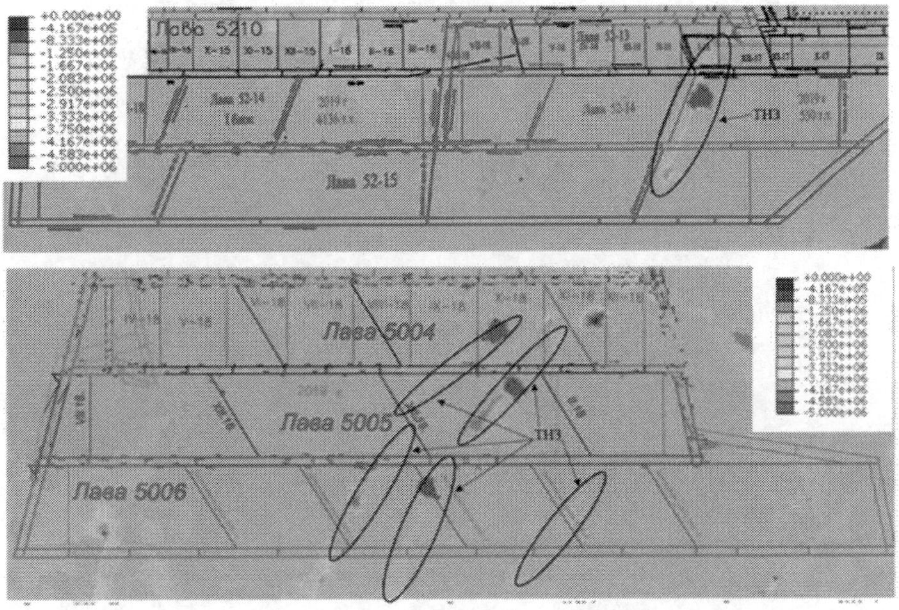

A)

B)

Fig. 5. Isochromatic curves of the distribution of maximum principle stresses in coal-seams of the mine "n.a. V.D. Yalevskogo" (areas planned for stoping until 2019): A) seam No.52; B) seam No.50.

5 The allocation of tectonically relieved and stressed zones, and prediction of the geodynamic and geomechanical conditions

The prediction of the geodynamic and geomechanical conditions of minefield's areas, which are planned to be mined, is carried out on the basis of a comprehensive analysis of risks of hazardous geodynamic events according to the distribution and the rank of hazardous zones within the scope of the estimated area.

The main criteria for assigning risk zones to the existing plan of mining operations for coal-seams of the minefield were the results of mathematical simulation and the development of the geodynamic model, the results of the geodynamic simulation.

The general approach of identifying existing risk zones for the predicted areas of the minefield, which are going to be developed in the near future, consists of overlaying all existing forecast maps on a plan.

It is also worth noting that the actual segregation of the tectonically relieved zones (TRZ) and tectonically stress zones (TSZ) was carried out roughly. This segregation was based on the distribution fields of the principle stresses in the initial state of the rock mass, id est before the start of mining operations. A layout and general principles of the formation of TRZ and TSZ during stoping, which take into consideration the impact of the mining operations on the neighboring areas, were provided in the text above. Limits of the hazardous zones are most clearly identified using calculations.

However, the indication of the predicted risk zones was carried out on the basis of the geodynamic model for the initial stress fields due to the fact that the position and spatial configuration of the limits of the hazardous zones (LHZ) are unstable. The position and spatial configuration of the LHZ depend on the state of the mining and stoping processes at the current time. The results of the introduction of risk zones to the plan of mining operations for the seam No.52 are presented in the Fig. 6.

Fig. 6. Introduction of risk zones based on the data of mathematical modeling and comprehensive analysis using neural network to the predicted area of the seam No.52.

It is shown in the figure 6 that the main risk zones within the scope of the predicted area are located in the zone of influence of the tectonic faults, which form the TRZ (based on the distribution of the stress fields in the figure) in their upthrow side and the TSZ in their lying side.

6 Conclusions

It is necessary to implement measures to reduce geodynamic risks for effective and safe application of modern technologies for intensive coal mining.

In general, the process of identification and minimization of geodynamic risks involves:
- conducting of geodynamic zonation, identification of potentially hazardous zones and consideration of the obtained results for mine layout and planning of mining operations;
- implementation of geodynamic monitoring and routine analysis of geodynamic risks;
- carrying out measure to prevent accidents due to the manifestation of geodynamic hazards.

The method of geodynamic zonation, which was used to indicate geo-dynamically hazardous zones during intensive mining of flat-lying seams of Kuzbass, has positively established itself. In addition to morphostructural analysis, neural networks can be applied and calculations of the strain-stress state during stoping can be conducted for more accurate assessment.

The presence of tectonically relieved zones (TRZ) is the most significant geomechanical property at the currently prevailing mining depth of 200-400 m for planning technological solutions and measures to manage the state of the rock mass.

The locations and limits of these zones can be determined both by office studying of the geological and geomorphological information, and by using full-scale geophysical methods.

It is preferred to locate development workings, setup and recovery rooms outside the TRZ and the TSZ during mine layout with preliminarily allocation of these zones.

To reduce the harmful impact of the TRZ and TSZ on the stoping it is recommended to identify these zones in a timely manner at the planning stage, and to provide ground supports, which move together with the active post immediately after the formation of the outcrop with the pass of the mining machine, in the timbering plan and the plan of strata control in the face.

It is necessary to correct some parameters of the gas drainage process, as well as ground control management plan in the zones of TRZ and TSZ.

References

1. Fan Chen Anye Cao, Linming Dou, Guangcheng Jing, Geosciences Journal, **3,** 1–16 (2019)
2. V. Kecojevic, Z. M. Nor, Journal of Coal Science and Engineering (China), **15:1**, 1–6 (2016)
3. T.V. Gudkova, A.V. Batov, V.N. Zharkov, Model Solar System Research, **51:6**, 457–478 (2017)
4. M. Burša, K. Pěč, Gravity Field and Dynamics of the Earth, **1**, 207-244 (1993)
5. A.M. Suchowerska, *Geomechanics of singleseam and multi-seam longwall coal mining* (University of Newcastle, Newcastle, 2014)
6. S.S. Peng, *Coal Mine Ground Control* (West Virginia University, Redmond, 2008)
7. A.M. Suchowerska, R.S. Merifield, J.P. Carter, International Journal of Rock Mechanics and Mining Sciences, **61**, 306-320 (2013)
8. Australasian Ground Control in Mining, *A case study on mine subsidence due to multi-seam longwall extraction.* URL: https://www.ausimm.com.au/publications/epublication.aspx?ID=12459
9. Ki-Dong Kim Saro Lee, Hyun-Joo Oh Jong-Kuk Choi Joong-Sun Won, Environmental Geology, 50:8, 1183–1191 (2006)
10. *Instruction for safely mining development at mines, extracting coal seams, liable to rock bursts* (MSHA, Moscow, 2000)

Modeling of processes and technologies of the Concentrating Raspadskaya Factory for assessing the impact of dilution of raw coal on the technological performance of enrichment products

Vladimir Udovitsky[1,*], *Vladimir* Kandinsky[1], *Artyom* Begunov[1] and *Larisa* Plotnikova[1]

[1]T.F. Gorbachev Kuzbass State Technical University, Department of mineral resources processing, 650000 Kemerovo, 28 Vesennya st., Russian Federation

Abstract. The paper presents a brief literature review on the impact of the dilution of mined coal from complex structural deposits on losses during selective and gross excavation of layers. The increase in dilution leads to a reduction in coal losses during mining. The effect of dilution on the loss of coal with beneficiation was established. A method is proposed for calculating sieve and fractional composition of coal depending on the ash content of the diluted coal using the example of sample No. 9 of the reservoir developed at Mine cut Raspadsky CJSC. The effect of ash content of diluted coal, enriched according to the technology of the Raspadskaya Processing Plant, is shown on the technological indicators of the enrichment products in the practical balance. It has been established by computational experiments that when the ash content of ordinary coal changes from 17.8% to 25.2% and the condition that the concentrate should have an ash content in the practical balance of not more than 6.8%, its output decreases from 84.4% to 73.3%, i.e. 11.1% of concentrate is irretrievably lost with rock shipped to the dump; the ash content of the rock decreases, which can lead to its ignition at the dump. An increase in rock yield, in turn, will lead to additional transport costs and even greater environmental damage.

1 Introduction

Dilution - loss of mineral quality during the mining process. It is expressed in a decrease in the content of the useful component or the useful component in the extracted mineral in comparison with their content in the mineral array (balance reserves). When assessing the quality of combustible minerals (coal, shale), the indicator is used - the ash content of coal (inorganic mass content, expressed as a percentage of the initial mass of the product). Dilution is the main feature characterizing the quality of the minerals mined; closely related

* Corresponding author: uvi@kuzstu.ru

to the loss of minerals. Usually, the greater the dilution, the smaller the loss during production [1].

The publication [2] shows the options for the dilution of coal mined: in the case of gross excavation, due to intra-layer rock layers; with selective - due to mining of coal plots by contact with the roof and soil. Previously, the authors of the publication [3] also indicated that when developing complex structural deposits, it is almost impossible to exclude losses and dilution of the mineral. This study shows the effect of the dilution of mined coal on the technological performance of products obtained at the processing plants (for example, Raspadskaya Processing Plant).

2 Materials and Methods

2.1 Initial data

The initial data for the calculation of technological processes and enrichment schemes are the sieve and fractional compositions of the sample number 9 of the layer of JSC Mine cut Raspadsky (Figure 1).

Size of the class, mm	Sitovyj composition		Fractional composition										T, %
			1.3 - 1.4		1.4 - 1.5		1.5 -1.6		1.6 - 1.8		> 1.8		
	γ, %	A^d, %	γ, %	A^d, %	γ, %	A^d, %	γ, %	A^d, %	γ, %	A^d, %	γ, %	A^d, %	
13-150	33,20	36,09	17,86	5,70	1,49	18,80	0,73	31,20	1,39	46,20	11,72	83,70	9,89
1.3 - 13	46,50	8,97	42,73	3,80	0,47	22,00	0,37	33,30	0,28	45,90	2,65	82,80	1,48
0.13 - 1.3	16,70	5,80											
0 - 0.13	3,60	19,90											
Total:	100,00	17,84											

Fig 1. Sieve and fractional formulations of sample No. 9 of the reservoir being developed at Mine cut Raspadsky CJSC.

The last column of table 1 shows the values of enrichment, T for machine classes, calculated according to the method GOST 10100-84. Obviously, the enrichment class more than 13 mm "difficult", and 1-13 mm - "easy".

The density of rocks of the Kuznetsk suite according to [3] is presented in table 1.

Table 1. The density of the rocks of the Kuznetsk suite

Species:	Density, g/cm^3
shale	2.58÷2.62
mudstones	2.62÷2.65
sandstones	2.65÷2.75
Natural burnt rocks	2.2÷2.6

In the development of reservoir deposits by selective or gross methods, the amount of crushed rock falling into the coal seam varies from 0 to 90 %. The resulting mixture is recycled to factories or concentrators.

In this work, the technological enrichment scheme of the Raspadskaya Processing Plant is adopted (Figure 1).

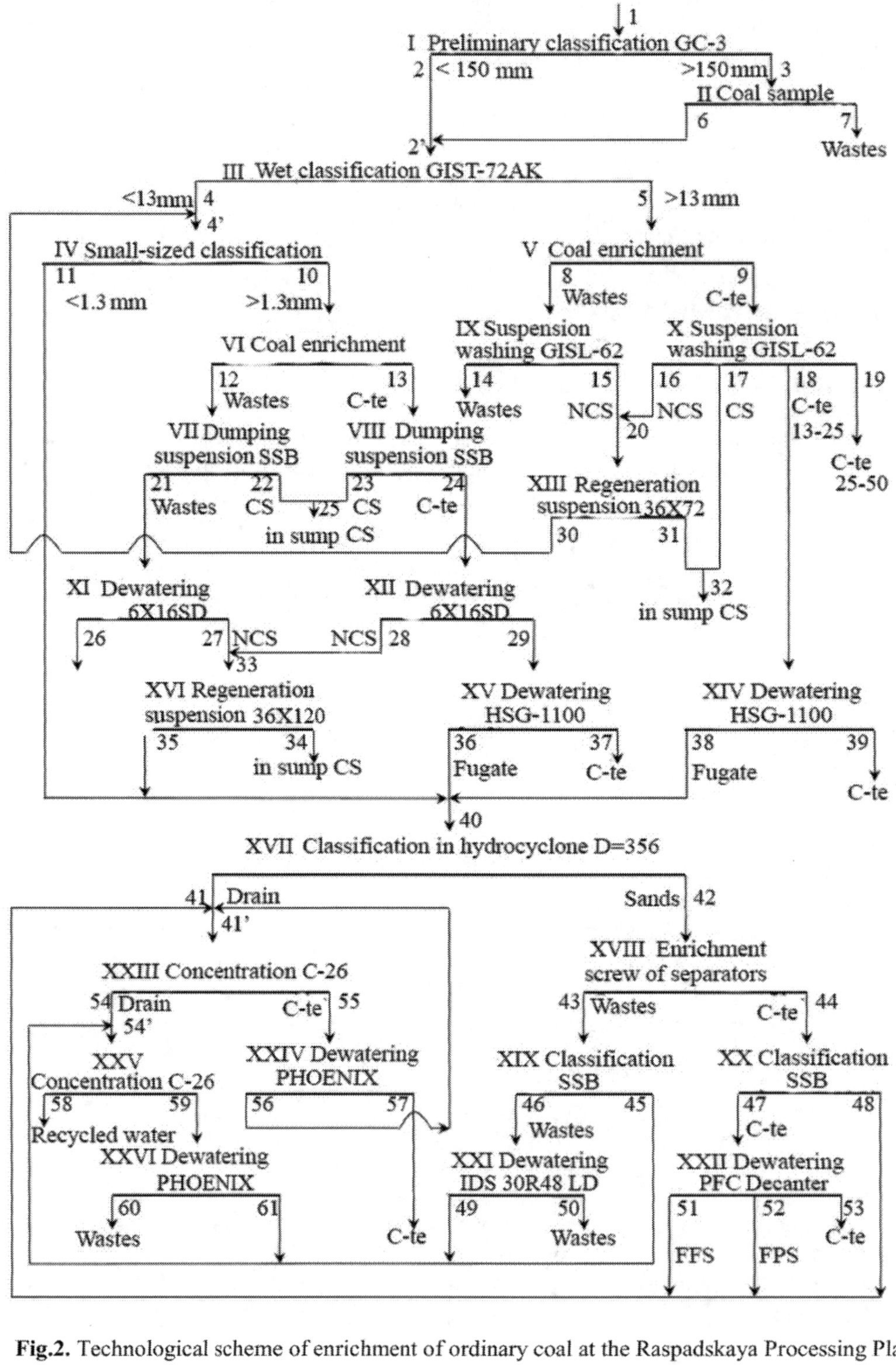

Fig.2. Technological scheme of enrichment of ordinary coal at the Raspadskaya Processing Plant.

The calculation of the yields and ash content of enrichment products in gravitational vehicles is performed according to known formulas, published in [4, 5, 6, 7]:

$$\gamma = \sum_{i=1}^{n} \gamma_i \cdot F(x_i); \tag{1}$$

$$A^d = \frac{\sum_{i=1}^{n} \gamma_i \cdot A_i^d \cdot F(x_i)}{\sum_{i=1}^{n} \gamma_i \cdot F(x_i)}; \qquad (2)$$

where n is the number of fractions; $F(x_i)$ – is the value of the Gauss probability integral (the calculation of the integral values for *Excel* and *TurboPascal* is described in detail in [6, 7]); γ_i, A^d_i – yield and ash fraction; x_i – is the deviation of the average density of the ρ_{cp} fractions from the planned separation density ρ_p:

$$x = \pm(\rho_p - \rho_{cp})/E_p \cdot 0.675, \qquad (3)$$

the plus sign in the formula (3) is taken in determining the extraction of fractions in concentrate and the minus sign in waste.

2.2 Mathematical models

Dilution of the coal produced occurs due to the ingress of rock into it, the density of which varies within 2.2÷2.75 g/cm3 (Table 2). Consequently, it can be assumed with sufficient confidence that during gravitational enrichment processes of ordinary coal, the rock will turn into a fraction with a density of more than 1.8 g/cm^3.

Depending on the amount of this rock, the sieve and fractional compositions of raw coal change, affecting the processing technology, as well as quantitative and qualitative indicators of the coal products produced. The mathematical model for calculating sieve and fractional compositions of diluted coal is represented by formulas (4) - (7):

Reservoir share in charge

$$\gamma_{plast} = 100 - \gamma_{por}, \%; \qquad (4)$$

γ_{por} – amount of collapsed rock, %

Output of rock falling into i - machine class

$$V_i\,por = D_i \cdot \gamma_{por}/100, \%; \qquad (5)$$

D_i – is part of the collapsed rock of its total number γ_{por}, falling from into i – machine class.

Output and ash content of the class, taking into account the breed that fell into it

$$\gamma_i = \gamma_i\,sit \cdot \gamma_{plast}/100 + V_i\,por, \%; \qquad (6)$$

$$A^d_i = (\gamma_i\,sit \cdot \gamma_{plast} \cdot A^d_i\,sit + V_i\,por \cdot A^d_i\,por)/\gamma_i, \%, \qquad (7)$$

γ_i *sit*, A^d_i *sit* –yield and ash content i of machine class by sieve composition, $A^d_i\,por$=92÷96 % – ash content of the collapsed rock.

According to the publications [4, 5, 6], as well as the current regulatory document VNTP 3-92 in the Russian Federation, the clogging of coal enrichment products is accepted in heavy-medium separators for a particle size of 13–150 mm

$$E_p = 0.015 \cdot \rho_p + 0.02, \text{g/cm}^3; \qquad (8)$$

in two-product heavy hydrocyclones

$$E_p = 0.03 \cdot \rho_p - 0.015, \text{g/cm}^3. \qquad (9)$$

The calculation of the scheme (Figure 1) was performed using methods that allow to determine the separation density at which the highest yield of the total concentrate of the required ash content is achieved. Techniques are published in [8, 9, 10].

Assumptions adopted in the calculation of indicators of enrichment end products:

• the ash content of the total concentrate should be constant and equal to 6.8 % at its highest yield;

• Dilution leads to an increase in the ash content of the reservoir (row "Total" of Table 1) from 17.8 to 25.2 % (Table 3) with intermediate values: 19.1, 20.5, 22.7, 23.9 % (Figure 4).

3 Results and Discussion

Using a specially developed program in *TurboPascal* to calculate the Raspadskaya Processing Plant presented in Figure 1, the results of the practical balance of the enrichment products for five dilution options for sample No. 9 of the reservoir developed at Razrez Raspadsky CJSC are obtained.

The program calculates separation densities in a heavy-medium separator (operation V on the diagram) and a heavy-medium hydrocyclone (operation VI) to achieve the maximum yield of the total concentrate.

The total concentrate consists of dehydrated concentrates of a heavy-medium separator with a particle size of 25-150 mm (operation X, product 19) and a particle size of 13-25 mm (operation XIV, product 39); screw separator concentrate after dewatering (operation XXII, product 53) and press filter cake (operation XXIV, product 57).

The practical balance is calculated for the ash content of the total concentrate equal to 6.8% in all options for dilution of sample No. 9.

Figure 3 shows the final fragment of the program, showing the practical balance for sample No. 9 with an ash content of 17.8 %. The initial data for the calculation is taken from Table 1.

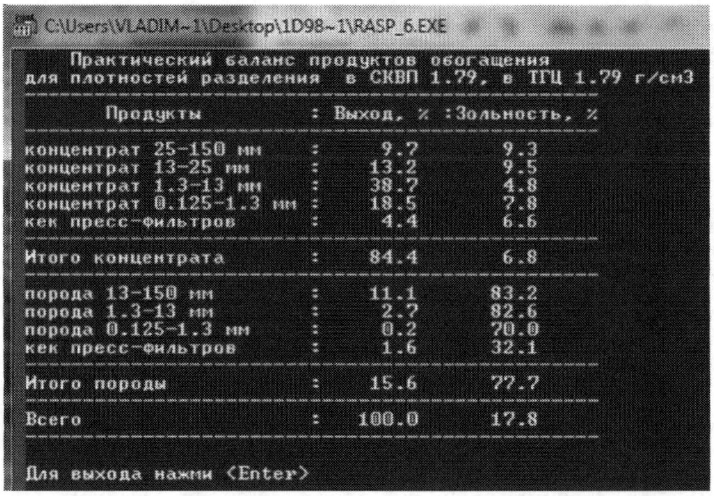

Fig. 3. A fragment of the program with a practical balance of enrichment products, calculated for a sample of ash content of 17.8%

Using the formulas (4) - (7), the sieve and fractional compositions of sample No. 9 diluted from ash content of 17.8 % (table 1) to 25.2 % (table 3) were calculated on a computer.

Size of the class, mm	Sitovyj composition		Fractional composition										T, %
			1.3 - 1.4		1.4 - 1.5		1.5 -1.6		1.6 - 1.8		> 1.8		
	γ, %	A^d, %	γ, %	A^d, %	γ, %	A^d, %	γ, %	A^d, %	γ, %	A^d, %	γ, %	A^d, %	
13-150	33,90	42,61	16,17	5,70	1,36	18,80	0,68	31,20	1,25	46,20	14,44	86,40	9,93
1.3 - 13	45,90	16,13	38,65	3,80	0,41	22,00	0,32	33,30	0,25	45,90	6,27	89,70	1,45
0.13 - 1.3	16,50	13,50											
0 - 0.13	3,70	29,40											
Total:	100,00	25,16											

Fig. 4. Sieve and fractional compositions of sample No. 9 of the reservoir being developed at Mining cut Raspadsky CJSC; dilution to 25.2 %

Tables 1 and Figure 4 draws attention to the enrichment index T, the value of which practically does not change as a result of dilution.

For the sieve and fractional compositions (Table 3), a practical balance of enrichment products was also calculated (Figure 5).

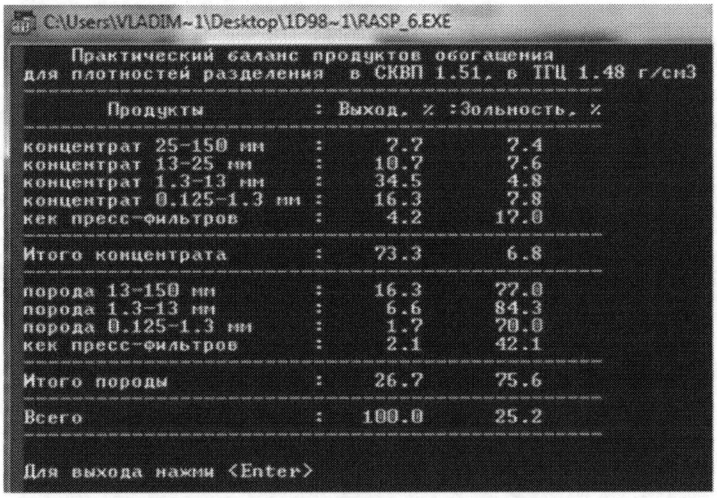

Fig. 5. Fragment of the program with a practical balance of enrichment products calculated for an ash content of 25.2 %

When the sample was diluted, the yield of total concentrate decreased from 84.4 % (Figure 2) to 73.3 % (Figure 3), i.e. by 11.1 %.

For intermediate values of ash content of diluted sample No. 9, similar calculations were performed. The results obtained are placed in table 2.

Table 2. The effect of dilution on technological indicators concentrate and waste.

Ash content diluted samples,%	Concentrate		Waste	
	Yield, %	Ash content, %	Yield, %	Ash content, %
17.8	84.4	6.8	15.6	77.7
19.1	82.8	6.8	17.2	77.8
20.5	80.8	6.8	19.2	77.7
22.7	77.6	6.8	22.4	77.4
23.9	75.7	6.8	24.3	77.0
25.2	73.3	6.8	26.7	75.6

4 Conclusion

Computational experiments have established:

1. An increase in the dilution of coal mined from 17.8 % to 25.2 % reduces losses during selective and gross excavation.

2. In case of fulfillment of contractual obligations by the concentrating plant to observe the constancy of the ash content of the concentrate being shipped (in our example no more than 6.8%):

• increases the loss of concentrate output from the beneficiation plant from 84.4. % to 73.3 %, i.e. by 11.1 %;

• increases the rock yield from 15.6 % to 26.7 %, i.e. 11.1 % of the ash content of 6.8 % will permanently go to the dump together with the rock;

• an increase in breeding yield will lead to additional transport costs associated with its transportation;

• A decrease in the ash content of the rock falling into the dump may lead to its ignition and additional environmental damage.

References

1. *Mineral dilution.* URL: http://www.mining-enc.ru/r/razubozhivanie/
2. V.L. Gavrilov, S.A. Ermakov, D.V. Hosoev, GIAB, **11** (2010)
3. S.A. Ermakov, D.V. Hosoev, V.L. Gavrilov, E.A. Hojutanov, Gornaya promyshlennost, **6** (2012)
4. K. Tromp, *Neue Wege fur die Beuerteilung der Aufbereitung von Steinkohlen* (Gluckauf, Berlin, 1937)
5. A. Terra, *Revue de lindustrue minerale* (Pellette, Paris, 1939)
6. A. Terra, *Sens de la courbe de partage anamorphosee et de lecart probable dans le controle des lavoirs* (Revue de lindustrue minerale, Paris, 1955)
7. J. Winckler, G. Mustafi, *Mining Challenges of the 21st Century* (APH Publishing Corp., New Delhi, 2000)
8. V.I. Chudovitsky, *XX International Mineral Processing Congress* (DTAH, Aachen, 1997)
9. V. Udovitsky, V. Kandinsky, A. Begunov, E3S Web Conf., **15**, (2017)
10. V. Udovitsky, V. Kandinsky, *E3S* Web Conf., **41**, 01011 (2018)

Design optimization of tumbling media for processing mining machinery parts

Vladimir Lyukshin[1], *Dmitry* Shatko[1,*], *Pavel* Strelnikov[1]

[1]T.F. Gorbachev Kuzbass State Technical University, 650000 28 Vesennyaya, Kemerovo, Russia

Abstract. Traditionally, in any country, mining industry is a strategic sector of the economy, for the development of which a large amount of resources necessary for the production of modern mining equipment, machinery and mechanisms are allocated. In recent years, a steady trend of technical re-equipment of mining enterprises is noted. This measure allows improving labor productivity and increasing the depth of mining due to the modernization of the existing mining equipment and the introduction of the new one; key attention being paid to the quality and reliability of the equipment produced. The current situation of the domestic mining machinery manufacturing market clearly demonstrates a strong dependence on foreign suppliers. In this case, a competent approach to the issues of import substitution in the industry, providing for the development and production of competitive domestic mining equipment that meets all the requirements and expectations of consumers, is urgently needed. As a rule, manufacturers of mining equipment use the entire machinery manufacturing cycle, ranging from blanking and to product assembly. At all stages of manufacturing, it is necessary to follow part processing technology, providing for the assurance of the required surface roughness, surface layer properties and other quality indicators. All this can be provided by a promising method of part processing – tumbling.

1 Introduction

Essentially, tumbling is mechanical action of abrasive media on a surface of a part, as a result of which each part is 3D processed all over. Tumbling allows polishing, grinding and finishing the surface of a work piece after cutting, casting, forging, welding and machining operations [1].

Casts, forgings, weld billets, which must be descaled, deburred, descuffed and given the desired roughness, are widely used in the mining equipment part production [2, 3]. The requirements for casts, forgings, cut and machined parts are specified in the industry standard "Products of mining machinery manufacturing. General specifications". All these requirements are successfully met by tumbling.

When tumbling, quartz sand, basalt or granite are used as natural tumbling media, and fragments of grinding wheels are most often used as artificial ones. The conducted

* Corresponding author: shdb.tm@kuzstu.ru

researches and the analysis of scientific literature and practice show that the tumbling media described above have some low operational capabilities [4, 5].

At the same time, works aimed at improving the grinding performance through the use of grains of a given shape, their orientation in the tool body, the development of new tool designs and their varieties, are known [6, 7].

Therefore, tumbling media, whose shape ensures maximum processing performance, are increasingly used in order to intensify the process and improve the quality of the processed workpieces. Consequently, the work aimed at the development and manufacture of tumbling media of a given shape is highly relevant and has good future prospects.

The tumbling technology performance depends on a number of factors, among which are the method and the processing mode, as well as design and operational features of tumbling media, which may be in the form of balls, cylinders, parallelepipeds, cones, tetrahedra, etc [8-10].

The shape of the tumbling media is selected in accordance with the required quality of a surface being processed and the shape of a part. It was experimentally established that it is advisable to use media that have the shape of a cone for polishing, removing small burrs from complex parts, as well as processing grooves, slots and holes. It is better to use media that have the shape of a sphere for polishing, removing large burrs on simple parts. The cylindrical shape of tumbling media proved to be the best for polishing and removing small burrs on simple parts.

In addition to the shape of tumbling media, the processing performance is also influenced by the material of which this media are made. For example, abrasive material, granite, fragments of grinding wheels are used for roughing. Steel balls, polishing lime, fine cut chrome-tanned leather, sawdust, etc. are used for polishing.

2 Methods and research

Considering the aforesaid, a tumbling body, which has wide versatility in processing parts of various shapes, was designed. The design of this body is shown in Fig. 1.

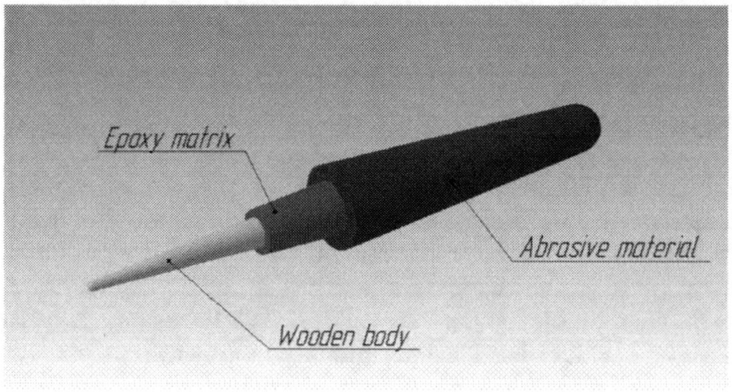

Fig. 1. Needle-shaped tumbling body.

The presented design has the following dimensions: length 10-25 mm, width - 3 mm, height - 3 mm.

The proposed needle-shaped tumbling media can be successfully applied in processing parts that have internal corners, small bending radii, narrow pockets, deep holes and other hard-to-reach places.

Tumbling bodies contain in their structure two different materials. Wood chips, which are waste production of chipboard, are used as core elements. Epoxy resin is used for binding a wood core and grinding grains. The proposed design is made of readily available materials and has a low cost.

The ratio of materials used to manufacture tumbling bodies: abrasive grain - 70%; wood chips - 25%; resin - 4%; curing agent - 1%.

Tests were carried out using a specially designed barreling unit with a horizontal axis of rotation. The unit is equipped with an electric motor with a rotational speed of up to 1500 rpm. The speed control panel is connected to the engine. The movement is communicated by means of a driving belt from the engine to the shaft, which, in turn, ensures the rotation of the barrel. To increase the efficiency of the operation, ribs are installed inside the barrel. They which do not allow the abrasive to roll down the wall, but force it to mix and move randomly.

To measure the wear of the proposed thumbing bodies, an LSE 3210 electronic scale was used; a measurement accuracy being of up to 0.01 g.

The experimental conditions were determined experimentally and include the following parameters:

- processing time is 180 min;
- the barrel rotation frequency is 80 rpm;
- the material of tested samples is 09G2S grade alloy steel.

In order to obtain the objective test data, a center-stiffener L-type part was selected. The sample has hard-to-reach angles at the points of junction and has dimensions of 70x70 mm.

The software "Image Pro Insight" was used to efficiently process, measure and analyze images for further evaluation of the amount of the ground material.

The experiment method contains the following steps:

1. Application of various kinds of dirt (paint, varnish, rust) to the sample.
2. Part processing.
3. Sample photographing.
4. Analysis of the ground material using software.

3 Results and discussion

The following experimental task was set: to investigate how well various kinds of dirt (rust, varnish, paint) are removed from a workpiece over a certain processing time.

In addition, the tumbling body durability, i.e. the ability of a resin to bind a wooden body and an abrasive grain, is also of great importance.

Below, in tables 1, 2, 3, there are data on the results of part processing using tumbling media of the proposed design, made of regular alumina abrasive grains of various grain sizes (No. 20, No. 40 and No. 100).

Table 1. The results of processing parts using tumbling media made of 13A20 abrasive grains.

Removed material, %	**Test number**				
	1	**2**	**3**	**4**	**5**
Rush	97.854	98.134	98.5318	97.456	98.143
Varnish	70.568	71.156	69.946	70.478	70.456
Paint	60.468	61.156	60.466	60.556	60.756

The average values of the removed area of dirt were: rust – 97.999%; varnish – 70.464%; paint – 60.666%.

Table 2. The results of processing parts using tumbling media made of 13A40 abrasive grains.

Removed material, %	Test number				
	1	2	3	4	5
Rush	100	100	100	100	100

Varnish	100	100	100	100	100
Paint	100	100	100	100	100

The average values of the removed area of dirt were: rust – 100%; varnish – 100%; paint – 100%.

Table 3. The results of processing parts using tumbling media made of 13A100 abrasive grains.

Removed material, %	Test number				
	1	2	3	4	5
Rush	100	100	100	100	100
Varnish	90.555	90.176	91.024	90.501	90.712
Paint	85.155	86.057	85.537	86.156	85.415

The average values of the removed area of dirt were: rust – 100 %; varnish – 90.564 %; paint – 85.534 %. Further tests of the epoxy ability to bind a wooden body and an abrasive grain were conducted. This parameter is extremely important, as it shows the ability of a body to resist the loads that act on it during processing. During the experiments, the fused abrasive alumina with grits No. 20, 40, and 100 was tested. The tumbling media, which were never used in the work, were taken. First, abrasive bodies was weighed, and then loaded into the barrel along with the part to be processed. Every 2 hours of processing, all the abrasive material that was separated from the wooden body was weighed, after which the operation was repeated. The durability of tumbling bodies is presented in Table 4.

Table 4. The durability of tumbling bodies.

Time, h	Grain size		
	20	40	100
2	2.22	2.17	2.28
4	5.81	5.55	5.87
6	7.35	7.14	6.95
8	20.54	21.11	20.34
10	55.37	53.55	55.11

After analyzing the data obtained, it can be concluded that the durability of grains No. 20, 40 and 100 is the same, taking into account the error of 5%.

The diagram of the dependence of abrasive grain binding durability on its grain size and the time of work is shown in Fig. 2.

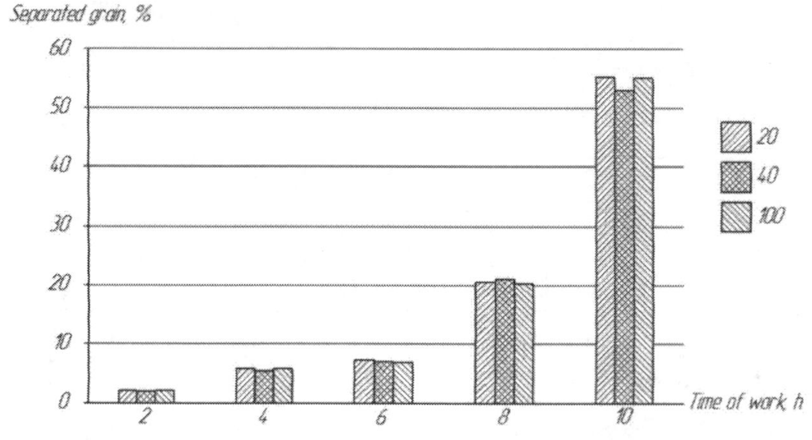

Fig. 2. The diagram of the dependence of abrasive grain binding durability on its grain size and the time of work.

4 Conclusions

Thus, the process of tumbling with the use of new media of a more rational design can provide a more efficient processing of mining machinery parts and thereby contribute to the achievement of their higher quality.

References

1. J. Wang, W. Li, Mater. Sci. Forum, **874**, 213-218, (2016)
2. W. Li, H. Zhu, J. Wang, C. Huang, AMM, **483**, 177-181, (2014)
3. J. Valíček, M. Harničárová, A. Panda, I. Hlavatý, M. Kušnerová, H. Tozan, M. Yagimli, V. Václavík, Adv. Struct. Mater., **61**, 111-120, (2016)
4. M. Tamarkin, E. Tishchenko, O. Rozhnenko, Russ. Eng. Res., **33**, 302-305, (2013)
5. M. Tamarkin, A. Tikhonov, E. Tishchenko, Russ. Eng. Res., **34**, 175-177, (2014)
6. D. Shatko, V. Lyukshin, V. Bakumenko, Mater. Sci. Eng., **22**, (2016)
7. D. Shatko, V. Lyukshin, P. Strelnikov, L. Samorodova, E3S Web Conf., **15**, 02012 (2018)
8. M. Harničárová, J. Valíček, M. Kušnerová, J. Krmela, A. Panda, Materialwiss. Werkstofftech., **50**, 635-645, (2019)
9. M. Tamarkin, V. Butenko, A. Isaev, E. Muregova, MATEC Web Conf., **226**, (2018)
10. M. Tamarkin, E. Tishchenko, I. Vyalikov, Russ. Eng. Res., **35**, 740-744, (2015)

Air pollution by coal dust as a factor of ecological compatibility for coal mining enterprises

Tatiana Korchagina[1,*], *Anatoly* Bykov[2], and *Evgeny* Schastlivcev[2]

[1]LLC "Mining Engineering Institute of Siberia" of JSC UHC "SDS-Ugol"
[2]Kemerovo Branch of Institute of computing technologies SB RAS, 650025, Kemerovo, Russia

Abstract. Coal mining enterprises (there are more than 130 in the Kemerovo region), are powerful sources of pollutant emissions (P) into the atmosphere. They account for almost a third of the total mass of the cumulative annual emissions in the region. A significant number of coal mining enterprises are located in the vicinity of the boundaries of residential settlements of Kuzbass. At the same time, emissions of coal mining enterprises are mainly carried out from unorganized sources of air pollution (SAI), which do not have treatment facilities. All this leads to pollution of the surface air layer by the coal mining industry. Currently, the category of the enterprise is determined only by its economic activity associated with the production of a product. In particular, all coal-mining enterprises, including the extraction and enrichment of coal, anthracite and brown coal, are classified as category 1 enterprises, i.e. enterprises with significant negative environmental impact (NEI) and related to the areas of application of the best available technologies (BAT). However, this approach to the definition of the category does not take into account the individual characteristics of the enterprise and the level of ecological compatibility of its production.

1 Introduction

Coal mining enterprises have a number of characteristic production processes that generate identical emissions into the atmosphere [1-6]. Its include drilling and blasting operations, coal loading and transportation, dusty coal warehouses and dumps, ventilation installations, quarry transport. In all these processes, coal dust (code 3749) is the most characteristic and unique P for coal mining enterprises. Therefore, we assume that the pollution of the atmosphere by coal dust can be an indicator of the level of ecological compatibility in the coal mining enterprise as a whole. For the integral evaluation of all the characteristics of atmospheric pollution with coal dust offers the option S_p (ha/tons) representing the area S (ha/kt) the zone of influence of emissions of the plant of coal dust divided by the annual output M (kt) coal company. To determine this indicator, it is enough to include coal dust in the composition of the calculated SV and determine the area of the zone of influence.

* Corresponding author: t.korchagina@sds-ugol.ru

According to [2] the zone of influence of emissions at the enterprise for a given P is defined as an area within which C_s>0.05 Mpcr, where C_s (mg/m3) is the maximum (in terms of wind speeds and directions) single concentration of this P, and PC_s (mg/m3) – the corresponding single maximum permissible concentration (for coal dust PC_s = 0.3 mg/m3). The distribution of Cs on the territory is determined by the calculation of air pollution on the basis of the normative methodology MRR-2017 [5] using unified programs for calculating air pollution [2].

The meaning of the indicator of ecological compatibility of the S_p is that it characterizes the size of the area with significant pollution by coal dust per 1000 tons of coal production. When determining the size of the zone of influence is taken into account the power of SAI emissions of the enterprise, their parameters and location on the industrial site, climatic parameters of the territory and the coefficient of relief. This takes into account almost all the parameters that determine the process of air pollution with coal dust. It is also taken into account how from an environmental point of view (dust suppression, humidification) the various stages of mining are organized.

Naturally, the proposed indicator S_p can be considered informative, if the inventory and the task of the calculated parameters of coal dust for different enterprises are carried out correctly and uniformly. The basis for the inventory of emission sources of coal mining enterprises is the industry methodology [6].

Note one important feature for determining the zone of influence associated with the setting of the settling coefficient of particles for the calculation of [5]. As a rule, in most calculations, project developers do not pay attention to the justification of the settling coefficient F, which takes into account the rate of settling of P (gases and aerosols, including solid particles) in the atmospheric air. However, as figure 1 shows, at a small distance (up to 1.5 km) from an unorganized areal source, the results of calculating the maximum surface concentration vary significantly.

2 Materials and Methods

Let us consider in more detail the procedure for selecting the settling coefficient F before carrying out the calculated estimates of atmospheric pollution by the method [5].

In the framework of the method [5], a dimensionless coefficient F is used in the calculation of atmospheric pollution, taking into account the rate of subsidence of P (gases and aerosols, including solid particles) in the atmospheric air. When developing projects, in most cases, task F is used on the basis of table 1 taken from Annex 2 to the methodology [5]. According to this table, all dust particles released into the atmosphere without treatment have a settling coefficient F=3. In fact, when particle size information is available, the value Of the F coefficient is set individually for each source, depending on the dimensionless ratio V_g/U_m, where Vg is the gravitational settling velocity of the particles.

Table 1. Dependence of the coefficient F on the degree of purification.

Degree of purification, %	Recommended sedimentation coefficient *F*
Over 90	2,0
75 to 90	2,5
Less than 75 or no cleaning	3,0

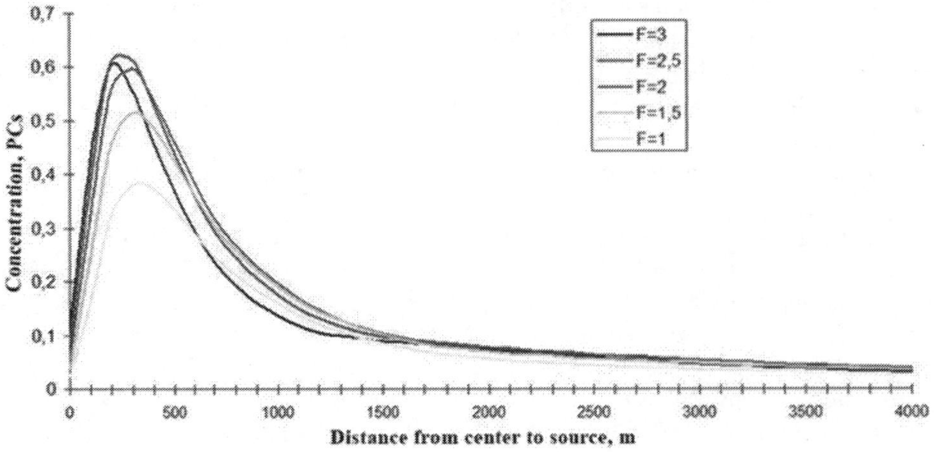

Fig. 1. Effect of the settling coefficient F on the graph of the surface concentration of coal dust (PCs = 0.3 mg / m3) along the axis of the torch from the coal warehouse height and size.

U_m – dangerous wind speed at which the surface concentration for a given source reaches a maximum (p. 5. 10. from [2]). In this case, the recommendations of Annex 2 of [2] can be summarized in table 2.

Table 2. Dependence of the coefficient F on the ratio Vg/Um.

V_g/U_m	Sedimentation coefficient *F*
≤ 0,015	1,0
0,015 < V_g/U_m ≤ 0,03	1,5
> 0,03	2,0-3,0 taking into account the degree of purification given in table 1

If the diameter of the particles D does not exceed 100 microns, the settling rate of the particles Vg can be found [7] using the Stokes formula:

$$V_{go} = \frac{10^{-8} \cdot g \cdot Ro \cdot D^2}{18\eta}, \tag{1}$$

where V_{go} is a gravitational subsidence velocity, cm / s;
g ≈ 981 is an acceleration of gravity, cm / S2;
R_o is a particle density, g / cm3;
η is a dynamic viscosity of air, g/cm·s;
d_g is an aerodynamic diameter of particles, microns.
In the future, instead of V_{go}, we consider Vg = 0.01 V_{go}, which has a dimension of m/s.

The value of η varies slightly within the range of air temperature changes typical for atmospheric pollution calculations (from -20°C to +20 ° C) [9]. This change in practice can be neglected and, taking into account that most calculations in accordance with [5] are carried out on summer conditions, it is possible to assume a dynamic viscosity of the air η = 1,8·10 - 4 (g/cm·s), which corresponds to a temperature of +20°C.

3 Results and Discussion

Most sources of air pollution by dust particles in coal mining enterprises are unorganized, and their emissions do not have an initial flare rise due to the initial velocity and overheating of the gas-air mixture. In particular, the source describing the technological explosion is approached in the calculation by an unorganized source formed after the explosion cloud rises to a height where it has lost its initial dynamic characteristics. Any unorganized source, both linear and areal, has a dangerous velocity U_m = 0.5 m / s, which is independent of the deposition coefficient F.

The density of coal varies from 1.20 to 1.5 g / cm3, and the density of the rock from 2.5 to 4.5 g / cm3. From tables 3 and 4 it is possible to determine the value of the settling coefficient F for coal particles of different diameters and densities.

Table 3. Ratio of deposition velocity V_g (m/s) to hazardous wind velocity U_m =0.5 m/s for an unorganized source depending on particle size D (μm) and coal density R_o (g / cm^3).

D	R_o=1,2	R_o=1,3	R_o=1,4	R_o=1,5
1	0,00008	0,00009	0,0001	0,00011
2	0,00031	0,00034	0,00036	0,00039
3	0,00069	0,00074	0,0008	0,00086
4	0,0012	0,0013	0,0014	0,0015
5	0,00186	0,00202	0,00217	0,00233
6	0,00267	0,00289	0,00312	0,00334
7	0,00362	0,00392	0,00422	0,00453
8	0,00471	0,00511	0,0055	0,00589
9	0,00595	0,00645	0,00695	0,00744
10	0,00734	0,00795	0,00856	0,00917
11	0,00887	0,0096	0,01034	0,01108
12	0,01054	0,01142	0,0123	0,01317
13	0,01235	0,01338	0,01442	0,01545
14	0,01431	0,01551	0,0167	0,0179
15	0,01642	0,01779	0,01916	0,02053
16	0,01867	0,02023	0,02178	0,02334
17	0,02106	0,02282	0,02458	0,02634
18	0,0236	0,02557	0,02754	0,02951

Table 4. Ratio of deposition velocity V_g (m/s) to hazardous wind velocity U_m = 0.5 m/s for an unorganized source depending on particle size D (μm) and rock density Ro (g / cm3)

D	Ro=2,5	Ro=3,0	Ro=3,5	Ro=4,0	Ro=4,5
1	0,00018	0,00021	0,00025	0,00028	0,00032
2	0,00065	0,00079	0,00092	0,00105	0,00117
3	0,00143	0,00172	0,00201	0,0023	0,00257
4	0,00251	0,00302	0,00353	0,00403	0,00451
5	0,00388	0,00469	0,00547	0,00625	0,00699
6	0,00556	0,00672	0,00784	0,00895	0,01002
7	0,00755	0,00911	0,01062	0,01214	0,01359
8	0,00983	0,01186	0,01384	0,01581	0,01769
9	0,01241	0,01498	0,01747	0,01997	0,02234
10	0,01529	0,01846	0,02153	0,02461	0,02753
11	0,01848	0,0223	0,02602	0,02974	0,03327
12	0,02196	0,02651	0,03093	0,03534	0,03954

13	0,02575	0,03108	0,03626	0,04144	0,04636
14	0,02984	0,03601	0,04201	0,04802	0,05372
15	0,03423	0,0413	0,04819	0,05508	0,06162
16	0,03892	0,04696	0,05479	0,06262	0,07006

Using data from the classification table for automatic partitioning of total dust emissions by fractional composition [3], we determined the dispersed composition of dust particles emitted into the atmosphere for the main processes of coal mining (table 5).

A comparison of table 5 with tables 3 and 4 shows that 98% of the particles are less than 10 microns in size when blown off the dumps, so the settling coefficient can be taken for coal as F=1.0 and for rock F=1.5. During blasting, 70% of particles with a diameter of less than 10 microns are carried outside the cut (large particles settle in the quarry), so it is advisable to represent the explosion emission in the form of two sources, where the first (70% of the emission) has a coefficient F=1.0, and the second (30% of the emission) F=2.0. For rock dust coefficient F should be taken 1.5 and 2.5, respectively.

Table 5. Division of total dust emissions by fractional composition.

		Particle size class, μm						
Code	Source type	0-5	5-10	10-20	20-30	30-50	50-100	>100
101	Blasting	50	20	15	10	4	0,9	0,1
201	Blowing from the dump	95	3	0.9	0.1			
301	Loading of coal	6	8	-	31	54	-	-
501	Crushing of rock mass	20	30	35	13	2-	0	-

The above estimates with the calculation of Vg according to Stokes law refer to the settling of particles in the resting air (in the research chamber or indoors). In the real atmosphere, dynamic and thermal factors are always present at the boundary with the underlying surface, where the propagation of P occurs, generating turbulence, under the influence of which the rate of subsidence can change. However, it is quite difficult to estimate the real limits of the change, because the data of standard meteorological observations are absolutely insufficient for the calculation of turbulence parameters.

Since 2012, the authors have been able to measure meteorological parameters using the acoustic mobile meteorological complex AMK-03. AMK-03 measures the following parameters: the projection of wind speed; wind direction; temperature; pressure; humidity; air density. In addition, the complex calculates the turbulence parameters: u* - friction velocity, m/s; L - Monin – Obukhov scale, m; H - vertical heat flow, cal/m2s. Thus, it is possible to specify the settling velocity Vg. First, use the real density and dynamic viscosity of the air in the Stokes formula (1), from which Vg is calculated. First, use the real density and dynamic viscosity of the air in the Stokes formula (1), from which VG is calculated.

Thus, the data of the mobile complex AMK-03, in combination with the laws of dependence of the parameters of the boundary layer of the atmosphere, allow reasonably enough to calculate the rate of deposition of particles of different sizes V_{dn} on the underlying surface. A special research module embedded in the ERA software package (www.lpp/EN), automatiseret calculation Vdn in the acquisition of archives of observations of metalocomplex AMK-03. When performing calculations for a specific time, it is possible to calculate the V_{dn} and set the coefficient F for these weather conditions.

The analysis of the calculations of dust particles deposition from atmospheric emissions with AMK-03 data (during which V_{dn} is calculated for each observation) shows that for

Kuzbass conditions V_{dn} may differ from V_g by 30% upwards and by 60% downwards. This can be taken into account in determining the actual subsidence factor F in tables 3 and 4.

Zone influence determination of coal mining enterprises on air pollution with coal dust was carried out on the basis of the new methodology MRR-2017 [5]. The deposition rates for sources associated with mining processes in table 5 have been introduced uniformly for all enterprises in accordance with tables 1 to 4.

By results of generalization of calculations of the indicator of S_p for the coal mining enterprises of Kuzbass it is possible to allocate the following categories of the enterprises which influence coal mining on atmospheric air.

Category 1 (low impact) $S_p < 1.5$;
Category 2 (moderate exposure) $1.5 < S_p < 2.5$;
Category 3 (severe impact) $2.5 < S_p < 3.5$;
Category 4 (hazardous effects) $3.5 < S_p$.

Enterprises of the 3rd and especially the 4th category use environmentally imperfect coal mining technologies. On them first of all atmospheric protection actions (dust suppression, humidification, the closed points of loading, etc.), at the organization of processes of production, transportation, storage and shipment of coal have to be introduced.

Let us consider some examples of calculating the S_P indicator based on data on sources of coal dust emissions for a number of open-pit coal mining enterprises in the Kemerovo region. Data on the first two sections are taken from departmental volumes of Maximum allowable physical impacts (MAPI) of the last years. For the rest of the summary volumes MAPI cities of Prokopyevsk, Kiselevsk, the Kemerovo and Mezhdurechensk developed in 2006-2008, OOO "Kuzbass environmental centre" (the authors express their gratitude to the Director of LLC T. G. Alekseichenko for the opportunity to use these data for scientific research). Data on coal production are taken for the years corresponding to the time of the inventory.

Table 6. Results of calculation of the indicator for some open-pit coal mining enterprises of the Kemerovo region.

Enterprise	Production M, thousand tons	Area of influence, S, ha	$S_P = S/M$, ha/ thousand tons
Mine Uzhnaya + Chernigovec	6174	2760 4970	1,26
MM Mayskoe	6654	13440	2,02
Oktyabrinskii	250	672	2,72
Vahrushevskoe	4115,7	5840	1,42
Prokopievskii	1678,4	3930	2,34
Kiselevskii	1905,8	4540	2,38
Tomusinskii	2421	5761	2,37
Kedrovskii	5040	12285	2,44

4 Conclusion

The results of the calculation of the S_P indicator are presented in table 6. It is easy to see that for 6 of the 8 enterprises considered, the values of S_p are quite close and lie in the range from 2.0 to 2.5, which classifies them as moderate impacts on natural complexes and the population.

References

1. A.A. Bykov, E.L. Schastlivtsev, Bulletin of the scientific center for safety in the coal industry, **1,** 128-134 (2014)
2. *Methodical manual on calculation, regulation and control of emissions of pollutants into the air* (Research Institute of the Atmosphere, St. Petersburg, 2012)
3. A.A. Bykov, E.L. Schastlivtsev, S.G. Pushkin, A.V. Smirnov, Bulletin of polzunovskii, **2**, 209-217 (2006)
4. *About the approval of criteria of reference of the objects having negative impact on environment to objects I and II, in III and in IV categories* (Government of the Russian Federation, Moscow, 2015)
5. *About the approval of methods of calculation of dispersion of emissions of harmful (polluting) substances in atmospheric air* (Government of the Russian Federation, Moscow, 2017)
6. *Industry methodology for calculating the amount of waste, captured and emitted into the atmosphere of harmful substances by coal mining enterprises* (Ministry of Energy o Russia, Moscow, 2003)
7. *Guidance on risk assessment for public health of exposure to chemicals polluting the environment* (Ministry of Energy o Russia, Moscow, 2004)
8. *Sanitary protection zones and sanitary classification of enterprises, structures and other objects* (Ministry of Energy o Russia, Moscow, 2014)
9. N.F. Tishchenko, *Atmospheric air protection. Calculation of the content of harmful substances and their distribution in the air* (Chemistry, Moscow, 1991)
10. M. L. Wesely, P.V. Doskey, J.D. Shannon, *Deposition Parameterizations for the Industrial Source Complex* (Argonne National Laboratory, Argonne, 2001)

Revitalization of Coal Brownfields in Solving Environmental Problems and Structural Development of Kuzbass Economy

Vitaly Zhironkin[1,*], and *Juraj* Janocko[2]

[1]National Research Tomsk Polytechnic University, 634050, 30 Lenina st., Tomsk, Russia
[2]Technical University of Kosice, Faculty of Mining, Ecology, Process Control and Geotechnologies, 04200 9 Letná, Kosice, Slovak Republic

Abstracts. The life cycle of mines lasts from several years to several decades. The closure of the mine occurs when its recourses are completely extracted, or mining is no longer profitable within its leasehold. Most regulatory authorities all around the world require the plan of mines and open-pits closure even before the mining begins, as it must be determined that the site would not pose a threat to the environment or society in future. Depending on the location, the site of mine or open-pit may be used in alternative way after closure or restored to the pre-operational condition. Local authorities increasingly require financial assurance that the funds needed to close mining enterprises will also be available in emergency situations. Special attention is paid to the environmental insurance intended to limit liability related to pollution elimination after mining activities at abandoned fields are over. In close connection with environmental insurance is post-mining – the development of alternative use of mining sites after the closure of mines and open-pits.

1 Introduction

In the Russian economy, the de-industrial trend actively manifested in the 1990s led to the mass closure of mines, machinery plants and agricultural enterprises. As a result, entire areas have remained abandoned, economically depressed and difficult to access for infrastructure for more than twenty years.

Kuzbass (industrial region of the Western Siberia) – the largest coal basin in the Russian Federation – has more than 50 mines having been closed since the 1990s. The territories of many cities of the Kemerovo region (Kiselevsk, Prokopievsk, Anzhero-Sudzhensk) consist in large part of abandoned mine fields, and their social-and-economic situation is permanently deteriorating [1]. In this regard, the problems of revitalizing the Brownfields in Kuzbass being solved by increasing open-pit coal mining is in force up to now.

* Corresponding author: v.zhironkin@inbox.ru

2 Materials and Methods

Different approaches to the analysis and evaluation of the abandoned mine or quarry fields, as well as to their definition, are determined by uneven state of the economy of various countries. For example, countries with high population density and competitive economy faced an urgent need to solve the problems with abandoned mine or quarry fields. In these countries the term "Brownfield" is understood as all abandoned areas, regardless their ecological condition: subsoil development sites, factory territories, agricultural land. These are the countries of the European Union, such as Great Britain, France, Germany, Austria, Belgium and Netherlands. On the contrary, low-density and highly competitive countries are focused on addressing environmental burdens or potential risks to human health, usually in urban areas. These countries view brownfields as abandoned territory only in the areas left by humans or contaminated with toxic substances. In the European Union such examples will be Denmark, Finland and Sweden [2-6].

Canada and the United States may be seen as countries having connected abandoned mining sites with environmental insurance.

In Russia brownfields represent deprivation epicenters of old industrial or agricultural territories with total depopulation (abandoned villages) or with structural degradation (mono-towns with closed plants and mines). Therefore, the issue of brownfields and their methodology is extremely relevant for Russia [6].

The etymology of the term "brownfield" is related to the color of land plots variously suitable for cultivation. While "greenfield" is an area or place with prosperous industry, i.e. a place which attracts investors in the construction of new enterprises, commercial and residential real estate, infrastructure then "blackfield" is a place abandoned by people affected by natural disasters, technological accidents, or influenced by deep economic depression (investors view investments within it as "toxic") [7].

So the brownfield is a territory which agricultural or industrial complex is almost destroyed, mainly for productive (depletion of mineral deposits or rural lands) or economic reasons (crisis of the national economy, fall of industry demand and prices, deep negative structural shift). The remains of production complex and infrastructure, settlements that could be brought back to life under certain circumstances, had been preserved [8]. Brownfields may have other economic potential, such as being used as tourism sites or national cultural heritage.

The problem of brownfields and their revitalization in the process of structural changes of national economy has become relevant for developed countries since 1960s. A common strategic vision for countries where mining was once highly developed is revitalization – transforming brownfields into socially, economically and environmentally sound areas through coordinated efforts by all levels of government, private investors and local people (Fig 1) [9].

In Kuzbass, brownfields were formed in the 1990s as a result of the decline in coal production and the closure of mines in the main coal clusters of the region:

- Northern Kuzbass cluster (Anzhero-Sudzhensk, Berezovsky, Kemerovo),
- Central Kuzbass cluster (Leninsk-Kuznetsky, Belovo, Prokopievsk, Kiselevsk),
- Southern Kuzbass cluster (Novokuznetsk, Mezhdurechensk, Myski, Osinniki), as well as in the ore cluster of Tashtagol destrict (ore mines in Kaz and Temir-Tau are closed).

 https://doi.org/10.1051/e3sconf/201913402002

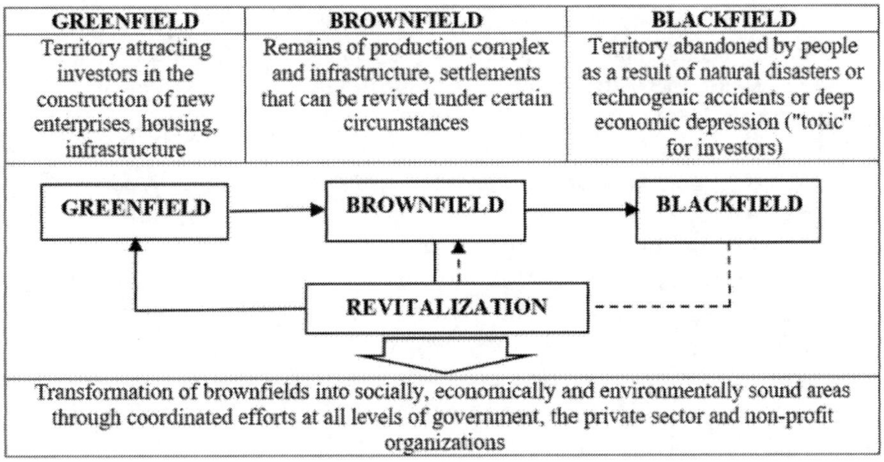

Fig. 1. Revitalization in the circle "Greenfield, Brownfield and Blackfield".

The main peak in brownfield development was in 1993-2002, although many mines were officially closed by the late 2000s. On the Kuzbass map (Fig. 2), they are the clusters of closed mines (brown circles) in the North, Center and South of the region.

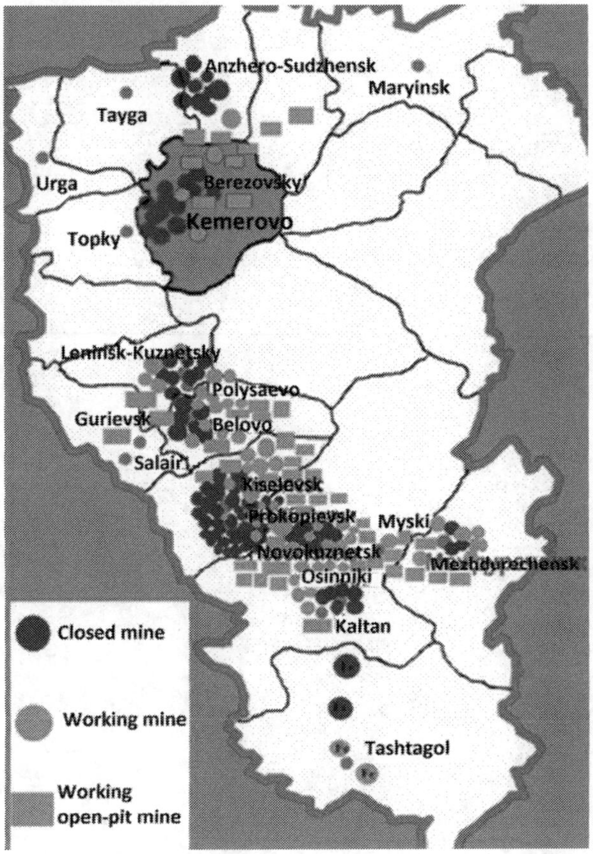

Fig. 2. The map of Brownfields and Greenfields in Kuzbass (drawn by author).

From 2003 up to now, the process of revitalization of Kuzbass brownfields takes place as opening of new coal open-pit mines and reconstruction of some mines (Fig. 2, green squares and circles on the map). New greenfield of Kuzbass coal industry are presented by working mineral extracting enterprises [10].

Most of the closed mines formed brownfields in the North of Kuzbass (22 closed coal enterprises versus 13 currently operating ones), while in the Central Kuzbass 54 mines and their segments were closed, and now 52 enterprises are operating. Half of them are open-pits, including working out old coal mine fields [11]. In the South of Kuzbass after the closure of 19 coal enterprises mining is carried out by 31 mine and open-pits [12]. Two iron mines (in the towns Kaz and Temir-Tau) were closed in Mountain Shoria, and two operate to nowadays (Tashtagol and Sheregesh) – Tab. 1.

Table 1. Sources of Brownfields and Greenfields in Kuzbass coal clusters

Coal mining clusters of Kuzbass (Cities)	Closed mines and Sections	Working mines and open-pit mines
North Kuzbass (Anzhero-Suzhensk, Kemerovo, Berezovsky)	22	13
Central Kuzbass (Leninsk-Kuznetsky, Polysaevo, Belovo, Kiselevsk, Prokopievsk)	54	52
South Kuzbass (Novokuznetsk, Myski, Kaltan, Mezhdurechensk) and Mountain Shoria iron ore cluster	19	31
Total	95	96

Dynamics and prospects of coal production in Kuzbass are reflected on Fig 3. It shows us that the period of brownfields development occurred in 1993-2003, when a large part of the Kuzbass mines was closed or suspended.

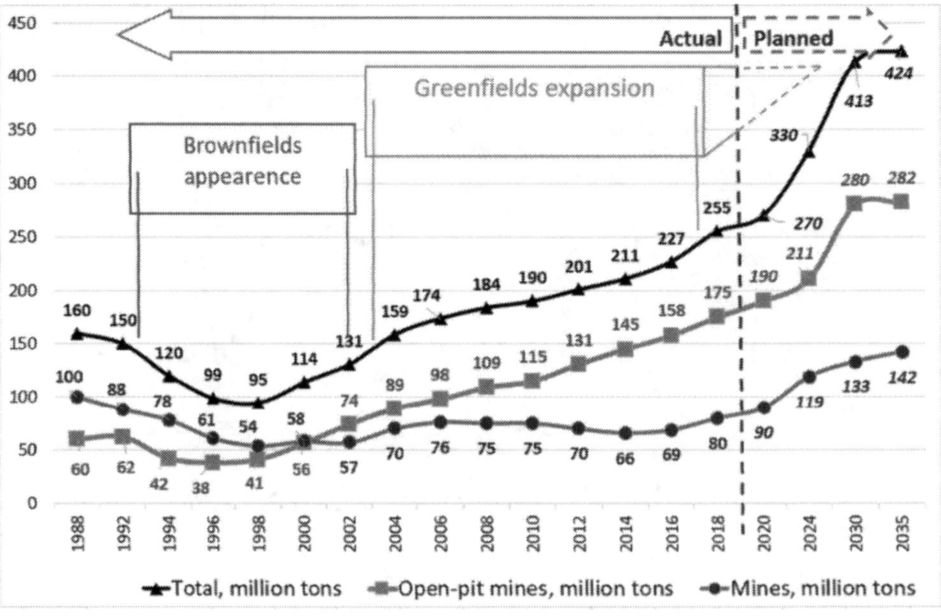

Fig. 3. Coal mining trends in Kuzbass (drawn by author).

Later, from 2003, the growth of domestic and global demand for coal, as well as the co-operation between coal producers and the state-owned railway company (established by the State Council of 2002 in Mezhdurechensk) allowed to increase coal production to the level of the USSR (160 million tons annually) already in 2004, and reach 250 million tons by 2019. At the same time, the main increase in coal production is made by the coal open-pits – new and reconstructed ones [13].

Thus, the greenfields – the new coal enterprises – arise on the sites of brownfields in Kuzbass. As a result, in the coal towns of the Center and South of Kuzbass the economic activity is being revived, new housing is being built. By contrast, in Anzhero-Sudzhensk (the North Cluster), the economic depression is intensifying as many mines are still closed.

At the same time, the present greenfields of Kuzbass are very similar to brown fields, as they are clusters of coal enterprises, just working, not closed (Fig. 4 and 5).

Fig. 4. View of brownfields in Kuzbass coal mining clusters.

Fig 5. View of greenfields in Kuzbass coal mining clusters.

The appearance of brownfields in 1990s had slightly reduced the burden on the environment, while their revitalization since the mid-2000s and the growth of greenfields, while bearing some social and economic benefits, does not improve, but deteriorates the regional environment – both in terms of land disruption and water and air pollution (Fig 6) [13, 14].

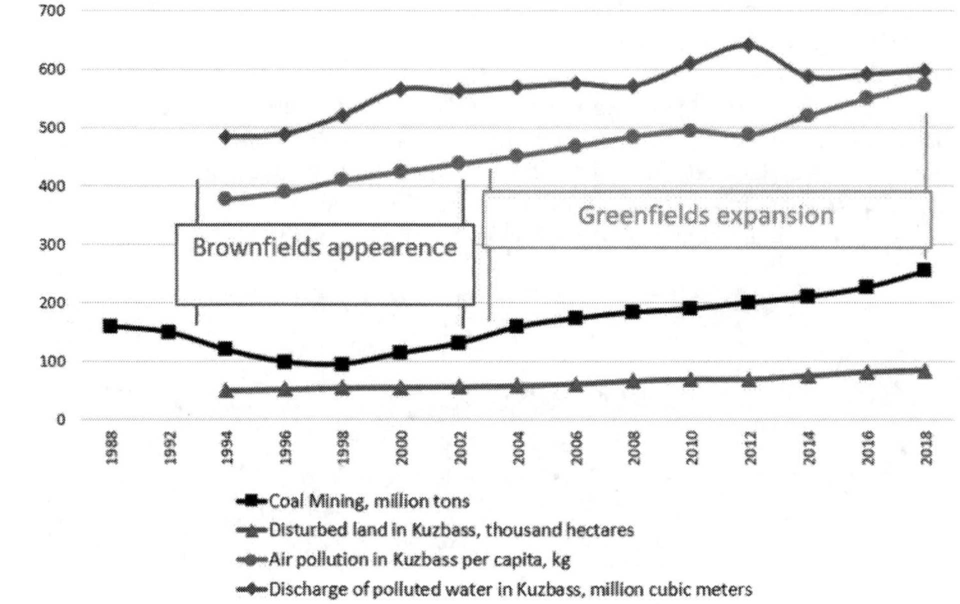

Fig. 6. Environmental damage compared to the volumes of coal mining in Kuzbass (drawn by author).

The diagram on Fig. 7 reflects the uncharacteristic process for modern greenfields depopulation process.

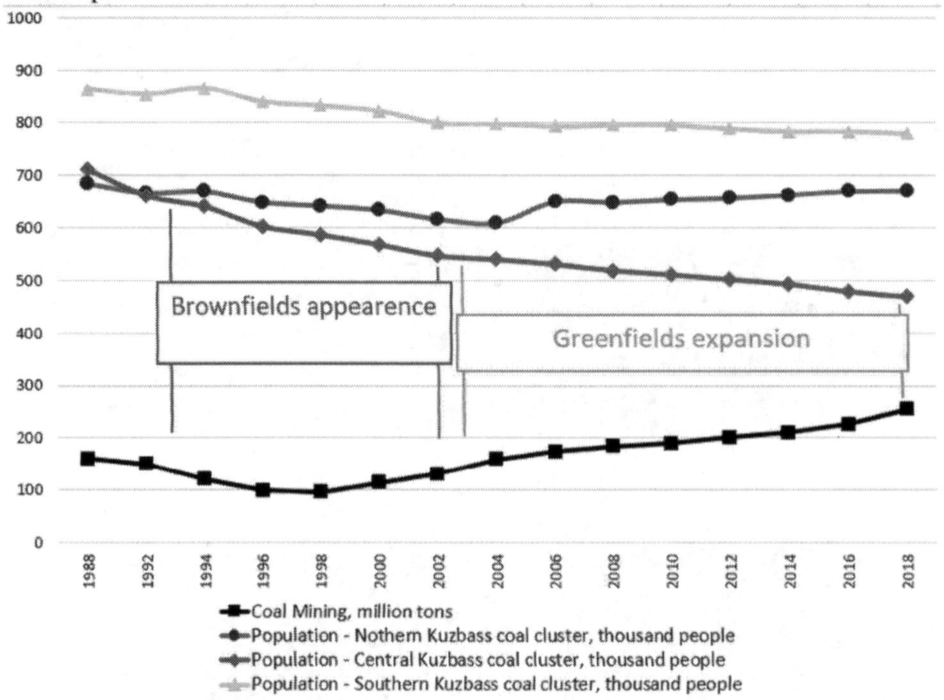

Fig. 7. Population growth/decrease compared to the volumes mined coal in Kuzbass (drawn by author).

Over the last 25 years the population of Kuzbass has decreased by more than 250,000 people, especially in the South and in the Center of the region (the North is held by the inflow of people to Kemerovo). And these are the Center and South of the region where the most of new coal enterprises are created [15].

There has been no diversification and positive structural shifts in Kuzbass economy during brownfields revitalization, and many old problems remain. The transition from underground to predominantly open-pit coal mining, the prevalence of investment in the coal industry, negligence to the environmental problems and small business potential largely equate the Kuzbass greenfields with European brownfields.

There is a clear lack of a positive link between the growth of coal production and the development of entrepreneurial conditions for small businesses in the region. Since 2006, the number of individual entrepreneurs in Kuzbass has been steadily decreasing, and since 2010 – their sales turnover too [16] (Fig 8).

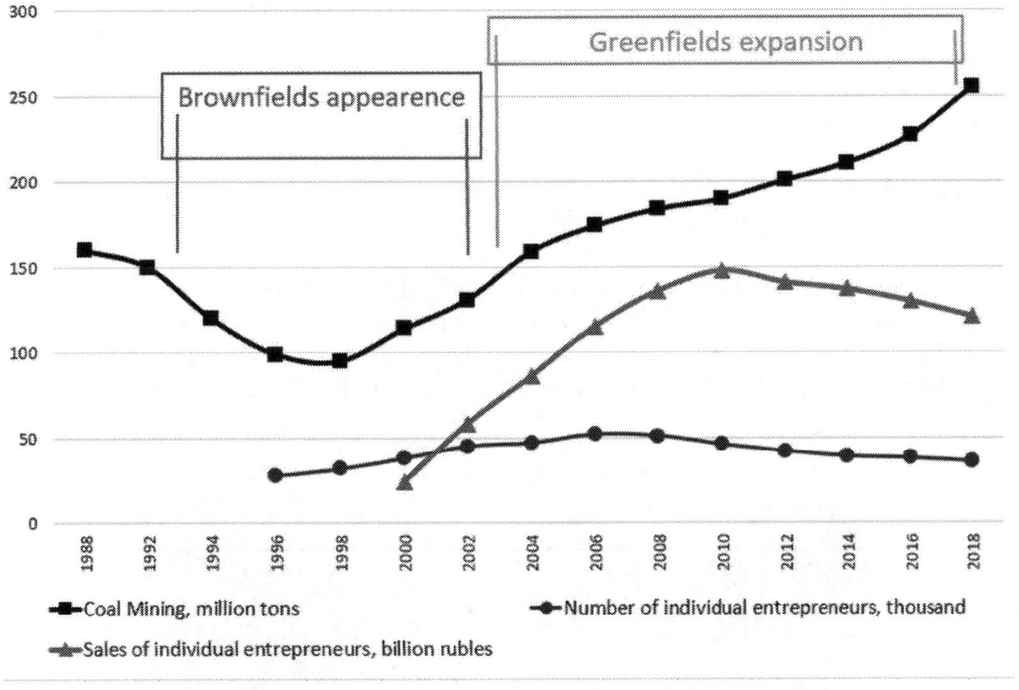

Fig. 8. Entrepreneurship dynamics compared to coal mining volumes in Kuzbass (drawn by author).

3 Results and Discussion

All mentioned above suggests that the existing mechanism for Kuzbass coal brownfields revitalizing is radically different from the European Union, as it does not improve the environment and does not allow the brownfield population to raise incomes outside the coal industry.

In this regard, it is appropriate to speak about the need to develop new mechanism of revitalization, the transition to greenfields of a new type.

It was clear that the creation of greenfields of the new type in Kuzbass, such as modern processing clusters of mechanical engineering, radio electronics, food industry, seems to be rather difficult in future, as it was not reflected in strategic documents (such as the Strategy for Social and Economic Development of Kemerovo Region until 2035). The machinery enterprises of former USSR, located in the towns Urga, Anzhero-Sudzhensk, Prokopievsk, Kemerovo today are also a kind of brownfield: they are closed, and technologies are irretrievably lost.

Another way of revitalization is implementation of new technologies for coal mining, for example, underground gasification of coal and methane extraction from coal seams did not take hold in Kuzbass.

The remaining promising way is the post-mining, that means the development of alternative activities in the brownfield territories.

The European Union countries are developing environmental insurance legislation as legal and the financial basis of post-mining. It allows the restoration of brownfield environment in future while mineral resources are being exploited today.

Environmental insurance makes it possible to place some of the revitalization activities of future investors on the acting mining companies [17, 18]:

A) liability insurance of mining companies for damage to land sites and their improper reclamation;

B) property insurance of residential areas and peasant farms in case of damage caused by accidents at mining enterprises;

C) insurance of costs of future investors in recovering brownfield's environment based on the "stop-loss" principle which sets the upper limit of their compensation;

D) personal insurance of life, health and property of mining enterprises employees.

Despite the variety of types of environmental insurance, it means the insurance of the liability of active investors to future ones who will convert brownfields and improve ecology.

Post-mining as a most promising way of brownfield revitalization means the development of alternative energy, ethnic and industrial tourism, and the re-establishment of agriculture after the restoration of disturbed lands. All this is successfully implemented in the Ruhr Basin in Germany, in Spanish Catalonia, where mines and open-pits were closed (Fig. 9 [19]).

Fig. 9. Conversion of coal industry clusters and development of alternative types of employment of the population: 1) development of alternative power engineering (wind, solar, thermal); 2 reconstruction of agricultural industry; 3) development of ethno-graphic tourism based on old mines, open-pits.

4 Conclusion

The main task assigned to the revitalization of brownfields is to integrate the restoration of production and economic activities on the depleted sites and old industrial areas at a new technological level. It also means reducing damage caused by mining to the environment.

By the example of Kuzbass, one can clearly see the negative results of the attempt to revitalize brownfields (abandoned mines closed in the 1990s as a result of the financial crisis in Russia) in the way of a transition from open to underground coal mining, with an increase in its volumes. These negative results include continuing environment deterioration, an increase in depopulation and a reduction in the attractiveness of the region for private investors (outside the coal industry).

A genuine solution of the brownfields problem in coal clusters of Kuzbass requires involving all participants such as business, authorities, the population, public organizations and trade unions in the process of revitalization. One of the main conditions for solving the issue of coal brownfields is the development of post-mining, as well as the establishment of an effective environmental insurance system. The latter will reduce the investment costs of future investors in restoring ecological balance and improving social infrastructure in the region.

References

1. N. N. Golofastova, V. G.Mikhailov, I. V. Seredyuk, Economics and Innovation Management, **1,** 66-75 (2017) DOI: 10.26730/2587-5574-2017-1 -66-75
2. Environmental Protection Agency (EPA), *Brownfields Assessment Grants FactSheet* (Environmental Protection Agency, Washington, 2009)
3. Environmental Protection Agency (EPA), *Potential Insurance Products for Brownfields* (EPA, Washington, 1996)
4. U. Ferber, *Brownfields Handbook, Lifelong Educational Project on Brownfields; Tech. Rep. Leonardo Da Vinci Pilot Project CZ /04/B/F/PP-168014* (Silezian University, Gliwice, 2006)
5. B. H. Fisher, *Brownfields Redevelopment and Gentrification: A Socio-Economic Evaluation of the EPA Brownfields Pilot Program* (Iowa State University, Ames, 2011)
6. J. Janocko, M. Cehlar, Z. Simkova, Economics and Innovation Management, **3(3),** 32-45 (2017) DOI: 10.26730/2587-5574-2017-3-32-45
7. H. Pavolová, L. Domaracká, N. Mitterpachová, *The Impact of Environmental Burden on the Reuse of Brownfields in Slovakia* (Univerisities of Košice Publishing, Košice, 2009)
8. M. Cehlár, L. Mihok, Theoretical and Empirical Researches in Urban Management, **8:4**, 60-72 (2013)
9. S. Zhironkin, S. Demchenko, G. Kayachev, M. Ryzhkova, O. Zhironkina, E3S Web Conf., 41, 04011 (2018)
10. O.O. Garshin, Z.A. Startseva, **2:5**, 33-41 (2019) DOI: 10.26730/2618-7434-2019-2-33-41
11. T. Gvozdkova, M. Tyulenev, S. Zhironkin, V. A. Trifonov, Yu. M. Osipov, IOP Conf. Ser.: Earth Environ. Sci., **50:1**, 012010 (2017)
12. *Mining Encyclopedia*, URL: http://miningwiki.ru/wiki

13. *Coal Mining Department of the Kemerovo Region Administration. Section: Mined Coals,* URL: http://www.ugolprom-kuzbass.ru/industry/indicators/mining/
14. T.V. Kiseleva, V.G. Mikhailov, Economics and Innovation Management, **4**, 70-78 (2018). DOI:10.26730/2587-5574-2018-4-70-78
15. *Ecology and Natural Resources Department of the Kemerovo Region Administration. The annual report,* URL: https://ako.ru/upload/medialibrary/7ff/doklad_2017.pdf
16. *Territorial branch of the Federal Service of State Statistics in the Kemerovo region. Section: Entrepreneurship*, URL: https://kemerovostat.gks.ru/folder/38691
17. M. Cehlar, J. Janocko, Z. Simkova, T. Pavlik, M. Tyulenev, S. Zhironkin, M. Gasanov, Resources, **8:1**, 21 (2019)
18. M. Cehlár, L. Domaracká, I. Šimko, M. Puzder, *Mineral Resource Extraction and its Political Risks; Environmental Science, Policy and Management* (ESPM Workshop, Tatranská Štrba, 2015)
19. J. Kretschmann, A. B. Efremenkov, A. A. Khoreshok, IOP Conf. Series: Earth and Environmental Science, **50**, 012024 (2017)

Development and debugging of MPI programs on the environmental research computing portal in the mining region

Andrey Vlasenko[1,1], *Igor* Sotnikov[1], *Eugeniya* Prokopenko[2], and *Anton* Demidov[3]

[1]Kemerovo State University, 650000 Kemerovo, Russia
[2]Kuzbass State Technical State University, 65000 28 Vesennya st., Kemerovo, Russia
[3] RMIT University, Melbourne, Australia, 3024

Abstract. The article is devoted to the development of automated debugging software for parallel programs used in the analysis of the impact of mining on the ecology of the region using the MPI communication interface. The system combines the approaches of static analysis of the source code and automated control of correctness at runtime. Each approach implements a separate component of the system. An innovative idea embedded in the system is to determine the necessary checks in the patterns of erroneous behavior that the user can control by modifying the templates from the proposed set, adding their own or deleting ones that are not required. A system for describing patterns is described and several examples of formalizing errors are given. Some of the examples of errors are suitable for static analysis, while others are suitable for automated verification. The architecture and the general scheme of the automated debugging system on a computing cluster are presented. Another software project described in the article is an engineering computing portal. The portal is an expandable environment for a comprehensive study of the environmental problems of the region, integrating various services united by high-performance computing.

1 Introduction

In the modern scientific world, the accuracy of numerical models of the studied objects and phenomena continues to increase, which causes an avalanche-like increase in computations in HPC problems.

The creation of an information portal that combines information on mining and their further use into a single model is extremely important for the Kuzbass. The region produces more than 250 million tons of coal, 20 million tons of iron ore. Accordingly, the total emissions into the atmosphere exceed 1 million tons, into the water basin – 500 thousand tons [1].

In the mining region, in which enterprises that extract, enrich and process coal, iron ore, gold (Kuzbass, Western Siberia, Russia), it is important to create a single computing portal

[1] Corresponding author: vlasenko_a@list.ru

that links together all the scattered information about the environmental impact. This information comes from each mine, mine, processing plant, metallurgical and chemical plant. The analysis of this information will allow creating a computer model of the environmental impact of the entire industrial complex of Kuzbass, identifying the most dangerous sources of pollution and putting together administrative, economic and engineering environmental protection measures.

In turn, the accuracy of data processing directly determines the quality of decisions to improve the environmental situation, and increases the requirements for developers of the portal software complex [2].

It is the reason for the active growth in the scale of high-performance computing. This dynamic can be easily traced by regularly updated rating lists of supercomputers, for example, the Top500 world list and the Russian Top 50. The increase in HPC tasks is reflected in the increase in the number of threads and processes of the corresponding parallel programs.

Despite the fact that many researchers use off-the-shelf computing packages (OpenFOAM, ANSYS products, FlowVision, etc.), a significant share in the HPC segment is occupied by proprietary programs. This fact, in particular, is associated with the development of new numerical methods, the search for optimal software technologies for a specific task, etc. Various implementations of the MPI standard continue to remain the most common programming tools for supercomputers. One of the main problems with MPI programming is debugging. Even in the case of sequential programs, debugging sometimes takes up to 2/3 of the total development time. And when compiling parallel code, the developer may encounter many additional semantic errors caused by the interaction of asynchronous processes and the use of MPI technology. Clause 3 describes some of these errors, and a more detailed look at their classification can be found in [3, 4]. The observed tendency to increase the size of the tasks being solved only aggravates the situation with debugging.

To detect logical errors in parallel programs, such dialog debuggers as Rogue Wave TotalView [5] and Allinea Distributed Debugging Tool (DDT) [6] are traditionally used. However, as the number of processes in programs increases, the usefulness of dialog debuggers decreases. After all, to observe the behavior of each of one hundred processes, executing the program in steps, is an incomparably more difficult task than the behavior of ten. In practice, users debug the task on "small dimensions" (this term refers to the number of processes / threads, the number of iterations of loops, the dimensions of arrays, etc.), and then upload it to the target computing resource in full scale. But at the same time, new errors may appear that did not arise on "small dimensions". In this case, dialog debuggers are almost useless.

The debugging system described in the article is able to function as a standalone software product. However, work is currently underway to integrate the system into the engineering computing portal [7, 8]. The purpose, structure of the portal, as well as the integration scheme of the debugging system are given below.

2 Materials and Methods

To detect logical errors in parallel programs, such dialog debuggers as Rogue Wave TotalView [11] and Allinea Distributed Debugging Tool (DDT) [9] are traditionally used. However, as the number of processes in programs increases, the usefulness of dialog debuggers decreases. After all, to observe the behavior of each of one hundred processes, executing the program in steps, is an incomparably more difficult task than the behavior of ten. In practice, users debug the task on "small dimensions" (this term refers to the number of processes / threads, the number of iterations of loops, the dimensions of arrays, etc.), and then upload it to the target computing resource in full scale. But at the same time, new errors may appear that did not arise on "small dimensions". In this case, dialog debuggers are almost useless.

The debugging system described in the article is able to function as a standalone software product. However, work is currently underway to integrate the system into the engineering computing portal [7, 8]. The purpose, structure of the portal, as well as the integration scheme of the debugging system are given below.

When constructing a system for automated debugging of MPI programs, the authors find it most productive to use a combination of several approaches, since various methods are suitable for searching for logical errors of various types. Suppose that the program contains deadlock, caused by the fact that two processes launched blocking communication functions, of which they must be participants together. Fig. 1 shows a fragment of such a program where the zero process calls the function of sending a message to the first (MPI_Send), and the first, among others, calls the group operation MPI_Gather. In this case, the zero process must also be a member of MPI_Gather.

The standard way to obtain information from the running processes of an MPI application is to use the profiling interface described in the MPI standard. This interface is used by almost all tools of automated correctness control, both developing projects and completed ones. Among them are Umpire, Marmot, MUST [9], Intel Trace Analyzer and Collector, which is the successor to the Intel Message Checker project [6]. The basic principle of using a profiling interface is as follows. The developer of the tool creates a library of MPI functions that is statically linked to the user program. While the user application is running, calls to MPI functions are intercepted by this library. Thus, the tool receives all the information about which process at which moment called the function and with what actual parameters. The tool processes the received data at its discretion, and in order for the action that the user expects to call this function to be performed, it is necessary to call the corresponding PMPI function at any place in the handler with exactly the same arguments as the original function.

For these reasons, when constructing our own automated debugging system, 2 methods were used: static analysis and automated correctness control.

3 Results and Discussion

The automated debugging system has a distributed architecture and is designed to work on a computing cluster (Fig. 1). On the head node (it is assumed that it is the compilation node), the source code of the user program is processed by the preprocessor, which is the first component of the system. The preprocessor introduces some additional operators and performs static analysis. To do this, from the presented set of templates (the path to the template directory is one of the input parameters of the preprocessor), he selects suitable for static analysis, which is determined by the number of processes installed in the unit, and the presence in the template text of certain keywords, which include, in particular, pseudo-function "SIZE". The preprocessor parses these templates and passes the user code, during which checks are performed for the occurrence of template situations. Information about the matches found is recorded in the resulting error file, where the name of the template is entered (field "Name"); names of source code files where an error was detected, with line numbers; The text of the lines of code that caused the error.

At the start of its work, the analyzer server parses text template files, filling out a list of templates with global errors and determining those MPI functions that need to be profiled based on the contents of the templates. The server analyzer passes the list of these functions to MPI processes. It also transfers those of the disassembled templates where the situation is described, for the determination of which information from only one process is sufficient (local errors). During their work, processes call MPI functions, which are intercepted by the profiling library. If the called function is included in any patterns for local errors, then the service flow enters information about the parameters of this operation into special structures and performs an analysis. If a function is included in at least one of the templates describing a global error, then its parameters and code line number are passed to the analyzer server. That server stream, to which this MPI process is assigned, parses the received line, enters data into special program structures and performs analysis for compliance with the current situation patterns with global errors.

Computing services are provided directly through the portal using the CompServiceGen component. In addition to services, users and experts interact through the portal, cooperate with specialists, and search among ready-made solutions. Administrators are provided with such functions as managing portal users, registering and assigning rights to access computing resources and services.

To conduct their own computational experiments on high-performance resources, the Onlide subsystem is used. This subsystem is a development environment with which you can create your own code, compile and run it on available high-performance resources.

Computing services, subsystems, as well as external client systems interact with the portal API, which is displayed in the Portal API component. The following functions are implemented through the portal API:

1) User authorization (Auth API). Data about users, as well as about user groups, computing resources and software is stored in an LDAP database (LDAP DB).

2) Management of user file storage (Storage API). The repository contains input files and calculation results.

3) Interaction with software on computing resources (Software Manager API). The Software Manager API provides the following functions:

- start and a stop of implementation of programs on remote computing resources in the form of the identified tasks;
- tracking of a condition of the started tasks (in turn, it is carried out, it is complete);
- obtaining information on the used computing resources (the RAM, loading of the CCP) and about the started tasks (date and time of start of a task, date and time of start of the program, date and time of end of a task, restriction of resources).

The execution of the above functions is delegated to a specific Software Proxy implementation, depending on the computing resource. There are currently two versions of its implementation.

In the first version, the interaction is performed by setting up an SSH connection with a remote computing resource and running a special Python script to be placed on that resource. This script is responsible for running the specified program and tracking the resources used during its operation.

In the second release, the interaction is performed by establishing a TCP connection with the remote agent being deployed to the desired resource. The agent starts once and processes startup requests. This version is implemented primarily for Windows computing resources, due to the problem of running the SSH server on this operating system (OS).

The calculation services are described in the business process simulation language BPMN. In Figю 1, the BPMN Engine is a business process execution environment. For software BPM, this environment provides its own set of APIs (Engine APIs). The CompServiceGen component is responsible for generating a Web interface to start and track the status of the business process.

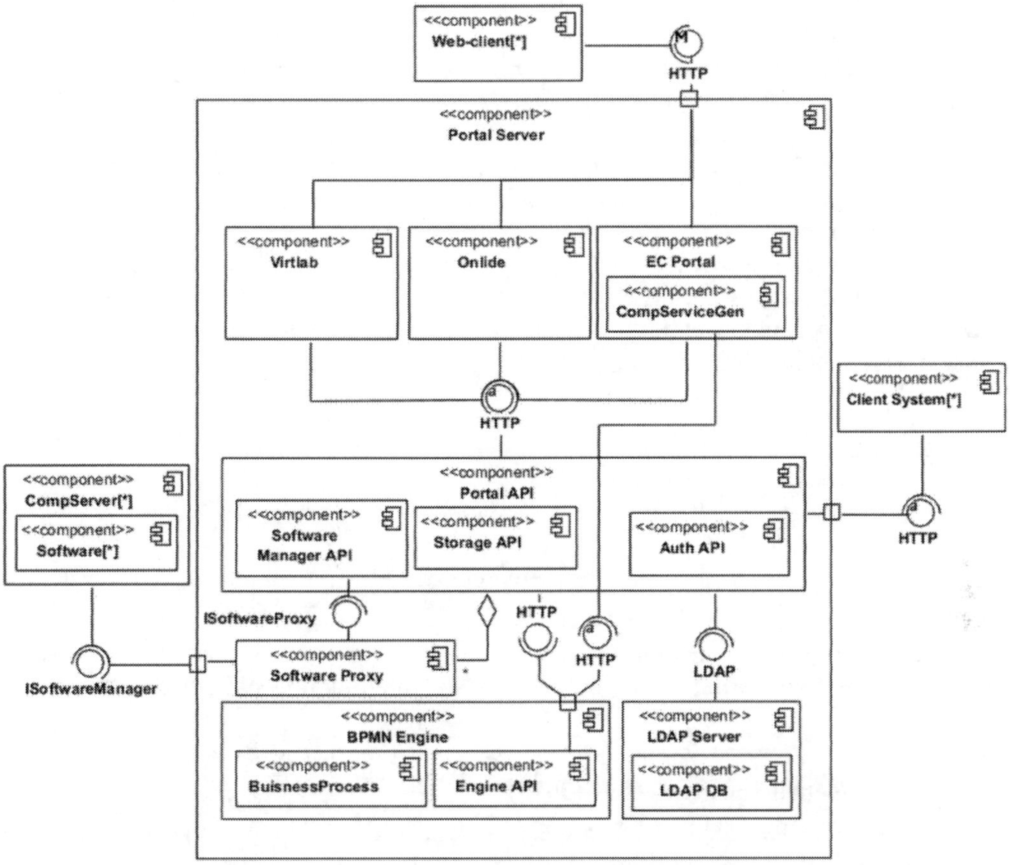

Fig. 1. Logical structure of the TPI in the form of a component diagram (UML2).

Camunda is selected as the execution environment for business processes. Camunda supports most of the elements that make up the business process, as specified in BPMN 2.0. In particular, they include an element that sends HTTP requests (for example, API requests) and an element that allows you to call another business process (sub-process). Based on the first element, two sub-processes are described.

The first sub-process is responsible for communicating with the user 's file store. The repository contains input files and results of service execution (video files, images, model 3D, etc.). As input parameters, the element takes the path to the file or directory and the action to perform on it (create, delete, move).

The second sub-process is responsible for interacting with software on compute servers, namely, running and checking status - the program is running or has already been completed.

4 Conclusion

Currently, environmental protection in a region with intensive mining is impossible without the creation of computer models of the impact of industry on the environment. These models should be presented in a single information portal, and information should be processed with maximum accuracy.

The system presented in this paper is intended to help researchers developing MPI applications in such a time-consuming and fairly routine part as debugging. The obvious advantage of the system is that the user does not need any preliminary work to obtain the result. All you have to do is specify multiple parameters and one command to run your program under system control. At the end of the system operation, the user has a ready list of logical errors with indication of code lines that caused the found errors. The automated debugging system has been tested on application parallel applications. In particular, it was used in writing the tutorial. With the help of the system, the authors of the manual found errors in the sorting programs of the array using the method of chet-odd permutation and block multiplication of the matrix by the vector. The current integration of the system into the engineering computing portal will further simplify the researcher 's interaction with the system and increase the number of users.

References

1. M. Tyulenev, S. Markov, E. Makridin, Y. Lesin, V. Gogolin, E3S Web of Conferences, **105**, 02022 (2019)
2. A. Medvedev, I. Kislyakov, Ye. Prokopenko, M. Semenkina, K. Brester, E3S Web Conf., **105**, 03020 (2019)
3. J. Desouza, B. Kuhn, B. Supinski, Proceedings of the second international workshop on Software engineering for high performance computing system applications, **1,** 02101 (2005)
4. A.M. Gudov, S.Y. Zavozkin, I.Y. Sotnikov, IOP Conference Series: Earth and Environmental Science, **115**, 1 (2018)
5. A. Gudov, V. Perminov, Y. Filatov, L.H. Un, S. Zavozkin, I. Grigorieva, I. Sotnikov, CEUR Workshop Proceedings, **1,** 61-73 (2017)
6. J. Protze, T. Hilbrich, M. Schulz, B.R. de Supinski, 43rd International Conference on Parallel Processing Workshops, **1**, 206-215 (2014)
7. S. Siegel, Proceedings of the 14th European PVM/MPI Users' Group Meeting, **1**, 13-14 (2007)
8. A.Yu. Vlasenko, A.M. Gudov, Journal of Computer and Systems Sciences International, 56:4, 708-720 (2017)
9. L. Williams, *Upgrade Adds Muscle to Debugger* (Oak Ridge National Laboratories, Oak Ridge, 2010)

Ecologization of the Development Management of the Mining Regions of Russia

Irina Kolechkina[1,*] *Irina* Verchagina[1], *Elena* Eltsova[1], and Mariana Petrova[2]

[1]T.F. Gorbachev Kuzbass State Technical University, Belovo branch, 65264466 Ilyica st. 32, Belovo, Russia
[2]St. Cyril and St. Methodius University of Veliko Tarnovo, Veliko Tarnovo, Bulgaria

Abstract. The article presents the results of the analysis of the development strategies of the mining regions of Russia for the environmental measures. The objectives of the strategic development of the regions, the inclusion of environmental development measures in the management system are studied, the content of the strategies of mining regions in relation to eliminating existing environmental damage and the implementation of measures to conserve forests and natural systems is studied.

1 Introduction

Environmental aspects related to the performance of mining enterprises have been the focus of attention of researchers since the second half of the last century. It is noted that efforts to improve the environmental situation often do not lead to the results that were expected [1], which requires changes in approaches to environmental management

Modern realities require the application of an increasingly in-depth, integrated and detailed understanding of the territorial management of environmental development, the ecology of the management system is particularly acute in mining regions, due to the significant damage that the mining industry causes to the environment. Russia is a state whose export potential is largely determined by the export of extracted minerals and minerals that have undergone primary processing. Mining leads to an inevitable increase in anthropogenic impact on regional ecosystems, and hence on global environmental processes. The high importance of mining for the economic development of individual states (including Russia) and the world community is beyond doubt.

In the global practice of strategic planning, such concepts as "ecosystem management" and "sustainable development" are becoming increasingly important. The concept of "sustainable development" in relation to the mining industry in the international community begins to take shape at the end of the XX - beginning of the XXI century. The relevance of this area has been confirmed by the international community. In September 2019, Chile hosted the 6th International Congress on Environment and Social Responsibility in the Mining Industry, Sustainable Mining 2019, which was attended by more than 275 participants from 16 countries [2]. (Russian representatives did not participate in the

*Corresponding author: 40882kip@mail.run

conference). The work of the congress was devoted to pressing environmental and social issues in the global mining industry. The conference examined the key areas of development of the mining industry, formed by the international community in recent decades.

The development of a territory development strategy is the prerogative of its highest authorities. It is at the strategic level that the mission and the general goal of the development of the territory are formulated, which are then detailed in tasks and implemented through specific programs, events, allowing to achieve results. The presence and nature of events in strategic documents allows us to judge the level of interest and involvement of the authorities in solving certain issues.

2 Materials and methods

The need for greening management, especially in mining regions, has determined a research interest in regional strategies for socio-economic development in terms of the presence / absence in these documents of modern approaches to solving environmental problems. It seems appropriate to study the content of strategic documents of the socio-economic development of the mining regions of the Russian Federation with a view to assessing the content of elements of environmental management of the development of territories in them.

The selection of regions specializing in mining was carried out by us in accordance with the Strategy for Spatial Development of the Russian Federation [3]. The named document highlights promising economic specializations of the regions of the Russian Federation. Moreover, in 41 regions out of 85, mining was named as a promising specialization. In addition, for the five constituent entities of the Federation, mining is an unpromising economic specialization critical for the region's economy. Thus, the object of the study was the strategic documents of the socio-economic development of 46 constituent entities of the Federation, specializing in including mining.

In this regard, the purpose of our study is the level of environmental friendliness of strategic management in the constituent entities of the Russian Federation, classified as extractive. The scope of analysis included strategic documents of 45 constituent entities of the Russian Federation (the Kursk region did not present an actual strategy on the Internet).

As criteria for the analysis of the environmental friendliness of strategic management, we took the following: K1 - the presence of environmental goals among the main strategic guidelines for the development of the region; K2 - the availability of measures to ensure environmentally-oriented economic growth and the introduction of environmentally effective innovative technologies; K3 - the presence of measures to reduce the negative impact on the environment; K4 - the presence of measures to preserve the environment.

3 Results and discussion

The study showed the following. Goal-setting is the foundation of strategic management. Criterion K1 allows you to study the priorities of regional development, promising areas of effort. By the nature of the strategic goals, one can judge the nature of the activities undertaken in the region in the next few years. The analysis showed that as part of the main strategic goals, only 23 producing regions have environmental goals and development priorities. That is, 51% of the producing regions do not currently focus their further development on the priority solution of environmental issues, on "green" standards in the management of the territory. It should be noted a significant difference in approaches to the formulation of strategic environmental goals in the regions.

It should be noted a significant difference in approaches to the formulation of strategic environmental goals in the regions. So, 17 regions have identified environmental goals as development priorities. For example, the mission of the Republic of Buryatia is formulated as a trinity of the natural environment, people and the economy. In one region - the Astrakhan region - for strategic purposes, a narrow region was identified - the reproduction of natural resources of the Volga-Caspian basin. In five regions, environmental issues are an integral part of the strategic objectives of improving the quality of the living environment for humans.

Ensuring environmentally oriented economic growth and the introduction of environmentally efficient innovative technologies (criterion K2) in most constituent entities of the Russian Federation involves the development of the production of environmentally safe and biologically high-grade agricultural products considering regional specifics (animal husbandry, crop production, etc.). In addition, the development of recreational and health tourism can be attributed to the K2 criterion. The border regions are planning the development of tourism as the core of an export-oriented health cluster (Kamchatka Territory, Magadan Region, the Republic of Karelia and others), the internal regions are mainly focused on the development of domestic tourism (Irkutsk Region, Altai Territory, the Republic of Buryatia and others). Strategic development activities of the territory management system aimed at using environmentally friendly innovative technologies are contained in 24 (53%) strategies of mining regions. At the same time, only in three strategies - 7% (Arkhangelsk region, Nenets Autonomous Okrug) the corresponding criteria for evaluating the implementation of these activities are provided. It should be noted that in 13 regional strategies (29%) specific actions and areas of action were named, 8 regions - (18%) limited themselves to a description of areas of work for the greening of territorial development management. In almost all regions, the development of an environmental monitoring system in one form or another is envisaged.

Measures to reduce the negative impact on the environment (criterion K3), in most strategies, are reduced, firstly, to the restoration of disturbed ecological systems, and secondly, to the efficient management of waste. The strategies of 13 regions (29%) include measures to reduce accumulated harm, and 26 strategies (28%) - to ensure the efficient management of waste. Most strategies include measures to reduce harmful emissions and improve the quality of drinking water. Almost all activities of this unit have a mechanism for monitoring their implementation through the formation of a system of relevant indicators. It should be noted that in some regions there are no indicators and indicators of waste management, the prospective development of non-waste production.

Environmental conservation measures (criterion K4) in regional strategies include the conservation and creation of new specially protected natural areas (18 regions or 40%) and the conservation and restoration of forests (11 regions or 24%). All events of this block are confirmed by indicators. It should also be noted that in the constituent entities of the Russian Federation that have unique ecological objects or a special territorial location, considerable attention is paid to the problem of restoring the damaged natural ecological systems of these natural objects. This distinguishes the strategies of the Kamchatka Territory (focusing on the resources of the Pacific ecosystem), Astrakhan Oblast (Caspian), Murmansk Oblast (Arctic zone), Irkutsk Oblast (Baikal). At the same time, insufficient attention is paid to the problem of reclamation of disturbed space, bioremediation of contaminated territories (in the strategies of the Orenburg and Kemerovo regions these areas are identified as separate types of activity). The border subjects of the Russian Federation are oriented towards participation in international cooperation programs in the field of environmental protection. So, the Astrakhan region - programs for the protection of the Caspian, Magadan region - in the programs of the Arctic Council and the Council of the Barents / Euro-Arctic Region (BEAC).

Analysis of strategic planning in the mining regions of Russia with a further review of the practice that has developed in the world community.

In the 1990s, wider information access and increased social activity of local communities in various countries led to widespread allegations of human rights violations in the areas where mining enterprises are located [4]. Social and environmental issues have led to clashes and conflicts between mining companies and local communities (including casualties). The need to solve these problems entailed several international initiatives in the field of mining. One of the first steps was taken at a conference organized by the World Bank in 1997 on the problems of implementing projects in the mining industry. Canadian mining expert James Cooney introduced the concept of a "social license to operate", the essence of which is that a company receives "approval" at various levels (from acceptance to permission) of the local community for the company's activity[5].

Another step towards sustainable interaction between mining enterprises and local communities was the MMSD project "Mining, Minerals and Sustainable Development" (MMSD), implemented with the participation of the independent international organization International Institute for Environment and Development. The key objectives of the mining industry, according to the MMSD project, are: 1. the viability of the industry, 2. control, use and management of land, 3. minerals and economic development, 4. local communities and mines, 5. mining, minerals and the environment environment, 6. integrated approach to the use of minerals, 7. access to information, 8. artisanal and small-scale mining, 9. sector management [6].

In 2001, the International Council for Mining and Metallurgy (ICMM) was created [7]. The Council unites 26 mining and metallurgical companies and 35 regional and raw material associations present in 59 countries. Russian companies do not participate in ICMM. The activities of ICMM are focused primarily on solving environmental and social problems, labor safety problems.

In 2003, the world community launched another initiative in the extraction of mineral resources The Extractive Industries Transparency Initiative (EITI) and sustainable development of underground mining regions. The Extractive Industries Transparency Initiative (EITI) is the global standard for managing oil, gas and mineral resources. He seeks to address key management issues in the extractive sectors. The EITI standard requires information on the entire value chain of the extractive industry from the point of extraction to before revenue flows through the government and contributes to the economy. This includes how licenses and contracts are distributed and registered, who are the beneficial owners of these operations, what are the tax and legal arrangements, how much is paid, how much is paid, where the income is distributed, and its contribution to the economy, including employment. The EITI standard is implemented in 52 countries. Each of these countries is obliged to publish an annual EITI Report with disclosure of information on: contracts and licenses, production, revenue collection, income distribution and social and economic expenses [8]. The Russian Federation has not joined the EITI standard. In recent decades, the concept of ecosystem management has also been introduced in UN documents, primarily in the United Nations Environment Program (UNEP). In the UNEP Medium-Term Strategy 2010-2013 ecosystem management has become one of UNEP's six priority areas [9].

Currently adopted IFC [10] activity standards, which are an international benchmark for determining environmental and social risks, are widely used in the activities of various subjects of socio-economic relations and environmental and social issues management. IFC standards cover the following aspects: 1) assessment and management of social and environmental issues; 2) staff and labour; 3) prevention and elimination of negative impact on the environment; 4) health and safety; 5) the use of land for industrial purposes and the relocation of people in connection with this; 6) conservation of biodiversity and rational use of natural resources; 7) indigenous peoples; 8) cultural heritage.

4 Conclusion

The key areas of international development of the mining industry and mining territories have become: environmental safety, financial and technological transparency of companies, resource conservation, company interaction with local communities.

According to the analysis of the level of environmental friendliness of strategic management in the constituent entities of the Russian Federation classified as extractive, we can conclude that the general pattern is an increase in the number of environmental parameters in the strategies of the "latest generation", in regional programs drawn up for the period up to 2030-2035. At the same time, in the analyzed strategies of the mining regions, there are no development indicators that are priority for the international community: financial and technological transparency of companies, resource saving, interaction of companies with local communities.

References

1. D.E. Pataki, Front. Ecol. Evol., **3**, 57 (2015)

2. M. Tyulenev, Y. Lesin, E. Tyuleneva, E. Murko, E3S Web Conf., **15**, 02003 (2017)

3. *Government of the Russian Federation of February No. 207-r 13* (Russian Gazette, Moscow, 2019)

4. U. Gankhuyag F. Gregoire. *Managing mining for sustainable development* (2018)

5. K. Moffat, A. Zhang. *The paths to social license to operate: An integrative model explaining community acceptance of mining* (2018)

6. S. Markov, J. Janočko, M. Tyulenev, Y. Litvin, E3S Web Conf. **105**, 01021 (2019)

7. M. Tyulenev, S. Markov, E. Makridin, Y. Lesin, V. Gogolin, E3S Web Conf., **105**, 02022 (2019)

8. Extractive Industries Transparency Initiative (EITI, Brussels, 2019)

9. J. Ahern, S. Cilliers, J. Niemelä. Landsc. Urban Plan., **125,** 254–259 (2014)

10. IFC Policies-Standards (EITI, Brussels, 2012)

Uranium ore mineralization at the Western Carpathians / Slovakia (conditions and social aspects)

Stanislav Jacko[1,*], *Kristína* Bátorová[1] , *Barbora* Ščerbáková[1] and *Peter* Blišťan[2]

[1]Institute of Geosciences, Faculty BERG, Technical University of Košice, 04001 Košice, Slovakia
[2]Institute of Geodesy and GIS, Faculty BERG, Technical University of Košice, 04001 Košice, Slovakia

Abstract. Uranium ore mineralisation occurrences at the Western Carpathians/Slovakia are deposited within Permian volcano-sedimentary sequences. The mineralisation principally consists of two genetic types, the stratiform one and U-Mo vein mineralization linked to andesite volcanites. Geological exploration of uranium mineralization in Slovakia is subject to rigorous assessment of Ministry of Environment and municipalities. The limited energy sources and dependence to foreigner, open space for prospection of new energy sources (reneweable, uranium or traditional from coal, oil, gas).

1 Introduction

Slovak territory covers the main part of the Western Carpathians forming the central segment of the European/ALCAPA arcuate belt. Similarly as other geological units of the Earth also structure of the Western Carpathians has been formed by several and multistage orogenic cycles. The Paleozoic evolution terminated in the Carboniferous and Permian (360 - 250 Ma.) by significant Hercynian orogenic cycle. Folded, metamorphosed and by granitoid bodies penetrated rock complexes had produced new mountain belts and their clastic material successively filled post-orogenetically modelled depressions. Basal parts of the depressions usually fill dark grey, bituminous Carboniferous sediments, i.e. the product of a humid and warm climate. They are sometimes interlayered with basic extrusive volcanic rocks. Most of pit-coal basins in the world are filled by Carboniferous stratas.

The onset of dry, semi-arid to arid climate in the Permian period was reflected in contrasting i.e. reddish coloration of the sediments and volcanites filling either top of the Carboniferous basins, or new depressions formed on crystalline basement respectively. of Permian arid climate reflected also into regular evaporites presence in Permian successions. Some of them (e.g. Germany, Poland) are major European producers of potassium and rock salt. Accompanying sediments are mainly formed by weathered material with high content of mica and feldspar transported to Permian basins from uplifted and eroded blocks of crystalline basement. Therefore these sediments contain increased phons of radioactive

* Corresponding author: stanislav.jacko@tuke.sk

elements, as well as plant cuttings. The mentioned sedimentary filling together with the significant volume of Ca-alkaline volcanites (mainly rhyolites, andesites and lesser possibly andesite-basaltic rocks) are the primary (stratiform) source of the mentioned mineralization in Slovakian Permian rock sequences [6, 7]. Hence the Permian stratas of the Western Carpathians contain elevated uranium mineralization.

2 Origin of uranium mineralization

Creation of uranium mineralization in Slovakia associates with two possible processes. The first relates to acidic volcanism, which is evidenced by the U-Mo stratiform deposits related to metaryolites and meta - andesites tuffs and tuffs. Such type of U- deposits are known from the Novoveská Huta near Spišská Nová Ves town and Jahodná area near Košice. The second type probably relates to the sorption effect of a coal flora during diagenetic-metamorphic processes. The last genetic type forms relatively small uranium deposits mainly from the Švábovce, Spišský Štiavnik, Kraviany and Vikartovce villages and Považský Inovec Mts. [2, 4].

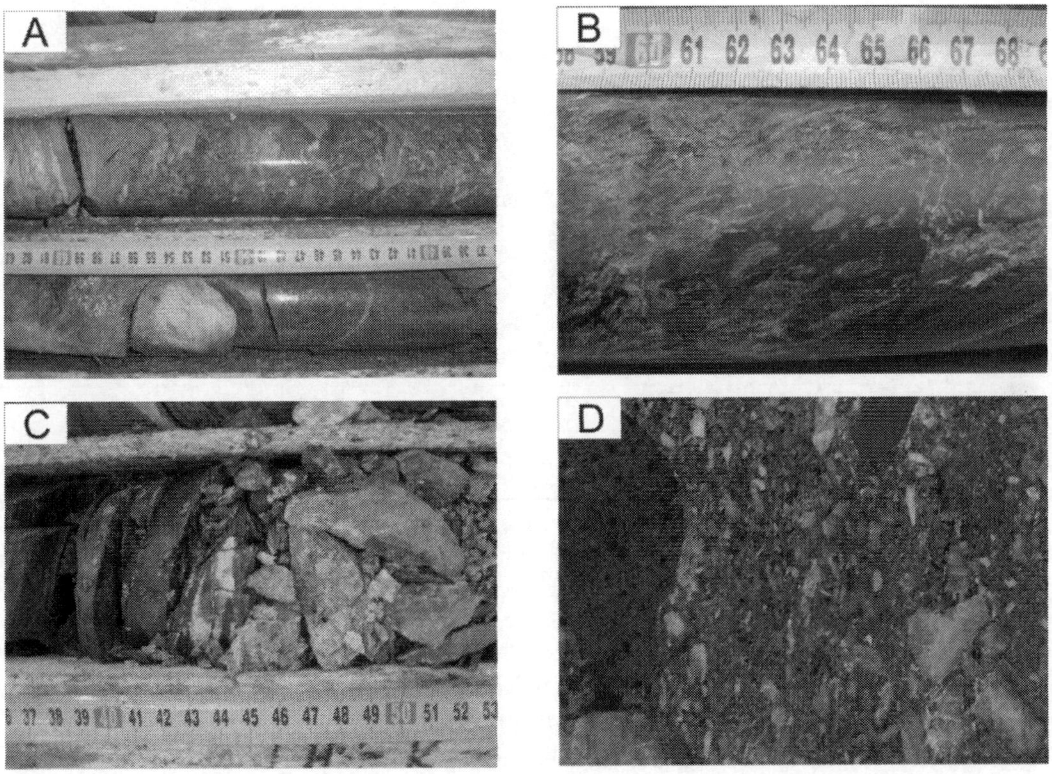

Fig. 1. Uranium mineralization varietes: a) In the Jahodná ore deposit near Košice A – stratiform type, B – vein type, C – as a fault filling. b.) Novoveska huta ore deposit D - breccia

Uranium mineralization performs in several mineral forms, but mostly as the uraninite and coffinite minerals forming a relatively lesser extent in the uranium ore deposit than the U-Ti oxides (especially orthobrannerite). More abundant uranium ore deposits are usually accompanied by increased molybdenite impregnations. Separate U-Mo mineralization forms either a filling in younger tectonic structures or an independent deposit segments. A younger i. e. the Alpine remobilization significantly contributes to all known U-mineralization localities. A good example of these relationships demonstrate the mentioned

Novoveská Huta and the Košice - Jahodná deposits [5, 10]. Discussed evidences suggest that we can´t currently expect a relatively homogeneous distribution of uranium mineralization (Fig. 1.). This is also reflected in the calculation of deposit reserves. Geological conditions of the U- mineralization in Slovakia are generally favorable in Permian sequences and - potentially, at their footwall in several areas containing at least rudimentary occurrence of uranium minerals (Fig. 2.).

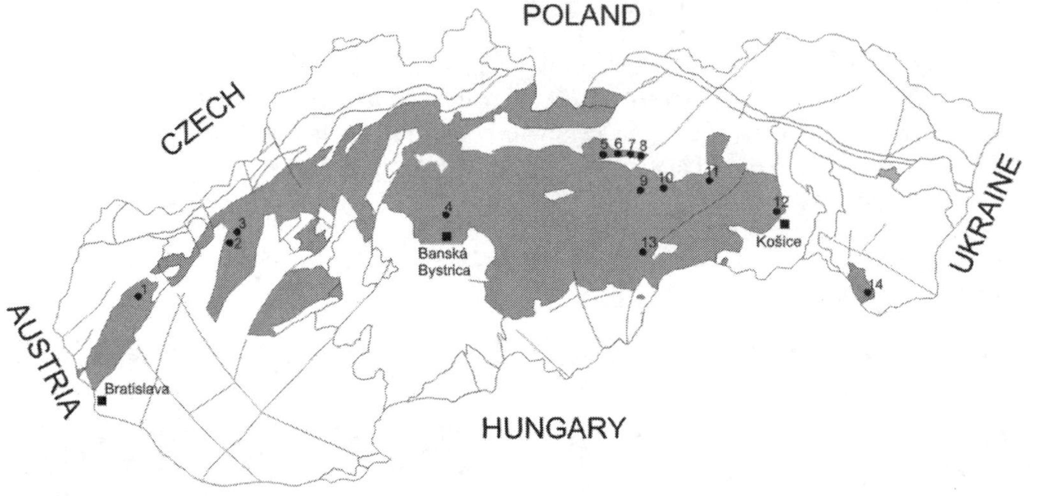

Fig. 2. Geological schetch of pre – Tertiary units and localities with uranium mineralization in the Western Carpathians (part Slovakia): 1 – Malé Karpaty, 2 – Kálnica, 3 – Selec, 4 – Špania Dolina, 5 – Vikartovce, 6 – Kravany, 7 – Spišské Bystré, 8 – Spišský Štiavnik, 9 – Novoveská Huta, 10 – Markušovce nad Hornádom, 11 – Spišské Vlachy, 12 – Košice-Jahodná, 13 – Gočaltovo, 14 – Zemplínske vrchy [1, 3].

3 Prospecting conditions

Prospecting history of uranium mineralization extends into the middle of the last century, when post-war Czechoslovakia utilizes energy and resource security. Since 1947 to 1992 y. exploration activities have been realized by the State Uranium Prospecting Company. The company has identified U- mineralization deposits occuring in the whole territory of the Western Carpathians [1]. In the nineties of the last century prospecting of uranium ores almost completely stopped. However industry boom and economic growth in the Slovakia started renewed interest in this ore material. After state companies privatization, exploration activities are carried out exclusively by private companies. Currently geological prospecting of U- mineralization being conducted at zero survey areas.

Companies providing exploration on the territory of Slovakia must full fill several criteria, which are listed in the geological Law No. 569/2007 [8]. Obtaining any prospection licence requires a favorable opinion either of the Regional Environmental Office or its subordinate offices. The request must also include statements of other organizations (eg Ministry of Health, Ministry of Defence, the District Mining Office, the Regional Monuments Board, etc.).

The geological prospection is realized on the base of the geological project where all geological operations have to be approved by the Ministry of Environment. The most common an usually the introducing case is geophysical prospection in orders to identify the geological structure, its basement and filling below the present surface. This non - invasive method has no negative impact on the environment and population. The second group of geological prospection methods relates to prospection drilling. Their purpose is to collect rock samples from different depths, which are then analyzed at certified laboratories. From the collected samples they determine the mineralization and other relevant parameters, depending on the purpose of exploration. Any exploration must be terminated by the final report, presented before the Ministry of Environment commission for reserves calculation. The permission is granted only in the case if the report contains all the requirements for declared economical reserves. Geological exploration calculated total reserves of the uranium deposits in volume 19 452t [9].

4 Social aspects of U-mineralization

Present discussions about uranium mineralization in Slovakia are mainly focused to security of exploration and economic reserves verification of this type of raw material for energy resources. This problem can be appreciated from two aspects. The first one, presented by environmental organizations, citizens and some municipalities resulted from premise that each survey will be followed by exploitation creating a potential risk to the environment, health and development of the region. The environmental impact to government forbid any mining activities related with extraction of radioactive ore deposits, without consensus of village residence leaving in the prospection area [8].

The second view presented mostly by professional geological community emphasizes the importance of each geological prospection for the state as the owner of the mineral resources and thus also for a particular community forming a part of the country population.

However Slovakia admittedly depends on an import of energy sources. The only perspective energy resource that we can economically evaluate is particularly the uranium. Its extraction and subsequent processing is quite difficult at current conditions. This is due to the lack of optimal technologies, impact studies, as well as long-term contracts to uranium import from Russia.

References

1. J.A. Arapov, G.V. Afanasjev, J. Badár, I.M. Bajuškin, J. Blažek, V.J. Bojcov, B. Bošina, B.V. Brodin, K.N. Černeceva, N.I. Česnokov, V.I. Dančev, K. Dědič, A.V. Djakonov, M. Doležel, V. Fiala, J. Habásko, A. Hájek, J. Halbrštát, J. Hrádek, J. Hřebec, K.E. Ivanov, A.M. Jakovjenko, V.S. Jegorov, M. Kolář, M. Kolek, J. Komínek, V.N. Kozyrev, G.A. Kremčukov, M. Lažanský, F. Lepka, V.I. Malyšev, I.G. Melnik, I.M. Meščerskij, I.A. Milovanov, V.P. Nikulin, V.P. Novik-Kačan, V. Nový, G.E. Ordynec, R. Petroš, F.K. Portnov, S. Prokeš, L. Prokop, K. Romanidis, K.T. Saveljeva, K.G. Stafějev, J. Steiner, F. Šorf, T. Veselý, S. Vilhelm, A.V. Zavarzin, G.D. Zinověv, N.S. Zontov, V.I. Žukova, Czechoslovak uranium industry, **1**, 7–365 (1984)
2. P. Baláž, and D. Kúšik, Št. Geol. Úst. D. Štúra, **5**, 163 (2011)
3. J. Daniel & I. Mesarčík, Enviromagazín, **13**, 10-11 (2008)
4. S. Ferenc, A. Biron, T. Mikuš, J. Spišiak, S. Budzak, J. Geosci., **63:3**, 277-291 (2018)
5. P. Grecula, A. Abonyi, M. Abonyiová, J. Antaš, B. Bartalský, J. Bartalský, I. Dianiška, E. Drnzík, R. Ďuďa, M. Gargulák, Ľ. Gazdačko, J. Hudáček, J. Kobulský, L. Lörinz, J. Macko, D. Návesňák, Z. Németh, L. Novotný, M. Radvanec, I. Rojkovič, L. Rozložník, O. Rozložník, C. Varček & J. Zlocha, Geocomplex, **1**, 3-834 (1995)
6. I. Rojkovič, Acta Geol. Univ. Comen., **7**, 117 (1997)
7. I. Rojkovič, J. Franců, J. & J. Čáslavský, Geol. Carpath., **43**, 27-34 (1992)
8. *Act. No. 569/2007 (Geological Law). Statue , part 237 (*Bratislava, 4, 2018)

9. *Slovak Minerals Yearbook* (State Geological Institute of D. Štúr, Bratislava, 2018)
10. R. Farkašovský, K. Bónová, M. Košúth, Geologos, 22 :**3**, 171-190 (2016)

The Use of Dust Suppression on Unpaved Roads at the Mining Allotment of the Kuzbass Mine

Tatyana Cherkasova [1,*], *Elizaveta* Cherkasova[1], *Anastasia* Tihomirova[1], *Alexey* Andryushkov[1], and *Oleg* Dormeshkin[2]

[1]T.F. Gorbachev Kuzbass State Technical University, 650000, Russian Federation
[2]Belarusian State Technological University, Minsk, Belarus

Abstract. Marker substances for the coal mining industry were identified as an information and technical reference book to the best available technologies for coal mining and processing. Inorganic dust when emitted into the atmosphere is defined as a marker substance for the coal industry. Inorganic dust emissions belong to the group of fugitive emissions which are carried out from large areas. For the solving of the mentioned problem on the territory of Mine no.12 in Kiselevsk city, it was proposed to use the Anti-dust and the Bischofit products.

1 Introduction

Air pollution in the area of coal mining enterprises depends on climatic and mining-geological conditions. Thus strong winds create conditions for the intensification of the dust entry into the surface layers of the atmosphere and the movement of dust in them. Even at a windspeed of 2 m/s, dry dust is blown away and carried over considerable distances. The construction of high dumps also contributes to the growth of dust emissions into the atmosphere since the wind speed increases as their height increases [1-7].

The solving of the relevant mine technical task aimed to increase the efficiency and safety of mining operations during the development of the mining allotment of the Mine no.12, LLC is economically and socially important and inextricably connected with the improvement of methods for fixing dusty surfaces at the objects of the mining complex.

The atmospheric air pollution by dust during mining is to the great extend caused by open fugitive dust emission sources, the main of which are dusty surfaces of warehouses, waste dumps, roads [8-15].

Hence, the goal of the study is to establish dust suppression efficiency when using the Anti-dust product of the brand D of the engineering company Brent, LLC and the Bishofit product of the JSC Kaustic on dirt roads at the mining allotment of the Mine no.12, LLC.

The research solves the following tasks:

- A comparative analysis of the concentration of dust blown from the surface treated with the Anti-dust product of the brand D and the Bishofit product, and non-treated surface;

* Corresponding author: ctg.htnv@kuzstu.ru

- Determination of the resistance to atmospheric precipitation of the surface treated with the Anti-dust product of the brand D and the Bishofit product;
- Determination of the optimal dosages of the reagent for using the Anti-dust product of the brand D on dirt roads at the mining allotment of Mine no.12, LLC taking into account characteristics of the treated surface and weather conditions.

2 Research method

The research object was the dirt road of the Mine no.12, LLC (Kiselevsk city, Kemerovo region). In 2013 it was decided to develop open-cut mining and cease dangerous supercategory underground mining. The annual production level of coal of grades K and T amounts to over 730 thousand tons.

The quality of atmospheric air is due to the ratio of certain substances in air composition. People's health, vegetation and wildlife state, strength and durability of any structures, buildings and structures depend on the air quality.

The main sources of air pollution are natural and man-made processes.

The formation of atmospheric air quality in the city of Kiselevsk is influenced by various factors, including the degree of industrialisation, the existence of highway systems with heavy traffic, as well as geographical location and climate pattern.

Within the framework of the study the industrial tests of the Anti-dust and the Bishofit dust suppression products were conducted on a dirt road section. The surface was treated by irrigation without prior preparation.

The dust suppression Anti-dust product has the characteristics presented in the Tables 1-2.

Table 1. Technical characteristics of the Anti-dust product of brand D.

No	Indicator	Brand D
1.	Appearance	Transparent liquid
2.	The content of active substances is not less, %	30
3.	Density, kg/dm^3	1,20 - 1,30
4.	Temperature range of application of the resulting solution, °C	-40…+40
5.	Temperature range of operation of reinforced soil, °C	-30…+40

Table 2. The field of utilization and consumption rates of the Anti-dust product of brand D.

	Field of utilization	Consumption per 1m^2 product+water, litre	Retreatment
1.	Roads, roadsides (horizontal section)	0,05+1	5-7 days
2.	Roads, erosion control on slopes (steep slope)	0,05+1	3-5 days
3.	Dedusting a surface	0,05+1	5-7 days

4.	Dedusting of coal during carriage	0,25+1	5 days
5.	Dedusting of coal upon storage in cones in open sections	0,25+1	3-5 days
6.	Dust suppression in places of coal and bulk materials transshipment	2 litres of 5,0% solution per 1 ton of coal	

The surface treated by irrigation does not require any preparation.

Depending on the specific goals and conditions various methods and options for surface treatment with the Anti-dust product of brand D are possible. The most common and convenient way is the irrigation method.

The irrigation method consists in uniform distribution of the previously prepared composition of the product over the surface using sprayers. As a result of irrigation the composition penetrates into the soil to a certain depth and binds it.

The Bishofit product absorbs moisture from the air and keeps it on the road for a long time. Having such a property as hygroscopicity (strong capacity to absorb moisture even from relatively dry air), Bishofit provides maximum moisture of the road surface. It does not allow dust to rise and hang in the air. This moisture-absorbing effect provides greater dust suppression compared to water. In addition, Bishofit extends the work lifespan of the roadway providing the soil stabilisation.

Dry (crystalline) magnesium chloride is dissolved in water in the ratio 1:1.

The consumption of the Bishofit solution – from 0,1 litre/m^2 depending on the type of the highway, traffic density and means of transport, dustiness of the area as well as the nature of the dust.

The Bishofit solution processing steps:

- Dissolution of the crystalline reagent on the mixing unit or directly in the tank of Multipurpose Road Maintenance Vehicle;
- Distribution of reagent solution to the surface counting from 0,1 litre/m^2;
-The highway is ready for use.

The duration of the reagent when using the specified proportions (period of use of the treated highway without dust):

- using grader equipment – 10 days;
- non-treated with a grader – 30 and more days.

The treatment of the tested sections was carried out with the use of the Anti-dust product of brand D of the engineering company Brent, LLC and the Bishofit product of the JSC Kaustic. The sections surface was treated at a temperature of +27°C at mainly clear weather with northwester at a windspeed of 2 m/s counting 0,3 и 0,25 litre per 1 m^2 of the Anti-dust and the Bishofit products respectively.

The concentrations of dust blown from the treated and non-treated sections surface were measured using the equipment that simulates the erosive effect of wind on the soil surface at the maximum possible wind loads of up to 25 m/s.

The installation diagram for the measuring of dust concentrations blown from the surface of treated and non-treated sections is shown at the Figure 1.

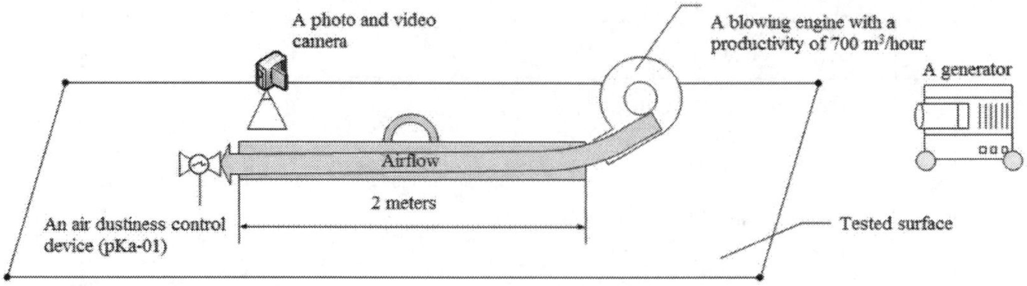

Fig. 1 The installation diagram.

Used equipment:
- A blowing engine with a productivity of 700 m^3/hour;
- An air conduit for creating a given airflow rate;
- An air dustiness control device (pKa-01);
- A photo and video camera;
- A generator.

An air conduit (2 meters long and 0.1 meters wide) was installed on the tested surface (for direct impact on a specific surface section and excluding external factors). A blowing engine was installed on one side of the air conduit; pKA-01 air dustiness control device and photo and video camera were installed on the other side of the air conduit.

3 Key findings

Table 3 presents the key findings of the study.

After measuring the concentration of dust at all sections, it was decided to stop the subsequent control of section no.2 due to the fact that the Bishofit product did not prove its effectiveness. Measurements were taken at section no.1 treated with the Anti-dust product, and at the non-treated section.

Retreatment of the section no.1 was carried out with the Anti-dust product of brand D at 5% concentration (0,05 litre/m^2). The product was applied to a damp surface at air temperature of +26°C in a cloudy weather with a north wind at a windspeed of 4 m/s.

Table 3. Dust Measurement Results.

Date	Weather			Dust concentration, mg/m^3		
	t - air temperature, °C	*V* - windspeed, m/s	Precipitation	Anti-dust	Bishofit	Non-treated
01.08.2019	29	3	clear	29,3	813,1	909,2
02.08.2019	29	3	clear	40,6	-	925,8
03.08.2019	31	3	clear	109,4	-	1335,5

By the end of the second testing day after the retreatment, the Anti-dust product began to lose its effectiveness and it was decided to treat the section with water in order to prolong its effect.

The following findings were obtained during the tests:
1. The Bishofit product did not prove the declared dust suppression charachteristics;
2. The Anti-dust product of brand D at a primary treatment effectively suppresses dust for four days;

3. After retreatment the Anti-dust product effectively suppresses dust for three days;
4. After watering the section treated with the Anti-dust product on the third day, the effective dust suppression is extended up to 12 hours.

4 Conclusions

Based on the results of the tests the declared dust suppression characteristics were confirmed and the optimal concentrations of the Anti-dust product of the brand D of the engineering company Brent, LLC were determined, which made it possible to reduce the amount of blown dust on various surfaces of road sections at the mine allotment of Mine no.12, LLC.

References

1. R. Zeng, X. Zhuang, N. Koukouzas, W. Xu, Int. J. Coal Geol, **61**, 833 (2005)
2. R.N. Peres, E.S.F. Cardoso, M.F. Montemor, H.G. Melo, A.V. Benedetti, P.H. Suegama, Surf Coat Technol (Part B), **303**, 372 (2016)
3. F. Khelifa, M-E. Druart, Y. Habibi, F. Bénard, P. Leclère, M. Olivier, P. Dubois, Prog Org Coat, **76**, 900 (2013)
4. B. Volarič, I. Corros Eng Sci Technol, **52**, 201 (2017)
5. R. D. Kiggins, *The Political Economy of Rare Earth Elements Rising Powers and Technological Change* (Palgrave Macmillan, London, 2015)
6. K. Binnemans, Y. Pontikes, P. T. Jones, T. Van Gerven, B. Blanpain, International Slag Valorisation Symposium, **3**, 191–205 (2013)
7. K. Binnemans, P. T. Jones, B. Blanpain, T. Van Gerven, Y. Pontikes, Journal of Cleaner Production, **99**, 17–38 (2015)
8. B. N. Kumar, S. Radhika, M. L. Kantam and B. R. Reddy, J. Chem. Technol. Biot., **86**:4, 562–569 (2011)
9. L.S. Wang, Z.Q. Long, X.W. Huang, Y. Yu, D.L. Cui G.C. Zhang, Hydrometallurgy, **101**:1–2, 41–47 (2010)
10. P. M. Reza, P. Norouzi, H. Rashedi, M. R. Ganjali, J Appl Electrochem, **40**, 1593-1603 (2010)
11. K. Shunsuke, K. Yuichi, K. Hiroshi, W. Kazuaki, Open Journal of Physical Chemistry, **3**, 69–75 (2013)
12. H. El-Didamony, M. M. Ali, N. S. Awwad, M. M. Fawzy and M. F. Attallah, J. Radioanal. Nucl. Chem., **291:3**, 907-914 (2012)
13. Tran X., Elsevier. ScienceDirect, **163**, 129-132 (2016)
14. F. Wojciech, M.M. Wiatros-Motyka, M. Wdowin, Environmental Science and Pollution Research, **22**, 9464–9474 (2015)
15. C.N. Lange, M.C Camargo, G. M. Figueiredo, B.A. Vasconcellos, R.B. Ticianelli, Journal of Radioanalytical and Nuclear Chemistry, **311**, 1235–1241 (2017)

Research of characteristics of PJSC "Koks" coal and coke dust fuel briquettes

Vyacheslav Solodov [1,2], *Tatyana* Cherkasova[1], *Nikolay* Kolmakov[2], *Sergey* Subbotin [1,2], and *Yuriy* Radchenko[3]

[1] T.F. Gorbachev Kuzbass State Technical University 28, Vesennyaya, Kemerovo, Russia, 650000
[2] PJSC «Koks» 6, 1st Stakhanovkaya, Kemerovo, Russia, 650021
[3]Belarusian State Technological University, Minsk, Belarus

Abstract. Known methods for the production of fuel briquettes are studied and an analysis of their qualitative characteristics is carried out in the present work. Based on the analysis of existing methods, pilot briquetting of a mixture of PJSC "Koks" coal and coke dust with various types of binder was carried out. Within the framework of studies, the main requirement for raw materials was the absence of demand for it in the market and in the technological processes of existing industries. It has been established that substandard urea is the best binder of the considered ones, since briquettes with its use have high strength and low sulfur content. Comparative briquetting by single and impulse pressing methods was carried out. It is shown that greater briquette strength is achieved by impulse pressing method. It has been established that the process of utilizing finely dispersed carbon-containing wastes is aimed not only at obtaining a new marketable type of fuel and at saving natural resources, but also at improving the environmental situation through the rational use of secondary material resources.

1 Introduction

Every year there is an increase in energy prices. The requirements for the quality of the final product are constantly increasing, therefore, a deeper cleaning and preparation of raw materials and intermediates is required, which in turn leads to the generation of additional waste. At the same time, there is a constant effort to reduce the environmental load in the world, where dumps and tailings are unacceptable. One of the rational solutions to these problems is the recycling of carbon-containing wastes of coal mining and coal processing enterprises to produce a marketable product in the form of fuel briquettes.

Briquetting is a process of mechanical conversion of coal or coke dust into lump fuel - briquettes having certain characteristics: shape, size and weight. Many methods for briquetting coke and coal dust, fines, sludges and other types of finely dispersed carbon-containing waste products are known. Having studied these methods, we can conclude that technologies for producing briquettes with the addition of a binder are the most efficient. As a result of adding a binder, the strength characteristics of the briquette are improved, but it is worth remembering that it is necessary to study in detail the properties of the binder in order to know the effects of the products of combustion of the binder on technology, hardware and ecology.

Previously, many researchers have repeatedly conducted tests to obtain fuel briquettes from various carbon-containing materials without the use of a binding agent. A positive briquette strength test result was achieved only in the case of additional heat treatment before or during the briquetting process. The briquettes produced without heat treatment had low strength indicators, which is unacceptable during loading and transportation of briquettes. In most cases, the sintering coal additive was used and heat treatment was carried out to a temperature of 250° - 400° C, in order to produce fuel briquettes which included sintered coal char. This method allows to produce fuel briquettes with acceptable strength, reduced yield of volatile substances and increased calorific value in comparison with briquettes made of coal or wood. But there are a number of drawbacks in the application of this method such as the need to capture the gas released in the heat treatment process, the high cost of equipment for the production, as well as the significant cost of energy to maintain the temperature during the heat treatment [1]. Briquetting methods with the use of binding agents are the most attractive, in view of the simplicity of the flowsheet and the equipment used. A rational approach to the production of fuel briquettes is the use of carbon-containing waste of coal mining and coal processing, as well as the chemical industry waste.

The most common types of binders are secondary polymers, oil refining wastes - paraffins and bitumens, wastes of coke and by-product process (for example, fusa), as well as the proposed binder in the form of a substandard urea product [2]. Since the mid-twentieth century Europe has accumulated rich experience in using various types of bitumen and oil refining waste as a binder for producing fuel briquettes [3-5]. But this binder is not widely used due to the high environmental load when burning fuel briquettes. Cement-based inorganic binders, using lime and other inorganic compounds, were also widely used [6]. Fuel briquettes produced using inorganic binders have high strength, but at the same time they significantly increase the content of ballast - ash - part of briquettes, which negatively affects both the economic and technical components of the process of using briquettes for energy purposes.

2 Materials and methods

PJSC "Koks" together with the Department of Chemical Technology of Solid Fuel of Kuzbass State Technical University conducted research on briquetting a mixture of PJSC "Koks" coal and coke dust using various binders (Tables 1 and 2). It was decided to use coal and coke dust as a basis for fuel briquettes because they are the only types of waste generated at PJSC "Koks" in Kemerovo. The coal component has a high reactivity when fuel briquettes enter the combustion zone, which will positively affect the initiation of the briquette ignition process. At the same time, coke dust has a greater calorific value, which in turn leads to greater efficiency and fuel saving. The experience of scientists who obtained positive results when using a mixture of sintering coal with non-sintering carbon-containing wastes was also taken into account [7,8]. Before briquetting, coke and coal dust were processed by the method of oil sintering to reduce ash content, and then they were dried. Substandard urea in molten form, together with coke dust, was fed into the mixing extruder shown in Fig. 1, after which the mixture was sent to the compression mould shown in Fig. 2, where the briquetting process took place. Briquetting in this case was carried out by repeated contact between the press and the surface of the briquette.

Fig. 1 The extruder-type mixing machine design.

Fig. 2 The compression mold and the stamp press.

Table 1. Technical characteristics of briquettes

Briquette type	A^d, %	S_t^d, %	V^{daf}, %	$Q_в$, kcal/kg
coal and coke dust + urea	5,4	0,05	1,3	7600
coal and coke dust + secondary polymers	6,4	0,8–1,0	1,5	6900
coal and coke dust + bitumen	6,2	0,1	1,7	7400
coal and coke dust + paraffin	5,8	0,5	1,3	7100
According to Specification 0761-004-43058211-2005 [7]	not exceeding 16	not exceeding 3	4,0	–

Table 2. Strength characteristics of briquettes.

Briquette type	Compression, kg/cm^2	Abrasion, > 25 mm lump content, %	Shatter, > 25 mm lump content, %
pressed coal and coke dust	10–14	74	84
coal and coke dust + urea	18–20	90	96
coal and coke dust + secondary polymers	50–60	99	99
coal and coke dust + bitumen	13–15	85	90
coal and coke dust + paraffin	14–16	88	94

3 Results and discussion

From the data obtained, it can be seen that the coal and coke dust briquettes have small strength characteristics, which limits their transportation over long distances. When a binder is added, the strength characteristics of the briquettes are noticeably improved, but these characteristics can change depending on the type and properties of the binder. Briquettes made with the addition of secondary polymers have the greatest strength, however, the presence of sulfur in them significantly increases this indicator in briquettes, which, if used, can be harmful to the environment. Briquettes using urea as a binder have the best characteristics.

The produced briquettes are presented in Fig. 3.

Fig. 3 Coal and coke dust briquettes with a substandard urea binder.

The dependence of strength characteristics on the type and concentration of the binder is shown in fig. 4. Based on the analysis of the dependence of the abrasion of briquettes on the concentration of the binder, it can be concluded that briquettes produced with the addition of secondary polymers have low abrasion indices, and with an increase in the concentration of the binder over 5% (8%, 10%, 15%), the abrasion indices did not actually change. The highest content of over 25 mm lump after testing was shown by briquettes with the addition of urea. A dynamic decrease in the abrasion index is observed in the interval between 2% and 5% of the urea content in the briquette, then the abrasion index decreases with an increase in the concentration of the binder, however, the abrasion index does not change even within the binder content from 10% to 15%. The abrasion indices of briquettes using bitumen as a binder are close to those of briquettes with paraffin. Briquettes using paraffin compared with the use of bitumen after testing gave a higher content of over 25 mm lumps. However, the dynamics of increasing the content of large lumps with an increase in the amount of binder between these two types of briquettes is close. Fig. 5 shows the dependence of the compressive strength of briquettes on the binder concentration.

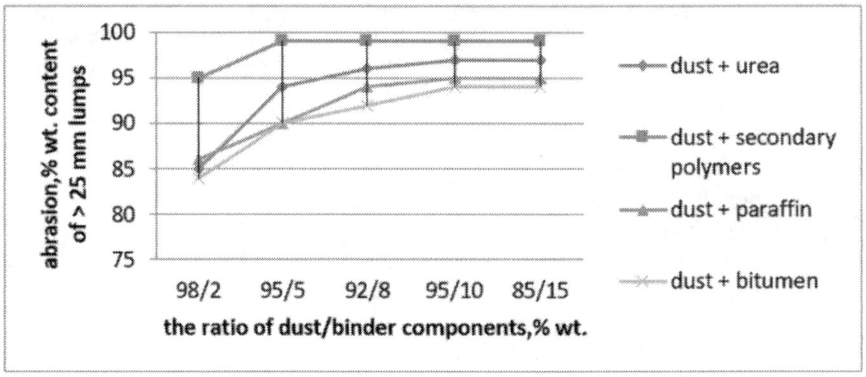

Fig. 4. The dependence of the abrasion of briquettes on the binder concentration.

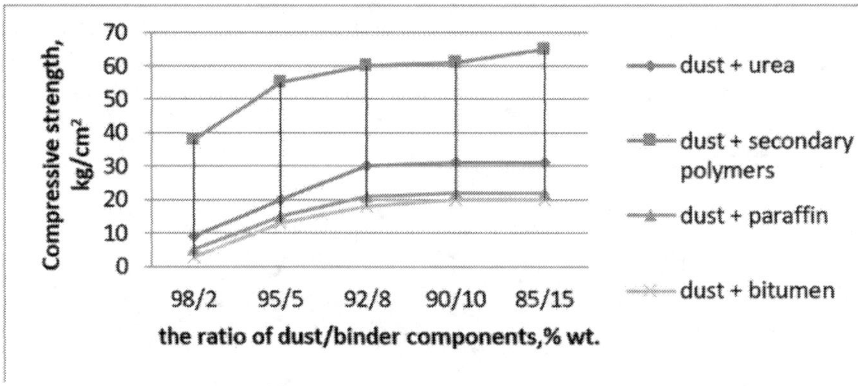

Fig. 5. The dependence of the compressive strength of briquettes on the binder concentration.

When analyzing the dependence, we can conclude that the greatest increase in the compressive strength index falls on the range of binder concentrations from 2% to 8%, then the indicator remains unchanged. In the case of using secondary polymers as a binder, after adding 8% of the binder, the strength continues to increase markedly.

As a result of the analysis of the dependences of the briquette abrasion and strength indices on the binder concentration, it can be concluded that the briquette with the addition of urea as a binder has optimal values.

Studying the strength characteristics of briquettes, the briquetting method should be taken into consideration, since this condition significantly affects the strength characteristics of briquettes. In practice, various types of pressing are known. The most popular and simplest type of pressing is the single-stage contact molding, which is considered the traditional method. Impulse compression is also known, characterized by multi-stage contacting.

Figures 6, 7 and 8 show graphical dependences of the strength characteristics on the type of compression with a final load of 15 t/cm^2.

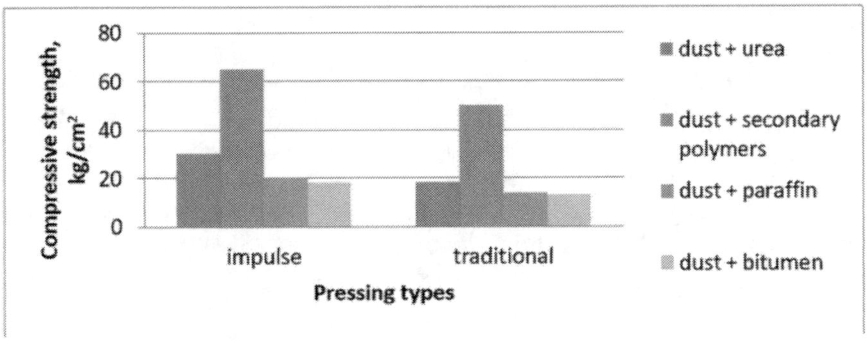

Fig. 6. The dependence of the compression strength of briquettes on the pressing type.

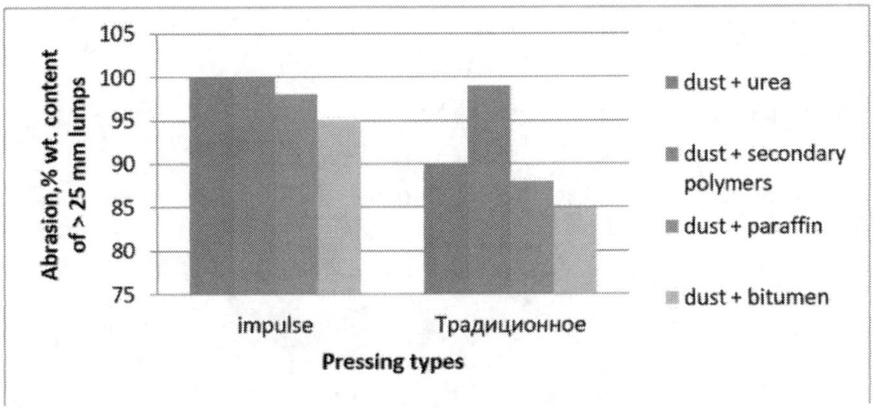

Fig. 7. The dependence of the abrasion resistance of briquettes on the pressing type.

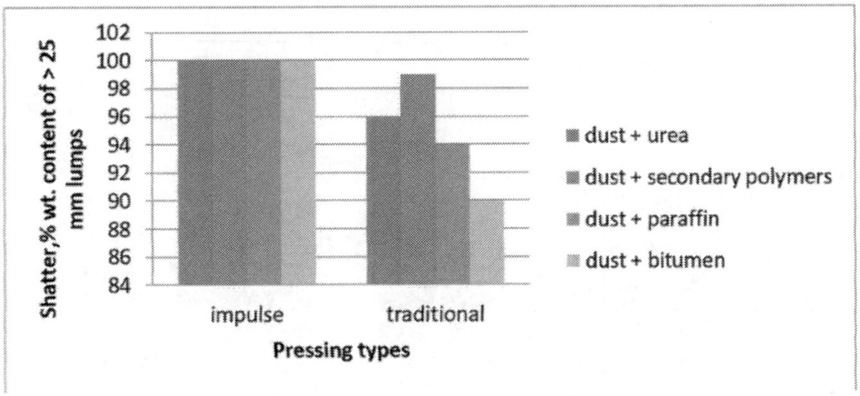

Fig. 8. The dependence of the shatter value of briquettes on the pressing type.

The analysis of the data shows that the multiple load on the fuel briquette (impulse briquetting) shows the best strength characteristics of the briquettes in comparison with the traditional pressing method.

4 Conclusion

According to the results of the research conducted, it can be concluded that the use of waste of coal mining and coal processing enterprises is not only a rational technological solution for non-waste production, but also socially significant and environmentally attractive. The process of utilizing finely dispersed carbon-containing wastes is aimed not only at obtaining a new marketable type of fuel and at saving natural resources, but also at improving the environmental situation through the rational use of secondary material resources.

References

1. G. Komarek, K. Komarek, Patent Bulleting, **293708**, 120 (1960)
2. I.Yu. Ryvkin, A.Ya. Eremin, Coke and chemistry, **10,** 36-44 (2000)
3. P. Zakar, J. Toth Brikettierung mit Erdolbitumen als Bindemittel, **2**, 1-10 (1958)
4. A. Věchet, J. Chvostek, Paliva, **41,** 350-353 (1961)
5. E. Gundermann, E. Knopfe, Freiberger Forschungsh, **238**, 95-121 (1962)
6. K. Macqueen, B. Ernest, Patent Bulleting, **864737**, 201 (1961)
7. I. Barbu, I. Stefanescu, N. Teodorescu, Cercetari metalurgice si miniere, **2**, 389-416 (1960)
8. I. Blum, I. Nistor, Studii si cer. cetari energ. Acad RPR, **10:3**, 563-575 (1960)

https://doi.org/10.1051/e3sconf/201913403001

Application of the "lean thinking" concept to the analysis of coal mining region's sustainable development objective

Vadim Mikhalchenko[1,*], and *Yuriy* Rubanik[2]

[1]T.F. Gorbachev Kuzbass State Technical University, Kemerovo, Russia
[2]Center for Contemporary Management Technologies, Moscow, Russia

Abstract. The article is devoted to the application of resource saving technologies as an important tool to achieve sustainable development at coal mining regions of the Russian Federation. The negative impact of coal-mining enterprises on the water and air basins of the regions, on withdrawal of fertile land continues to be unacceptably high. As a result the objective to preserve natural environment in these regions has gained high priority. Its solution requires development and implementation of fundamentally new, cost-effective and environment friendly underground and open pit coal mining technologies. It is shown that an effective solution of the problem, a radical contraction of resource consumption requires radical rethinking of the whole coal mining management and production system. The basic principles of "Lean thinking" methodology have been applied to structure the issue and to develop guidelines for coal-mining systems reengineering.

1 Introduction

Ensuring sustainable development in the process of economic activities aimed at meeting the needs and ensuring the high quality of human life, the world community has recognized a very acute and pressing problem of our time. At the end of the UN Conference (Rio de Janeiro, 1992), the heads of state and government of the world recognized that the development of modern society cannot be considered in isolation from the preservation of the environment, destroyed by man-made activities. According to the decision adopted at the conference, the Concept of Sustainable Development is raised to the rank of state and interstate policy. All governments of the world are recommended to ensure the fulfillment of the conditions of sustainable development in the implementation of national development programs.

On the eve of this conference, in one of the largest industrial centers of the Russian Federation - the Kuznetsk coal basin, a large-scale study was carried out on the ecological and economic assessment of the functioning of coal mining and processing technologies [1, 2]. A system analysis of the previous twenty-year period of economic activity of 78 mines, 25 cuts and 16 processing plants and the consequences of this activity in terms of its impact on humans, water and air basins, on land resources and the region's mineral resources, revealed an extremely high

* Corresponding author: v.mikhalchenko@mail.ru

level of destructive impact all listed areas of the environment. It was shown that further intensive development of the coal-mining complex will lead to an increase in environmental problems in the region and to change the current situation, greening of mining production is necessary, development and introduction of fundamentally new, not only economically more efficient, but also environmentally friendly coal mining technologies is needed.

The years passed, the socioeconomic transformations and the restructuring of the coal mining industry of the Russian Federation, which took place in the 90s of the 20th century, significantly changed the characteristics of the external business environment, the criteria and indicators of economic efficiency of enterprises in relation to the planned economy. The closure of unprofitable mines, the commissioning of new capacities in favorable geological conditions, an increase in the share of open coal mining, the introduction of modern powerful mining and transport equipment, increased concentration and intensification of mining operations brought the industry to a competitive level of efficiency and contributed to its integration into the global economic space.

However, these transformations almost did not affect the historical technological paradigm of the production systems of coal-mining enterprises. As a result, the environmental problems of a large mining region not only did not lose their acuteness, but became even worse.

In addition, the market principles of economic activity fundamentally changed the working conditions of business entities. Coal-mining enterprises, basically formed in stable and fairly well-predicted conditions of a planned economy, faced fundamentally new realities. New economic conditions are characterized by a high degree of variability and uncertainty of the market environment, which limits the ability of coal mining enterprises, which are characterized by high resource intensity and inertia of the production system, to fully realize the potential of their economic efficiency. As a result, the coal mining industry, despite the above transformations, continues to give up its position in the competition with alternative energy sources. Its share in the domestic market of hydrocarbons is reduced due to the expansion of the use of cheaper and more technologically advanced fuels. If at the beginning of the 90s the ratio "coal: oil: gas" in the fuel and energy balance of Russia was 26:33:41, then by this time the share of coal here has been almost halved [3].

It should be added that the modern period is characterized by an intensive search for solutions to create efficient power plants based on alternative energy sources to hydrocarbons. A number of forecast estimates indicate the closure of the "era of fossil hydrocarbons" [4, 5]. In accordance with the forecast of the famous futurologist Ray Kurzweil, technical director of Google, by 2028 alternative energy will be one of the key technologies for the development of all others. It will be approved as the dominant concept. Solar panels will become so efficient that they are enough to cover most of the energy costs of humanity [4].

In addition, following the results of the 2016 Paris Climate Conference, the protocol of which prescribes the next reduction of CO^2 emissions, the idea of creating in Siberia, where the main coal deposits of the Russian Federation, carbon-free zone are concentrated, is being actively discussed. This makes the coal industry of Russia even more acute the task of transformation in the light of new environmental requirements.

Thus, the above, speaks of the incredibly high competitive pressure faced by the coal mining industry in the global energy market, and in the coming years, taking into account the noted tendencies and the growth of environmental restrictions, will experience it to an even greater degree.

In the light of the above, it becomes quite obvious that the problem of sustainable development and improving the efficiency of coal mining production continues to be highly relevant and needs to be addressed.

2 Materials and methods

Successful solution of the problem under consideration requires a clear understanding of what principles, what concept should be used as the basis for building production systems for coal mining in modern conditions.

The world economy considers the problem of sustainable development in close connection with the problem of resource conservation, for the solution of which it is necessary to increase the efficiency of resource use based on the principle: spend less - produce more [6]. This approach is the dominant direction in the development of modern production systems in various sectors of the world economy.

It is believed that the basis of modern economical, resource-saving production systems and methods of organizing and managing production was made up by the achievements of Japanese manufacturers and, above all, the automotive giant Toyota [7, 8, 9].

Having become a pioneer in the development of new methods and technologies based on the concept of continuous improvement of processes and systems (quality concept) and their introduction into the structure of their enterprise, Toyota built the Toyota Production System (TPS) production system, achieved outstanding economic results and became of the leading leaders of the world economy.

The main idea of the "Toyota Production System" is a new way of thinking that provides a movement towards efficient resource-saving production, based on the ubiquitous and continuous elimination of any activity that consumes resources, but does not create consumer value. This system has become the object of careful study throughout the world in many industries. J. Womack and D. Jones summarized the results of these studies and formulated the main methodological approaches that ensure the use of TPS ideas in various industries and in all kinds of organizations in their work «Lean thinking: banish waste and create wealth in your corporation» [7].

In accordance with the recommendations of this study, the construction of resource-saving production systems is a process consisting of the following five steps:

1. Definition of use value;
2. Designing a value stream;
3. Organization of flow movement;
4. "Pulling" the product;
5. Continuous improvement.

At the same time, the technological scheme of mining enterprises is fundamentally different from the technological schemes of industrial production, where the idea of "Lean thinking" is being developed. It is a combination of basic and auxiliary production processes in combination with the workings necessary for their execution, mechanization and automation tools, which, with the rational organization of work, ensure safe and efficient development of the field. The technology of mining production assumes the continuous dynamics of the working zone of a coal-mining enterprise, when technological systems move along a complex constantly changing three-dimensional surface of the coal-bearing strata. In the process of implementing technological processes, significant amounts of mineral and host rocks are extracted and transferred (mass transfer), which requires the involvement of a huge amount of material resources (in the cost structure of mining products, material costs (past work) reach 50% or more total production costs). Moreover, these resources, from the point of view of creating consumer value, are used extremely inefficiently, and the intervention of mining production in the biosphere, both at the stage of past labor and living, and its active destruction, predetermine a high level of environmental pressures on the environment.

Thus, when considering the problem of sustainable development of a coal-mining region, success, to a decisive extent, will depend on the results of finding solutions that minimize the volume of mass transfer and resource consumption, and also increase the efficiency of their use, which, as a result, will create prerequisites for reducing the level of negative impact of mining on the biosphere.

The following are the basic principles for building production systems that implement the concept of "Lean thinking" in the aspect of their implementation in coal mining enterprises.

3 Results and discussion

Principle 1. Definition of use value. Consumer value is a key moment in the construction of production systems on the principles of "Lean thinking", this is the purpose and meaning of their existence in modern economic conditions. Creating maximum use value, i.e. the fullest satisfaction of the interests of the consumer is what serves as the main criterion for the optimization of all processes in the organization. The enterprise discards everything that is unimportant or secondary, and concentrates on the main factor causing its survival - the consumer. It is the consumer who satisfies his needs, pays them, and thereby finances your business [10].

This approach is radically different from the traditional vision of the organization. Only relatively recently, in the conditions of an unsaturated consumer goods market, the situation when demand exceeded supply was characteristic. For the specified state of the market, rather homogeneous products of a very limited range were produced - "mass production". In this case, the consumer was not a limiting factor: everything that was produced was consumed, and, as a result, the manufacturer dominated the market, and the production systems of the inception and flourishing of the industrial era were built on the principles of "PUSH" - pushing production. The main task facing the production systems of this period is to ensure maximum production volumes with minimum costs.

In the second half of the 20th century, the situation in the global economy changed dramatically. By this period, the problem of raising the standard of living was solved, markets were saturated with consumer goods with an acceptable standard level of quality, supply exceeded demand, competition between manufacturers in various market segments sharply intensified. The consumer has become a key market player, its limiting factor. New conditions transfer the attention of the manufacturer to the question: how and what do you produce? - the question of what and how you produce. This means that the consumer needs to offer not what your company can produce, but what it would like to receive. If he is not interested in what he is offered, then you lose the source of the working capital.

With regard to the products of coal-mining enterprises, the overall goal of production can be defined as the satisfaction of the consumer's need for additional energy to ensure the vital needs of a person (heat, light) and its production activities.

The main place in the set of indicators affecting the use value of coal is occupied by the heat of combustion, ash, moisture, volatile content, that is, components that characterize the ability of coal to concentrate and provide movement to any distances in the smallest possible amount of energy, as well as features of ignition and heat emission dynamics. At the present stage of development, these traditional indicators of coal quality retain their value, however, this group of characteristics is considered to be performed unconditionally, but the focus and attention of the consumer are also the volumes and reliability of supply, and the supplier's ability to predictably and with high accuracy ensure an agreed delivery schedule and product price at the point of delivery, and environmental performance of mining production, etc.

Thus, in order to achieve the highest production results, it is necessary to build production systems and manage the entire distribution chain, focusing not on one's abilities, but on the criterion of consumer value, to the fullest possible satisfaction of the needs and requirements of the end user.

At the same time, the consumer is absolutely not interested in the mining and geological conditions in which your enterprise carries out work, or the level of technical and technological perfection of your enterprise, nor indicators of equipment use efficiency, nor the geographical location of your enterprise, nor your level of costs. In modern economic conditions, the manufacturer must know exactly what should be done in order to fully meet the needs and expectations of the consumer. This requires him to continuously work with consumers and implement specific technical, technological, organizational and managerial decisions aimed at meeting the identified requests and expectations of the latter.

Principle 2. Designing a value stream. The value stream is a set of all actions that must be performed to bring a product with an established use value from the idea to the release of the finished product. In the process of implementing this stage of value creation, it is necessary, firstly, to know exactly what you are creating exactly what the consumer needs, and secondly, you do it as efficiently as possible from the standpoint of creating use value. At this stage, it is necessary to exclude from the process or minimize all operations that consume resources but do not create use value, i.e. it is necessary to ensure that the ratio of resources spent in a useful way (to create use value) to the total amount of resources used is as high as possible.

From this point of view, it is very important to understand - what type of production systems currently constitute the main potential of the coal mining industry, to what extent they meet modern economic conditions and efficiency requirements from the standpoint of "resource-saving production", and if not, what needs to be done to radically change the situation .

Technological processes implemented in real coal-mining production systems, from the point of view of participation in the creation of consumer value, can be attributed to one of three types [8]:

1. Operations or processes that can be defined as unconditionally involved in creating the use value of a product: for example, destruction of a seam, separation of coal from an array of rocks, its excavation, enrichment or sorting, and storage

2. Operations or processes that are not involved in creating use value, but they are inevitable due to the characteristics of currently used technologies and types of equipment. With regard to the mining industry, for example, all processes associated with the preparation, excavation, movement and storage of overburden can be attributed to this category of losses.

3. Operations or processes that do not create use value and are not technologically based. In the mining industry, such situations may arise as a result of, for example, erroneous technological solutions.

Analysis of efficiency indicators for the use of resources such as metal, electrical energy, diesel fuel, oils, and explosives showed that, for example, in open-cast coal, most of the resources are consumed by technological processes not involved in creating use value. As a result, the consumption efficiency of the above resources is 17%, 15%, 6%, 10% and 25% respectively.

In accordance with the classification given above, in the case under consideration there are losses of the second kind, that is, losses whose presence is predetermined by a constructive imperfection of the technological cycle. An example of a solution to a problem considered may be the reengineering of a production system. The abandonment of the currently widely used, in-depth longitudinal development system, when the entire volume of overburden moves beyond the boundaries of the quarry field and is placed on external dumps, in favor of development systems with internal dumping, for example, longitudinal longitudinal blocks transverse, will allow to increase the indicator of efficiency of use of resources, respectively, for these systems by 1.2; 1.3; 1.4 times.

It should be noted that this result was obtained solely by creating conditions for internal dumping with a corresponding change in the direction of cargo overburden, reducing the range of movement of quarry cargo and, consequently, reducing the amount of resources consumed by the technological chain, but maintaining the traditional structure of the latter.

Significantly higher efficiency (several times) can be achieved by implementing transformations of the traditional structure of the technological chain and building a new one, with a large number of operations involved in creating use value. An example of such a solution is the transition to open-pit coal mining [11]. In this case, there are options when underground mining is carried out mechanohydraulic method. In this case, the technological chain is built up, when most of the processes — opening and preparatory workings from the quarry space along the mineral reservoir, cleaning works, coal transport in the form of coal-water mixture and dehydration — are involved in creating use value. In this case, the coefficient of efficiency of resource use increases several times and reaches a value of 0.5, i.e., it increases 4.1 times.

Thus, the review allows us to conclude that a qualitative increase in the efficiency of resource use requires a transition to new principles for the formation of coal production systems. In particular, in the structure of the technological chain, the number of processes involved in creating consumer value should tend to the maximum and the absolute amount of resources consumed in the technological cycle to a minimum. The implementation of this principle will contribute to a significant increase in the efficiency of resource consumption, which, as a result, will ensure an increase in the competitiveness of coal-mining enterprises in the current economic conditions, as well as the fulfillment of sustainable development conditions.

Principle 3. Organization of flow movement. Most modern coal mining production systems are organized in such a way that they have no direct connection with their end user and the interaction is carried out through a chain of intermediate suppliers and consumers. Obviously, the effective functioning of such a production and logistics chain is possible only if each "internal" supplier within the framework of this system would satisfy the requirements of its "internal" consumer, but at the same time, they would ensure the needs and requirements of the end user. If this condition is not fulfilled at least in a single link, then the entire production and logistics chain of creation of consumer value may collapse. This, in turn, requires the joint and coordinated action of all components of the production and logistics chain. It makes no sense to improve the quality and cost indicators of the components of the system separately, if the entire gain is lost due to inconsistency of interaction at the joints. An attempt of this kind will be an example of sub-optimization, i.e. local (imaginary) optimization, which does not benefit from the point of view of the system as a whole. The considered approach to the management and optimization of the production and logistics chain of creating consumer value as a single system is a fundamental moment of "resource-saving production" based on the principles of "Lean thinking".

At the same time, to ensure the coordinated activities of all participants in the production and logistics system, the choice of management criteria is of fundamental importance. Until recently, as a criterion in the production systems of the coal-mining complex, the concept of costs is widely used - the cost of production.

However, in recent years, the concept of cost has been criticized [12]. To date, a sufficient number of examples of the negative consequences of its use have been accumulated associated with the local suboptimization of individual parts of business systems to the detriment of their effectiveness as a whole.

Significant progress towards the development of a set of indicators to ensure system optimization has been achieved in Goldrath's work [12]. As Goldrath has shown, the main problem of the cost-based management system is the impossibility of adequately accounting for the interaction of various elements of the system. This is due to the fact that its model is based on the assumption of the independent nature of the costs arising in various elements of the system. However, in fact, the level of costs, as a rule, is the result of the interaction of elements of the system and is determined by the circumstances of this interaction at the previous and subsequent stages of the technological cycle. According to Goldrata, in modern conditions, the achievement of system optimization requires the abandonment of the concept of costs and the transition to using the system of "global operational criteria". In accordance with [12], this system includes:

- the rate of revenue generation by the system;
- operating expenses;
- the average for the period level of equity capital.

The use of this system of indicators focuses the management's attention on organizing the optimal interaction of the processes of the managed system in order to identify and eliminate the bottleneck, that is, the resource that limits the rate of revenue generation by the system as a whole. Management of the limiting resource is aimed at both the growth of the system's performance indicators as a whole, and the reduction of operating expenses and the amount of capital linked in the system.

The Constraints Authority aims performers to focus on those factors that are most critical to overall success, that is, on the constraints of the system.

Thus, the secret of sustainable development is to create such a set of systemic factors that would provide the opportunity for system-wide optimization both within the enterprise and in the chain of intermediate partners of the logistics chain - in the interests of the end user.

Principle 4. "Pulling" the product. Depending on the mode of operation, production systems can be divided into "push" (PUSH) and "retractor" (PULL) type systems.

Until recently, the majority of production and logistics chains traditionally implemented in the coal mining industry belong to push-type systems. The logic of the functioning of systems of this type is determined by the requirement of the most complete use of expensive fixed assets. "The equipment should not be idle!" This is the meaning of the control system and organization of production of the "pushing" type. The reason for this decision was a seemingly completely "obvious" fact - the transfer of capital invested in fixed assets is possible only when these funds are working. In accordance with the above system of organization and production management, including methods of current and operational planning, the system of organizational policies of the enterprise (including wage systems) direct managers and employees to the fullest possible load of funds.

At the same time, coal mining production systems operate in conditions of a significant level of seasonal and conjunctive variability of demand and the said organization of work leads to the fact that the appearance of products at the output of the production and technological system is inconsistent with the volumes of its sales. As a result, in the production system, the level of work in progress and stocks of finished products in the warehouse increases periodically. This leads to the binding of significant amounts of working capital and the emergence of significant economic risks.

The general negative consequence of the "pushing" type of production is predetermined by the fact that the optimization of the system is carried out not by the system-wide criterion of economic efficiency, but by the totality of local efficiency criteria. Figuratively speaking, instead of optimizing the ability of the production system as a whole to make money (to ensure maximum efficiency of the rate of generation of cash flow with the minimum level of resources used and the capital linked in the system), the system is set up to "minimize costs", "evenness" or "continuity of production", "Permanent employment", i.e., local optimization.

The negative effects that the "pushing" production system has on enterprises have stimulated the development of "pull-in" type systems. The main feature of this type of system is the synchronization of production, i.e., the release of products demanded by the market in a mode consistent with the dynamics of market demand.

In this case, an extremely important task for working in a market economy is being solved - the response time of the production system to a consumer request is reduced. In the conditions of unstable demand, the systems of "intrusive" type make it possible to abandon the practice of outstripping production of large volumes of products based on unreliable long-term forecasts and to produce products only to order. A direct consequence of this approach is a sharp reduction in the level of work in progress, the release of working capital, a reduction in the need for storage space for finished products with a corresponding reduction in production costs and the entire set of production and economic risks.

Principle 5. Continuous improvement. "Resource-saving production" is a production system aimed at constant struggle with losses of various kinds. The solution of this problem requires constant monitoring of the state of the production system, control of variability (change) of the parameters of its processes, analysis of the collected statistics, development of improvement programs and their implementation.

The manufacturer focuses on the "causal factors" that limit the effectiveness of the system, and improving the processes, seeking to reduce the variability of parameters and characteristics of the latter, better use of working time (less breakdowns, setups, downtime), better use of materials (less scrap), reducing the cost of rework and control. This allows both to increase the degree of uniformity (quality) of products and to reduce its cost. It is on this basis that the efficiency of resource-saving production is formed.

At the same time, it is impossible to achieve this result without the creative participation in this action of those people who implement the entire set of production processes. This requires the development and implementation of appropriate organizational policies in the field of recruitment, training, motivation and personnel management, creating an environment of cooperation in the organization, maintaining a balance of interests of all participants in the process of creating use value.

Thus, it will be possible to achieve the goals of increasing the competitiveness of coal mining enterprises and the implementation of the conditions of sustainable development in the region only if all workers and employees of the coal mining complex are actively interested in this.

Conclusion

The considered approaches to the construction of production systems on the principles of "resource-saving production", tested in world practice in various industries, despite the significant specificity of the structure of mining production compared to the structure of production systems in the processing and processing industries, of course, adapted to the conditions of the coal-mining industry, but will require a fundamental change in the fundamental principles of design and management of industrial coal mining systems.

The review allows us to conclude that a qualitative increase in the efficiency of resource use through the implementation of a new approach to the formation of production systems and their construction on the principles of "resource-saving production" will contribute not only to a significant increase in the competitiveness of coal mining enterprises in modern economic conditions, but also to solve sustainable development of a large coal mining region.

References

1. V.V. Zhdamirov., V.I. Kuznetsov, V.V. Mikhalchenko, Coal, **11**, 3 (1991)
2. V.V. Mikhalchenko, R.M. Artyushkova, A.P. Kuzmin, Coal, **1,** 40 (1992)
3. Yu.N. Malyshev, V.E. Zaidenvarg, I.M. Zykov et al., *Restructuring of the coal industry (Theory. Experience. Programs. Forecast)* (Nedra, Moscow, 1996)
4. R. Kurzweil, *The Six Epochs. In The Singularity Is Near: When Humans Transcend Biology.* (Viking, New York, 2005)
5. D. Rifkin, *The Third Industrial Revolution: How horizontal interactions change energy, the economy and the world as a whole* (Alpina non-fiction, Moscow, 2014)
6. E. Weizsäcker, E. Lovins, L. Lovins, *Factor four: New report to the Rome Club* (Academia, Moscow, 2000)

7. J. Womack, D. Jones, *Lean manufacturing: how to get rid of losses and achieve prosperity of your company: Transl. from English* (Alpina Business Books, Moscow, 2004)

8. T. Ono, *The Toyota Production System. Departing from mass production: Transl. from English* - Institute of Integrated Strategic (Studies, Moscow,2005)

9. I. Macmillan, *The Japanese Industrial System* (Progress, Moscow, 1988)

10. G. R. Niv, *Dr. Deming's Space: Principles for Building a Sustainable Business: Transl. from English* (Alpina Business Books, Moscow, 2005)

11. V.V. Mikhalchenko, A.A. Atrushkevich, Patent # 2143075 RF, Bulletin, **35** (1999)

12. E. Goldrath, *Purpose: a process of continuous improvement* (Popurri LLC, Minsk, 2004)

Mechanisms for Harmonizing the Sustainable Development of Kuzbass as a Resource Region

Yuri Fridman[1] , *Ekaterina* Loginova[1] , and *Galina* Rechko[1, 2,*]

[1]Institute of Economics and Industrial Engineering of the Siberian Branch of the Russian Academy of Sciences, 630090, Novosibirsk, Russian Federation
[2]T.F. Gorbatchev Kuzbass State Technical University, 650000, Kemerovo, Russian Federation

Abstract. The article views the problems related to harmonizing the development of a resource area. Kemerovo Oblast - Kuzbass is selected as the object of this study. We point out that, in the absence of consensus between authorities and business concerning the prospective development of the region, its economy may decay even despite obvious success in the fundamental industry, namely coal mining. It is strategically important for Kuzbass to create an efficient infrastructure for transforming the development of the coal industry into the economy-wide development of the region. Taking into account best international harmonization practices, we propose a conceptual system of mechanisms aimed at coordinating the interests of the commodity-driven business and the resource area built on the 'common grounds' shared by the coal industry and Kemerovo Oblast. The concept in question relies on the need to introduce business models to the industry management that would harmonize with the interests of the territory. At the same time, the territory should start to build a new economy by creating new standards that would establish requirements for doing business.

1 Introduction

A major issue for the majority of Russian regions producing raw materials (resource regions) is the business and authorities disagreeing over how the area should develop. This causes it to decay even despite obvious success in fundamental industries (e.g. [1]), as is happening today in Kemerovo Oblast–Kuzbass, the center of the Russian coal industry. Unfortunately, neither at the national nor regional levels is there any holistic understanding of how to practically transform the successful development of the primary sector into economy-wide development. Among the reasons is the lack of tools for converting the effects of mineral resources development into economic system's productivity that would be understandable to all actors in the regional economic space.

In our opinion, a way out of this situation in Kuzbass is, first of all, to reset the entire system of relations between the coal industry and the area at the 'entrance' to the region. The second solution is to introduce principles of the harmonious development of a resource region in Kemerovo Oblast, which is crucial as part of improving the territory's

* Corresponding author: rgn.vt@kuzstu.ru

competitiveness. This purpose requires identifying 'common grounds' between the coal industry and the region, as well as setting out conflict resolution mechanisms, which will allow forming an effective model for the long-term development of Kuzbass.

2 Materials and Methods

International scholars began studying harmonization problems in the early 1990s. This period signified the emergence of the famous 'resource curse' theories [2, 3]. Then, thanks to the success stories featuring the resource economies in Canada, Australia, Norway, and others, works appeared proving that the resource curse is not inevitable [4]. For one, global experience shows that even now the mining industry is fully capable of stimulating growth in national and regional economies [5]. Chile may serve as an example: for twenty years this country remained the fastest growing one in South America.

Yet, there are no consolidated mechanisms for market self-regulation that would ensure the harmony of interests between business and resource regions. Each country solves these problems on its own. In Russia, almost every region, including Kemerovo Oblast, has gained experience in using different tools for interaction between business and the area. In most cases, they are based on informal agreements between regional authorities and business. But our studies (e.g. [6, 7]) and practices in Kuzbass show that, under certain conditions, authorities and business can find interest coordinating mechanisms to harmonize the development of the resource area [8] (Figure 1).

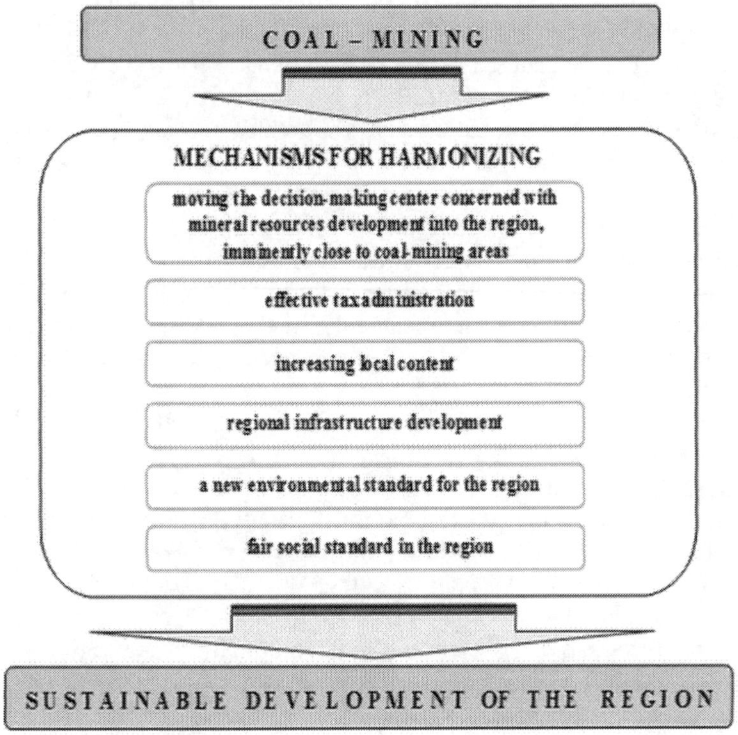

Fig. 1. Fragment of the harmonization model for the development of the primary industry and the area: harmonization mechanisms.

3 Results and Discussion

The advances of the coal industry in Kuzbass are obvious. Several factors have led to the development of this business in Kemerovo Oblast:
– In 1994–2016 the Russian coal industry underwent an avalanche of elaborate restructuring. Having the World Bank, the International Monetary Fund, and the Hoover Institute at their side, Russian experts carried out a plan to 'softly' eliminate inefficient coal capacity. Over ten years, the Russian coal sector experienced changes in the mining technology, the management system, and the funding vehicle;
– The vast majority of coal enterprises, despite the change of ownership, are managed by highly skilled professionals well aware of the particular characteristics of the Kuzbass fields;
– A big portion of coal-mining facilities in the region have fallen under the control of large vertically integrated companies (SUEK, EVRAZ, MMK, Mechel, UMMC), which has facilitated promoting Kuzbass coal in global markets and reduced the dependence of coal production on coal consumption decline within Russia;
– against the backdrop of the oil and gas booms, coal yielded relatively low profits; as a result, the Russian government simply did not pay much attention to it in terms of the tax burden;
– Although in the early 2000s Kemerovo Oblast announced the strategic significance of the coal industry, it put forward too lenient criteria for its development regarding several 'common grounds' simultaneously.

In the period between 2007 and 2017, the coal industry and the whole region showed very 'impressive' development results: coal production grew 1.3-fold, gross regional product decreased by 8 percent, real gross wages increased by 18 percent, real household disposable income dropped by 24 percent.

The issue of Kuzbass harmonization as a resource region is of particular relevance now, since the further growth of the coal industry in the region is becoming a nationwide project. For the next 15 years, Kemerovo Oblast will need to increase coal production by no less than 100 million tons against 255.3 million tons produced in 2018.

Kuzbass needs to create an effective infrastructure for transforming the development of the coal industry into regional development based on the principles of an inclusive economy [9] by involving all economic actors in the decision-making process, together with distributing risks and responsibilities among them. The *following mechanisms* should contribute to this:
– Moving the decision-making center concerned with mineral resources development into the region, imminently close to coal-mining areas.

Such an approach, in our opinion, will make it possible to consider the interests of the territory and its residents to the greatest extent. Formally, it can be expressed, for instance, in using the two-man rule when issuing mining permits. Kuzbass authorities insist on bringing this practice back to the coal industry. When looking at this issue more broadly and taking into account the experience of resource-rich countries, we are talking about introducing a strategic planning system in the development of the region's mineral and raw materials (which requires to create a long-term development program for the coal industry in Kemerovo Oblast). The system is supported by the region and subsoil users signing 'licensing agreements' even before they start coal mining. Such a document should contain mechanisms for transferring a part of the value obtained while developing a field towards the territory, specify how a subsoil user increases employment opportunities for the residents and provides comfortable living conditions inside the region, which should include improving social services and solving environmental problems associated with the mining industry;

– Effective tax administration.

At the moment, the main formal tool for transforming coal-generated financial resources into territorial development is taxes. The taxation of the coal industry in Russia is based on gross indicators (revenue, production, and export of coal products). This practice is typical of developing countries. It benefits from the relative ease of administration and guaranteed tax revenues to the budget. What is different is that the essential portion of these revenues consists of the income tax, whose value depends on fluctuations in world markets due to the export-oriented model of the coal business in Russia. We think the coal industry requires a change in fiscal management so as to stimulate business to invest, assume high risks of developing new deposits but also look to the long term, nonetheless. For one, Kuzbass may become a platform for a tax experiment on the financial result of coal enterprises. Moreover, it is equally important to resolve the issue of redistributing tax deductions from the coal industry in favor of the resource region itself;

– Increasing local content.

Synergy achieved from the fundamental industry development may be called the main condition for harmonizing the development of the resource area. This is due in no small part to the amount of local content in production and the involvement of local business in supply chains of goods and services for mining, the region's anchor industry. Local content in Kemerovo Oblast today is currently quite low, below 10 percent, according to our estimates. Within Kuzbass coal companies' annual purchases at about 400 billion rubles, products of regional origin amount to only 40–45 billion rubles. The amount of local content in the region is not governed by licensing agreements or regional legislation. However, large coal mining companies in Kuzbass use the world's best technology, while global leading mining equipment manufacturers strive to promote their products on the region's market. It creates prerequisites for the rapid growth of localization thanks to the developing service sector. Another promising trend is related to the fuel supply of coal mining enterprises: investment activity in oil refining should be accelerated to start producing diesel fuel for the coal industry. Besides, it is definitely important to adapt regional research centers to the demands of the modern coal business in mining, transportation, and processing of solid fuels;

– Regional infrastructure development.

High-quality infrastructure is traditionally considered mandatory for the efficient development of minerals, the investment attractiveness of the resource area (e.g. [10]) and its subsequent improvement. Due to the historical peculiarities of its exploration, Kuzbass differs from other Siberian regions by having a rather strong infrastructure in general (transport systems, power network, etc.). Nevertheless, when mining new deposits in Kemerovo Oblast, businesses face an increase in development costs, related to the preparation and operation of the infrastructure necessary, in particular, to supply energy to enterprises, transport coal by road and rail, etc. The high wear rate of most municipal infrastructure all over the region remains a problem. In the long term, infrastructure constraints might inhibit the development of Kemerovo Oblast. We think a way out may be to actively involve the commodity business in large infrastructure projects on the principles of public-private partnership and then to share infrastructure facilities created in so doing;

– A new environmental standard for the region.

Coal mining is among the top 10 industries with the greatest negative impact on the environment. With growing production and beneficiation, the technogenic load of the coal industry on Kemerovo Oblast will definitely continue to rise, which becomes a serious risk for the region and coal business. Kuzbass authorities have put forward an idea of establishing a regional environmental standard that, according to open sources, should include a sanitary protection zone (at least a thousand meters from open-pit coal mines), local authorities hearings when issuing permits for natural resource management, relocations from subsided surfaces, and other services. In our opinion, an environmental standard for Kuzbass should be an active and efficient mechanism for coordinating interests between the territory and coal companies. We believe that such a document should follow advanced environmental monitoring and contain not only standards for the regional coal industry but also incentives to practicing environmental consciousness and transparency, introducing the best available technologies, promoting environmental insurance, biodiversity offsetting, etc.;

– Fair social standard in the region.

Employment rate and wage conditions are important indicators showing how much the territory 'feels' the industry's presence. Increased labor productivity in the coal industry, on the one hand, leads to a reduction in employment; on the other, no adequate wage growth is observed in Kuzbass. The authorities are trying to influence the coal business by tightening social measures (for example, settling people in the sanitary protection zones of coal enterprises at the companies' expense). However, the efforts undertaken are clearly not enough: they do not solve the problem of encouraging human capital investment (education, health, etc.) in Kemerovo Oblast. It is indispensable to envisage an up-to-date fair 'social standard for Kuzbass' that would regulate the obligations of the base materials sector to the local community (wage levels, local content in employment, direct social obligations to the municipal entity where coal is being mined, satisfying the interests of small indigenous peoples, etc.). The fulfillment of these standards should also be part of a license agreement between the resource area and the subsoil user.

4 Conclusion

Thus, the proposed mechanisms for harmonizing the interests of authorities and business rely, on the one hand, on introducing business models in the Kuzbass coal industry management that cater to the needs of the territory. On the other hand, they come from formalizing and standardizing approaches towards interaction between Kemerovo Oblast and the resource business. In other words, relations between the region (government, society) and the industry (business) should base upon a clear set of rules for 'the game', which will help resolve conflicts and mitigate certain immediate risks inside the coal industry [11] and the resource area.

References

1. V. Basareva, N. Mikheeva, Regional Research of Russia, **8:2**, 184 (2018)
2. J. Sachs & A. Warner, National Bureau of Economic Research Working paper, **5398** (1995)
3. P. Van Der, Journal of Economic Literature, **49:2**, 326 (2011)
4. De Ferranti, G. Perry, D. Lederman & W. Maloney, *From Natural Resources to the Knowledge Economy* (IBRD, Washington, 2002)
5. V. Kondrat'ev, World Economy & International Relations, **60:1**, 41 (2016)
6. Iu. Fridman, G. Rechko, Problems of Economic Transition, **55:6**, 30 (2012)
7. Yu. Fridman, G. Rechko, E. Loginova, Journal of Mining Science, **51:5**, 924 (2015)
8. V. Kryukov, A. Sevastyanova, A. Tokarev, V. Shmat, Economy of region, **13:1**, 93 (2017)
9. *Fraser Institute Annual Survey of Mining Companies* (British Columbia Press., Vancouver, 2014)

Resource Regions as an Initial Base for Transformation of the Russian Commodity-Based Rental Economy Model

Ludmila Kusurgasheva[1,*], *Lydia* Lubyagina[1], *Pawel* Stefanek[2], and *Arunas* Lapinskas[3]

[1]Department of Economics, Institute of Economics and Management, T.F. Gorbachev Kuzbass State Technical University, 650000 Kemerovo, Russia
[2]KGHM Polska Miedź SA, Rudna, Poland
[3]Department of Economics, Saint-Petersburg Mining University, 199106 Saint-Petersburg, Russia

Abstract. The article analyzes the current commodity-based rental model of economical growth in Russia. In the conditions when the Russian economy faced serious challenges and risks both within the country and on the outer contour, the question of inertial development possibility and the need to modernize the existing model has become particularly relevant. The aim of the work is to demonstrate the inefficiency, the exhaustion of the commodity-based rental model in relation to growth of opportunities and the futility in relation to the capitalist world-system laws of the development. The article shows that at the present stage the transition from the commodity-based rental model of economical growth to a nationally oriented model of development is possible through the implementation of a new industrialization policy, whereby the resource regions should shift to a sustainable inclusive growth.

1 Introduction

The state of the Russian economy, despite the fact that official data showed some growth acceleration after their adjustment for 2018, continues to be assessed by specialists as stagnation. Overcoming this negative trend and putting Russia on the path of sustainable dynamic development is a task of paramount importance. Its solution is connected with the need to carry out systemic changes in society, which should alter, first of all, the current model of economic development in Russia. The latter is called in different ways; the most operational term is the "commodity based rental model", which determined the nature of economic growth dynamics in the post-Soviet period and which has now exhausted its potential. Naturally-geographical factors of sectorial labor division determine the localization of the raw material structure in a certain area - within the framework of resource regions. Therefore, it is quite logical to consider them as an initial base for transformation the Russian commodity-based rental economy model and their characteristics not only as a source of problems and issues related to the impact of natural wealth on economic growth and welfare, but also as a potential for transformation.

*Corresponding author: klv.eti@kuzstu.ru

2 Materials and methods

There is a vast literature on the relationship of the countries resource endowment and economic growth. Despite years of research in this area, there are still no concurrent views as to whether the resource wealth is a "curse" or a boon to the country. However, all research on the effects of resource abundance begins with a statement of an obvious fact that some countries with large reserves of natural resources, especially such as oil and gas, tend to have a lower level of economic development than countries without such resources. For the first time, the thesis on the negative impact of resource abundance ("resource curse") was defined in a paper [1] and has since received many empirical evidence [2–4], including the case of Russia [5–6]. Existing research focuses mainly on identifying the influence channels through which resource abundance undermine the development of a resource-dependent economy. Among the reasons of negative connection between the abundance of natural resources and economic growth, in particular are analyzed: 1) the "Dutch disease" effect, according to which the extractive industries boom has a dismal influence on other (non-resource) industries [7]; 2) decline of investment (including investment in human assets), associated with the volatility of commodity prices [8-11]; 3) rent-oriented unproductive behavior, soaring corruption, deterioration in quality of institutions, and a high probability of civil conflicts [12]. It should be noted that all these significant results were obtained using econometric methods and models. Standard cross-country models are also used when passing to the analysis of the resources abundance influence for particular regions within the national economic structure, showing similar effects at the regional or local levels.

The presence of a conflicting viewpoint, explaining resource abundance as a boon to the economy ("resource blessing") [13-14], although it cannot be the basis for conclusion of econometric models nonapplicability, indicates their insufficiency. The latter is connected not only with the fact that the model building results are largely determined by the choice of variables, but also by the complexity of the modeling object itself. The two main factors that impart complexity are, firstly, the fundamental uncertainty inherent in a market economy and, secondly, the historically determined place of the national economy in the capitalist world-system. Uncertainty as an objective characteristic of the real world, defined by the complexity and scale of information necessary to economic agents for decision making, as well as the fact that it does not yet exist, it has not yet been created (due to the absence of the fact itself, an event that can still occur as a result of unintentional actions), puts in doubt the realism of econometric models. The second factor necessitates the expansion of analytical base of research by combining various methods, united by a common research program. The systemic paradigm of J. Kornai [15] can act as such general program, according to which society is considered as a unified system requiring complex studies that do not become isolated within private disciplines (economics, political science, sociology). This justifies the use of the world-system approach to the study of the Russian commodity-based rental economy model transformation. World-system analysis allows us to identify and include in the analysis global and macroeconomic factors in the development of resource regions that are not taken into account in papers on econometric modeling of a resource-dependent economy [16].

3 Results and Discussion

The size of natural rent is the main characteristic of the entire Russian economy, which is why it is called rental. Almost half of the federal budget revenues are provided by rental incomes, which include, in most of their part, taxes and duties paid by oil and gas companies. Among the external signs of Russian economy rental nature, we can distinguish such features as: high dependence of the economy on foreign markets, appearing primarily in dependence of the general state of economy on hydrocarbon prices; structural degradation, appearing in the "weighting", the simplification of the economy structure and deindustrialization; technological gap with developed countries; permanent investment crisis; enterprises dependence on state orders and subsidies; consistently low share of innovation-active enterprises; human capital stagnation; high level of income inequality; low quality of institutions; rent-oriented behavior of economic agents, politicians and officials (high propensity for acts of corruption). The rental-commodity nature of Russian economy is also manifested in the increasing unevenness of the country's spatial development: the regions are actually divided into a developed "center" that plays the role of accumulating and distributing rental incomes and an undeveloped "periphery" consisting of resource regions that generate rent.

The economic system with such characteristics brings to the Dutch disease, which emerging mechanism and development is well known: in a favorable market environment, the commodity-based industries receive additional impulse for expanding their activities, increasing profits and wages, which leads to a decrease in the growth potential of the manufacturing industries and results in deindustrialization of national economy. This approach adequately describes the deepening of regions differentiation, although other factors come into effect at this level, in particular, the infrastructure, production and human potential – all accumulated in the previous period by individual regions.

The Russian economy shows all the symptoms of the Dutch disease. Its background, apparently, was formed as part of the planned Soviet economy, but its mechanism began to work in full force after price liberalization, liberalization of foreign trade and privatization. In other words, the Dutch disease was caused by the shock transition of Russia to a market economy. This raises the question: what were the reasons for choosing such option of transforming the Soviet economic system into a market one? The thesis that this choice was due to political and ideological factors can be considered as proven. The policy choices made by the country's leadership at that time do not fit the logic of purely economic feasibility and efficiency, but can be rationally explained in terms of the ruling classes political interests of the late Soviet society and the "center" countries - privatization, liberalization and "return to the civilized world". Interests, in their turn, are an expression of objective need; in this case, they expressed the need for unlimited expansion (external economic expansionism), which is objectively inherent in the capitalist world-system. On the other hand, the Soviet economy, although it was planned, it can hardly be defined as socialistic. There are strong arguments for the hypothesis of state capitalism. At least, in its depths, in view of the underdeveloped background for the transition to a new social system, a new class formation process was underway, which resulted in the formation of a party-economic nomenclature class, which was in fact a new owner of the production means and a "collective entrepreneur." Furthermore, there was a large segment of clandestine capitalism, the subjects of which were various kinds of "shadow merchants". These two classes were the social layer that was objectively interested in restructuring and changing the social system. The interests of both classes, with the flow of resources into the shadow economy, diverged increasingly from the interests of the population majority, but at the same time meet with the ruling class interests of developed countries. As a result, the country had shock 1990s reforms, Russia was reintegrated into the capitalist world-system [16] and, interrupted by the "socialist experiment," the dependence of the Russian economy on the "center" was restored with all the consequences.

Accordingly, the essence of the commodity-based rental model of Russian economy can be defined as rental capitalism of the world economy peripheral zone. The main relation in this system is the center's exploitation of the periphery by means of production, accumulation and redistribution of natural rent, produced in the resource-based industries of the peripheral economy. This relationship lays behind the dependence of peripheral economies on developed countries, which appears on the surface of phenomena as their "backwardness" and which is supported by various instruments. The dependence of the periphery countries is the real cause of the "resource curse" and the Dutch disease.

The qualification of Russian economy as rental characterizes its reproduction as a system based on rent. Rent is an additional income (superprofit), resulting from the use of a rare resource in production, that gives the owner of this resource the possibility of monopoly command, both on the terms of ownership and management with its use. The latter serves as the basis for distinguishing various types of rent: mine, land, administrative, intellectual, technological, insider and others. Of all rent types, only natural rent and its varieties are the result of productive activity, yet others are the result of redistributive processes and are essentially a source of fictitious capital. The Russian economy is not just rental, but commodity-based rental, as the main type of rent is natural rent. However, the specific nature of the reproduction process of Russian economy and its institutional originality highlighted above as features of the economy's rental nature, are determined not so much by its commodity-based rental nature, as by the peripheral status in the global economy.

The modern world economy is not a simple group of equal countries, albeit at different stages of development, but a unified system in which individual elements play different roles subject to their place. The dependent periphery role is the maintenance and reproduction of the capitalization system in the "center" countries. Within the framework of the capitalistic world-system there is no exit for dependent countries. Therefore, the current economic model in Russia is not only ineffective, but also non-sovereign, and thus futile.

The situation with inefficiency and futility of the commodity-based rental model would seem to be in contradiction with the fact that the growth rate of the Russian economy is quite high in the pre-crisis years of the first decade of the 21st century (approximately 7% per year). However, structural analysis of real GDP growth shows that during this period, the main contribution – almost half – to output growth was provided by external demand. Its significance after the 2008-2009 crisis, and especially after the well-known 2014 events, has seriously decreased. This further demonstrates the dependent nature of Russian economy on external factors and the need for systemic reforms that should change the fundamental society basis, including economy, politics and ideology. Such changes are impossible to apply, relying on neoliberalism as a theory and ideology of the state economic policy. It is impossible either by returning or restoring the command-and-control system in Russia.

Historical experience indicates that Russian economy showed growth rates higher than the world when it was regulated and socially oriented, and at the same time was aimed at ensuring national interests. The intention of building a nationally oriented model of socio-economic development in the country is being formulated on this basis. Its characteristic features are: the diversity of ownership forms, the combination of market self-regulation institutions and state planning, business initiative and social justice, the priority of the common good over private interests. Such model should be based on adequate productive forces. The latter implies a new industrialization policy aimed at a radical transformation of the technical and technological base through re-industrialization and the transition to new, progressive technological modes.

The place and role of resource-type regions in neo-industrialization is determined by the fact that new industrialization requires significant investments, the source of which, in the current situation, can be the natural rent in the first place. This implies the need for sustainable development of resource regions. The concepts of sustainability and associated with them inclusiveness, balance, and innovativeness of economic growth are extensively and widely discussed in the literature. Выделим главную проблему. We highlight the main problem. It lays in determining the priority of social, economic or environmental purposes in determining the development strategy of resource regions. The choice of strategy depends on a number of external and internal factors, among which an important place is occupied by regional policy, which is a part of the state macroeconomic policy. At present day the implementation of the neo-industrialization policy in Russian regions, where it is necessary to systematically create conditions for sustainable inclusive development largely depends on the actions of the government.

4 Conclusion

The analysis of the existing in Russia commodity-based rental model has shown its inefficiency, exhaustion from the viewpoint of growth opportunities and the futility of further development within the framework of the center-peripheral structure of world economic relations. Its formation was the result of two laws of nature: firstly, the objectively inherent capitalistic world-system needs for unlimited expansion and, secondly, the need to legalize the property rights of the party and economic nomenclature and the clandestine bourgeoisie of Soviet society. That is why the implementation of 1990s market reforms according to the Washington consensus, the state's withdrawal from the economy with a simultaneous bureaucracy increase in the power, the weakness of market institutions and the "failures" of the state that caused the effects of the resource curse and the Dutch disease, became a reality. The main direction of the commodity-based rental model transformation is the new industrialization, which should be the basis for the formation of a new, nationally oriented development model that meets the objectives of achieving above the world growth rates and changing the institutional structure towards ensuring the country's national economic security and sustainable inclusive development of resource regions.

References

1. R. Auty, *Sustaining Development in Mineral Economies: The Resource Curse Thesis* (Routledge, London, 1993)
2. J. Sachs & A.M. Warner, European Economic Review, **45:4-6,** 827 (2001)
3. M. Humphreys, Journal of Conflict Resolution, **49**, 508 (2005)
4. F. Van der Ploeg, S. Poelhekke, Journal of Environmental Economics and Management, **60:1**, 44 (2010)
5. C.G. Gaddy, B.W. Ickes, Eurasian Geography and Economics, **51:3**, 281 (2010)
6. A. Kudrin, E. Gurvich, Russian Journal of Economics, **1:1**, 30 (2015)
7. W. Max Corden and J. Peter Neary, The Economic Journal, **92:368**, 825 (1982)
8. Ph. Aghion, Ph. Bacchetta, R. Ranciere, K. Rogoff, Journal of Monetary Economics, **56:4**, 494 (2009)
9. S. Zhironkin, S. Demchenko, G. Kayachev, M. Ryzhkova, O. Zhironkina, Cluster E3S Web Conf., 41, 04011 (2018)
10. S. Zhironkin, O. Aleshina, V. Gorev, Y. Gunyakov, O. Zhironkina, E3S Web of Conferences, **105**, 04001 (2019)
11. M. Prokudina, O. Zhironkina, O. Kalinina, M. Gasanov, F. Agafonov, E3S Web of Conferences, **21**, 04003 (2017)
12. H. Mehlum, K. Moene, and R. Torvik, The Economic Journal, **116:508**, 1 (2006)

13. M. Alexeev, R. Conrad, The Review of Economics and Statistics, **91:3**, 586 (2009)
14. T. Gylfason, European Economic Review, **45:4-6**, 847 (2001)
15. J. Kornai, *The System Paradigm* (William Davidson Institute at the University of Michigan, Detroit, 1998)
16. N. Osokina, E3S Web of Conferences **41**, 04032 (2018)

Ecological Consciousness in a Shift of Scientific Paradigms

Mariya Yazevich[1,*], *Oksana* Kalinina[2], and *Olga* Zhironkina[3]

[1]T.F. Gorbachev Kuzbass State Technical University, Rector, 650000, 28 Vesennyaya St., Kemerovo, Russia
[2]Tomsk Polytechnic University, 634050 Lenin Avenue 30, Tomsk, Russia
[3]Kemerovo State University, 650000, 6 Krasnaya st., Kemerovo, Russia

Abstract. The analysis of the relationship between environmental conscious and scientific paradigms in different Western civilization development periods is presented in the article. The peculiarities of the influence of classical, non-classical and post-non-classical science on the types of environmental conscious characteristic of modern and postmodern societies are studied. The provision on the existence of the connection between the logical and philosophical attitudes of the New European (classical) science and the formation of the principles of anthropocentric environmental conscious that have ensured the formation of the anthropogenic civilization is proved. It is determined that within the framework of the non-classical paradigm the theoretical and philosophical foundations of the eco-centric model of environmental conscious begin to take shape. In the post-non-classical era the model of eco-centrism and worldview orientations directed to the development of the strategy for the safe interaction of nature and man is being formed on the basis of the theories of postmodernism and under the influence of modern cognitive practices. The influence of ontological and epistemological attitudes dominating within the framework of scientific paradigms on the formation of principles of nature management in anthropocentric and eco-centric systems of ecological conscious is traced.

1 Introduction

The aggravation of global environmental problems in the twentieth century has provoked a need for understanding the causes and sources of environmental crises. Awareness of the lack of prospects for the anthropogenic civilization and the need to develop a constructive strategy for the development of society made it necessary to create a new model of relations in the system of nature-society, ensuring their harmonious and holistic coexistence. At the same time, great importance in overcoming global contradictions and leveling negative trends from human activities is attached to the development of a worldview attitude capable of providing civilization with new guidelines in relation to the environment and the principles of effective nature management.

* Corresponding author: maria762003@list.ru

Understanding the relationship between environmental crises and the mental attitudes of society involves addressing the problem of human-nature interaction, as well as their conceptual interpretations, characteristic of certain epochs. Socio-natural relations are mobile, are transformed in the course of historical development and are conditioned by various causes: from individual and social needs, cultural traditions to specific cognitive practices, epistemological attitudes and ontological prerequisites that dominate in a certain epoch. In this regard, it should be noted that the formation of worldview orientations that contributed to the development of environmental crises is largely associated with the emergence of the phenomenon of scientific knowledge in the modern era in Western European society.

2 Materials and methods

For research of features of interrelation of ecological consciousness and scientific paradigms researches were used: J. E. Baugher, F. Capra, I. V. Chernikova, R. Rorti, V. S. Stepin, J. Habermas, M. Foucault, L. P. Pojman, P. Pojman, C. Weston and others. This made it possible to cover all the important stages in the development of science and environmental worldview, to reveal their correlation. For determination of features characteristic for different periods of science and types of ecological consciousness, and also identification of their General principles methods were applied: rational and logical analysis, structural and functional approach, conceptual analysis, historical and logical.

3 Results and discussion

The emergence and spread of science became the source of modern anthropogenic civilization and determined the formation of a certain type of ecological consciousness, which ensured the development of both an industrial boom in Western society and associated negative environmental processes. Environmental conscience is an intellectual and spiritual prerequisite for the formation of social practices that include environmental knowledge, concepts, categories, principles, ideas, theories, and an established model of ecological behavior and manifests itself in certain intentions, actions, providing ways to interact nature and man development of nature, conservation of the biosphere, use of resources, environmental safety, etc.). [1]

The genesis of the environmental conscience of anthropogenic civilization begins in pre-modern societies, but its active development falls on the epoch of modernity and is closely connected with the classical scientific paradigm. In the opinion of the V.S. Stepin, the reasons for the eco-crisis relations are not easy in the imperfection or mismanagement of technology, they are laid in the deeper foundations of modern society. This refers to the characteristics associated with the emergence and establishment of a specific new European mechanism for the development of culture, a mode of activity that includes, in particular, such an attitude towards nature, humanity, such a system of value determination of its behavior that, in appropriate socio-historical conditions, could not exacerbate environmental relations. [2]

The emergence of science occurs in the era of modern times in the conditions of transformation of traditional ideas about the structure of the world and society, the source of which were the logical and philosophical attitudes of the ancient and medieval world view. In the conditions of changing philosophical and ideological orientations (secularization processes, Reformation, the emergence of Protestant ethics and anthropology, the transition from cosmocentrism and theocentrism to anthropocentric ontology, the manifestation of symptoms of axiological anomie, the modernization of social practices, etc.), the classical model of science and the development of anthropogenic civilization attitudes. The basis for describing the reality of modern European science is the idea that explains the world order as a single mechanism consisting of moving material elements (atoms), obeying objective universal laws and principles, as well as devoid of hierarchy, ethical meanings and value measurement. All the differences in material nature (inherent in the ancient medieval view of the world) are absent in this picture of the world: any element of the created nature is in principle equal, more accurately equal to any other, and the whole creation, if it is considered separately from the Creator, is unified.

The comprehension of natural laws presupposed the discovery of cause-effect relationships in the phenomena of nature through a properly organized methodology (observation and mathematical calculation). At the same time, the prevailing Cartesian model of cognition prescribes the original ability of the human consciousness (subject) to think in accordance with the logic of the existence of the world (object). The main way to obtain a reliable knowledge of nature is experimental research, which is based on mind-controlled observation and measurement. This methodological setting stimulated not only the extensive growth of natural theories, but also the design of a new system of views on nature, where the world around is losing its "sacredness" and becomes a "great mechanism", a "machine", comprehensible through a set of measuring practices.

The unfolding of empirical epistemology based on the principles of logocentrism promoted the formation of subject-object relations, where the autonomous subjective consciousness cognizes fragments of reality through experience and logical procedures, creating theoretical constructions-projections that explain and model the surrounding reality in accordance with the specifics of one's own consciousness. [3] In this case, the subject opposed to the object becomes not just a participant in cognition, but the creator of reality, the basic ontological unit with the help of its own mind, providing the reality of the external world. He does not simply discover the laws of the universe, but encodes being, formalizes in theory, and initiates practices that make it possible to extract from nature a benefit on an "industrial" scale. This metaphysical setting becomes the basis for the anthropocentric picture of the world where, by setting the measure of what is, the person prescribes laws to nature, receiving the status of her master and the moral right to change reality according to her own understanding.

Of particular importance in the classical scientific paradigm is the idea of progress, on the basis of which an interpretation of natural laws is formed. Progress was understood as a movement similar to a mechanism that can be studied, predicted, managed, and consciously directed. The doctrine of progress justified the statement about the evolutionary development of the biosphere and determined the intervention of man as a necessary factor for the optimization and balance of natural processes. The idea of progress presupposed the activity of a sovereign subject endowed with reason and rights, carrying out activity in all spheres of the external world from epistemological practices to social modernization. As a result, progressism becomes the dominant conviction in the modernist picture of the world and prescribes the existence of a relationship: technological progress - material progress - progress in human rights. Thus, in the public consciousness, an instinct for subjugation, the transformation of nature, the subordination of it to human understanding was formed, and its justification was the idea of progress, which gained supremacy in the worldview of the Enlightenment.

It should be noted that the epoch of modern science is characterized not only by the fact that the values of society in relation to nature change (ethical and aesthetic views of the ancient and medieval installations are replaced by material and pragmatic ones), but also because in connection with such a transformation of public views, soon there was a need for conservation. But concern for nature does not arise with the purpose, for example, of preserving aesthetic characteristics or reverence for "God's creation," but for continuing the "productive" exploitation of nature and extracting the maximum profit from it. At the same time, nature conservation activities that were dictated by the principles of pragmatism were not excluded. So a person should save the natural environment so that future generations could use it.

Within the framework of the logical worldview of the classical scientific paradigm, an anthropocentric model of ecological consciousness is formed, which has conditioned the development of anthropogenic civilization and has become a source of global ecological crises. The main dominants of anthropocentric environmental conscience include a number of value orientation attitudes that determined the emergence of environmental contradictions.

One of the most important foundations of the anthropocentric worldview is the idea of progress that encompasses the spheres of production, consumption, social relations, the system of knowledge, moral norms, culture, the nature of man, etc. It is fixed and associated with the enlargement of forms of social cooperation, the development of production, and raw materials.

Another important priority of anthropocentric consciousness is the idea of the supreme value of man, which is the center of being (the basic ontological unit) and rises above nature. This provision is based on a myth according to which for man as the bearer of the mind there is no limit in the dissemination of activity. In this case, the very activity of a person in material reality is interpreted as the expansion of norms and rules into nature for its transformation in order to ensure its own security. Over time, as a derivative of material security, the concept of comfort is formed - physical, moral, social. In this idea there is a conviction in a certain evolutionary rule: the movement of a person in the direction of improving tools for his activities (progress) will provide him with a comfortable, conflict-free existence.

So in the system of anthropocentric attitudes, nature is perceived as an object of human activity, where the goal of its interaction with nature is the satisfaction of various utilitarian needs. In this logic, the notion of a "pragmatic imperative" is asserted, where it is right and normal to recognize that it benefits and maximizes the benefits to a person. [3] This position is strengthened with the help of utilitarian ethics, which received its natural spread in the conditions of the modernization and development of the market economy [9-12]. The active dissemination of "market ethics" and a passion for personal enrichment became "the driving force of the whole machine civilization" [4] However, ethical norms do not extend to interaction with the natural world, but the development of nature is conceived as a process that must be subordinated to the goals and tasks of man. Nature in anthropogenic culture is gradually becoming not a" temple, "but a" workshop ", an" environment, "a means of social progress. [4]

Thus, the new European scientific worldview became the foundation for the formulation of anthropocentric consciousness of anthropogenic civilization. The role of the "great modernizer" was assigned to the scientific and technological revolution. It was to create an "ideal world" where all the elements of nature would be completely curbed, the material universe is subordinated to man, which would determine the triumph of the all-ordering mind and control over the outside world.

However, the further development of social history and the crisis of the paradigmatic foundations of classical science have cast doubt on the principles of anthropocentrism in ecology. The ensuing era of postmodernism began with a disappointment in the scientific and technological revolution, where the destructive consequences of a number of new European landmarks that determined the nature of the environmental and social relations of the anthropogenic civilization are becoming evident. It turned out that tough industrial technologies and consumer attitude towards nature can destroy the ecosystem and thereby undermine the prospect of a person's continued existence on the Earth. In addition, it becomes clear that they are capable of enslaving the man himself, imposing his own mechanistic standard and norms of interaction on him. The rapidly growing influence of the industrial "megamachine" is becoming increasingly dangerous, trying to change the person himself, turning it into a functional element.

In the European system of understanding of the world since the end of the XIX century. the transformation of onto-epistemological and value systems, which caused shifts both at the level of the paradigmatic foundations of science, and in social practice, begins to be observed. In this period, a non-classical model of science is formed, where the notion of reality is based on the principles of relativism and complementarity, and also affirms the position about the dependence of the image of the world on the position of the observer in the knowledge of the microworld. The specificity of cognitive practices is determined by the features of the object of quantum-mechanical reality, which is a network of interrelated events, including the subject-observer as an element of the observable system. [5] Due to the transformation of subject-object relations, as well as under the influence of I. Kant's transcendental apriorism and practical dialectical materialism, an activity-based model of cognition is formed. In this context, a person acquires a new meaning, he is regarded as a subject of activity and an existential being responsible for his actions and constructing reality, including through science.

The formulation of new theoretical and practical problems, as well as the emergence of global environmental problems, led to the emergence of the foundations of ecocentric ecological consciousness. Among the most important priorities is the need to develop new principles of nature management, while searching for an alternative system for explaining the material world in the context of its relationship with the person and his activities. There is the emergence of a new type of scientific rationality, the criteria of which is the rejection of the cognition model as an act of contemplation of the subjective mind in the cognition of the world and the attainment of absolute truth. It is found necessary to take into account the influence of methods of cognitive activity in the description of the object, as well as the formation of epistemological practice on the basis of the principle of equality of various theoretical approaches to describing phenomena. Thus, the nonclassical picture of the world does not allow a "completely objective description of nature" [6], stimulates at the paradigm level the need to rethink the anthropocentric model of reality and initiates a gradual rejection of the central position of man in the material universe. Nature begins to be gradually perceived as equal to man and complementing him through the possibility of realizing his own activity. Recognition of the equivalence of nature determines the need for responsible attitude to their own projects and taking into account the consequences of their interference in the biosphere, including through science and technology.

In the mid-60's. there are a number of conditions that contribute to the crisis of the worldview attitude in Western society, which manifested itself in the statement of global problems, the deployment of environmental and demographic crises with irreversible destructive changes in the ecosystem. There comes an awareness of the need to transform not only the technical and technological processes, but also the mental and theoretical prerequisites that have determined the emergence of global contradictions. In these circumstances, there is a change in the paradigmatic foundations of science and the transformation of value orientation attitudes in the environmental conscience of civilization. The design of new worldviews is shaped by the ideas of postmodernism, post positivism, social constructivism. The world's picture is based on the ontology of polycentricism, which prescribes to the world the properties of mosaics and irrationality. [7, 8] Typical ontological categories are "simulacra", "hyper real", "fragmentation". [2] Cognition of the world is understood as a discursive practice, proceeding in conjunction with socio-cultural circumstances and relies on a system of conceptual-final agreements, and social norms are relative and lie in the plane of subjective preference, often unconscious.

Postmodernism contributed to the development of the principle of ecology, where society and man are placed in a single context with the world, nature and unite in a single system of geo-bio-sociogenesis. The subject-object principle of the New Time, when the subject - "Promethean man", elevating himself above nature and culture, which should be changed at his own discretion - is opposed by the principle of co-responsibility, co-ownership, co-dimension. Thus, postmodernism has debunked the myth of a man towering over the world and referring to him as a means. On this basis, numerous attempts have been made to revise the Western principles of social being and mentality. [1].

In many ways, under the influence of postmodernism, the processes of globalization, the ideas of multiculturalism develop, and the science and theoretical knowledge are orientated to the principles of the diversity of cognitive procedures and images of reality. In these conditions, science is experiencing a series of cardinal changes that allow us to fix a new stage in its development - post-non-classical. V.S. Stepin singled out the following signs: the change in the nature of scientific activity associated with the revolution in the means of obtaining and storing information (the computerization of science, the merging of science with the industrial sphere), the dissemination of interdisciplinary research and integrated research programs; increasing the importance of economic and socio-political factors and objectives; the discovery of self-organizing systems; inclusion of the axiological moment in the composition of scientific theories; the use in the natural sciences of the methods of the humanities.

As the dominant cognitive practices in post-non-classical science, a dialogue and evolutionary model of epistemology is distinguished. [4] On their basis, the principle of equivalence in the interaction of man with the outside world takes shape, the desire for a deep understanding of the essential links between nature and society. In the dialogue epistemological scheme, the subject enters with the object in a relation that Habermas called "communicative action". Cognition, therefore, is here a dialogue, an act of interaction between the world and man, representing a single whole and having an indissoluble connection with each other. In evolutionism, a person is seen as part of a developing nature, and his cognitive structures are commensurate with the world and are with him in complicity. Thus, a person becomes not just an explorer, an observer of the world, but a participant in the evolutionary formation of the universe, setting new trends in its development.

Modern science, focused on the study of self-developing objects, such natural complexes, including humans, like the biosphere and the noosphere, is forced to include in the internal context what it used to endeavor to eliminate - the person. In this model of cognition, the position of an outside observer does not exist, the "human-sized" object already includes the observer and his evolving mind. Thus, a person who knows the world, as well as knowledge, procedures and tools of nature research - is all part of a self-developing system and includes many perspectives and variants of development of events, for the consequences of which the person himself is responsible.

Standards and benchmarks of post-non-classical science have become a theoretical platform for the design of eco-centric consciousness, formed on the basis of some ontological principles of postmodernism, synergetic and evolutionism. In general, the existence of nature and man is conceived as a process of developing mutually beneficial cooperation. On this basis, nature is defined as an equal subject in interaction with man. In the process of nature management, it is important to maintain a balance between meeting human needs and the needs of the whole ecosystem.

There is a change in the axiological system of social consciousness, where the harmonious development of man and nature, connected with each other and representing unity in bio- and noosphere, is defined as the highest value. This sets new ethical guidelines, emphasizing the unity of moral norms and standards, which apply equally to the interaction between individuals in society, and to relations with nature. Central to the ethical paradigm of eco-centrism is the "ecological imperative", which affirms the principle: "only that which does not violate the existing environmental balance in nature" is right.

Particular importance is given to efficient use of natural resources and to reducing the anthropogenic load on the Earth, which is based on the need to reduce consumption and technological modernization. Environmental protection activities from the position of eco-centric environmental conscious are dictated by the need to preserve nature for its own sake, as well as for present and future generations living in harmony and balance with the outside world. Special importance is attached to science, which should create the basis for the process of building a new type of relationship between man, society and nature.

4 Conclusion

In general, the development of science and environmental conscious is a single interrelated process, the emergence of which occurs against the background of the dynamics of ontological, epistemological principles and values and ideological orientations characteristic of Western European society. Depending on the specifics of the paradigmatic foundations of science in a certain era, a model of a certain type of ecological consciousness was formed. Thus, classical science became the source of anthropocentric environmental conscious with an emphasis on human domination and the priority of its needs in environmental management. In the framework of non-classical and post-non-classical models of science in the late 19th-20th centuries. The structure of ecocentric environmental conscious with the justified system of reference points to the principles of optimal and equilibrium interaction with nature, ethical attitude to it and mutually beneficial cooperation with the environment is maturing, developing and developing. This trend is dominant in the modern worldview system of Western society, but it does not fully determine social practices and environmental management processes.

References

1. T. Anderson, *The origins of postmodernity. London: Verso, 1998J.Baudrillard, Simulacra and Simulations* (Stanford University Press, Stanford, 1988)
2. J. E. Baugher, *Ecological Consciousness, Moral Imagination, and the Framework for Strategic Sustainable Development* (Emerald Group Publishing Limited, London, 2016)
3. I. V. Chernikova, *Epistemology: the main problems and the evolution of approaches in the philosophy of science*, (Kuzbass University, Kemerovo, 2007)
4. F. Capra, *Uncommon Wisdom: Conversations with Remarkable People.* (Flamingo, London, 1989)
5. J. Derrida, *Globalization, Peace and Cosmopolitism in Negotiation: Interventions and Interview 1971-2001, trans* (Stanford University Press, Stanford, 2002)
6. W. Heisenberg, *Physics and philosophy. The revolution in modern science* (Ruskin House. George Allen and Unwin Ltd. Museum Street, London. 1971)
7. V. Zolotukhin, V. Gogolin, M. Yazevich, M. Baumgarten, A. Dyagileva, IOP Conf. Series: Earth and Environmental Science, **50**, 012027 (2017)
8. M. Foucalt, *The Archaeology of Knowledge and The Discourse on Language.* (Panteon books, New York, 1972)

9. E.A. Gasanov, T.S. Boyko, N.S. Frolova, Economics and Innovation Management, **3**, 4-8 (2018) DOI: 10.26730/2587-5574-2018-3-4-8
10. S. Zhironkin, S. Demchenko, G. Kayachev, M. Ryzhkova, O. Zhironkina, Cluster E3S Web Conf., **41**, 04011 (2018)
11. S. Zhironkin, D. Khloptsov, N. Skrylnikova, I. Petinenko, O. Zhironkina, E3S Web Conf., 41, 04010 (2018)
12. S. Zhironkin, O. Aleshina, V. Gorev, Y. Gunyakov, O. Zhironkina, E3S Web of Conferences, **105**, 04001 (2019)

The Solution of Environmental Problems and the Dynamics of Demographic Processes in Industrialized Regions

Anton Bel'kov[1,*], *Vladimir* Zolotukhin[1], *Natalia* Zolotukhina[1], *Natalia* Sedina[1], and *Marina* Kozyreva[1]

[1]T.F. Gorbachev Kuzbass State Technical University, Kemerovo, Russian Federation

Abstract. The article analyzes the relationship of solving environmental problems with the dynamics of demographic processes in industrialized regions. The attention is focused on the fact that social policy, development of social infrastructure, as well as its relevance depends on the demographic processes taking place at the regional level. Especially important to improve the legal framework for the development and disclosure of human potential, regional investment attractiveness, which will ultimately contribute to the stabilization of the demographic and socio-cultural situation both at the professional and daily levels. In this regard, attention is drawn to the fact that the disclosure of labor potential is impossible without taking into account the interests of the business community in educational technologies, in which, in the regional labor market, not only specialists with a narrow set of practical skills, but also with a wide range of fundamental training are in demand. The purpose of the development of the dominant regional segment of large business (coal companies, holdings, etc.) is the introduction of new technics, improvement of technologies, etc. in the field of mining and processing of minerals to level of competitive, environmentally «clean» product not only for foreign markets, but also for domestic consumers. The economic niche of small and medium-sized businesses are enterprises of social and infrastructure sectors of the economy. Related to this are issues of staffing, social and socio-cultural problems in the industrial development of the regions.

1 Introduction

The environmental situation in the Russian Federation in the today's modern social-economic situation can hardly be called prosperous. Regional specificity, especially in industrialized regions, is characterized by a number of serious problems, such as contamination of air, water, soil, radioactive contamination, etc. The Russian Federation has all possible sources of environmental pollution, such as electricity, ferrous and non-ferrous metallurgy, oil production and refining, chemical, gas and coal industries. Hence the emergence of various socio-economic risks and social tensions, the deleting of which is

*Corresponding author: kemerovo_text@mail.ru

associated with the resolution of the demographic problems associated with the Russian mentality [1], social-economic situation [2, 3], the solution of technological problems in the field of innovative development [4], as well as various methodological and organizational approaches to solving environmental problems [5].

2 Material and Method

In the framework of the study, the relationship of solving environmental problems with the dynamics of demographic processes in the industrialized regions is paid attention to the implementation of «cultural capital» (R. Solow) in terms of reducing social risks and overcoming socio-cultural contradictions (A. S. Akhiezer) in the framework of making compromise decisions to achieve regional socio-economic stability. Issues related to retraining of employees and transformation of their consciousness are considered within the framework of the system of transforming education (Bowd R., Gordon I., Elias D.), an important component of which is a critical rethinking of approaches to environmental consciousness. In the description of modern trends and principles of functioning of various aspects of the relationship of solving environmental problems with the dynamics of demographic processes in the industrialized regions used methods: rational and logical analysis, structural and functional approach, conceptual analysis, historical and logical comparison, generalization, method of hypotheses with their subsequent verification, as well as methods of correlation analysis.

3 Results and Discussion

Studies Yale Center for Environmental Law and Policy at Yale University periodically generate the Environmental Performance Index (last time 2018) as a combined indicator of the state's achievements in terms of the state of the environment and natural resources management. For Russia, it is only 63.79 points, which corresponds to 52 out of 180 places in the environmental ranking of States [6]. In various regions, in particular in the Kemerovo region, there are all the main components that have a negative impact on the environment industry. First of all, it is the dominant regional segment of large business (coal companies, holdings, etc.), a significant level of concentration of objects which is located in the South-Western and Western parts of the region. Infrastructure feature is the presence of high-class production hazards.

Air pollution adversely affects the ecological situation of the region as a whole, living organisms and vegetation in particular. The main source of pollution of the atmosphere with dust is the conduct of open development, sorting and transportation of coal. Gas changes in the composition of air occur mainly due to the receipt of harmful toxic impurities contained in the rock mass into the atmosphere during underground oxidation processes, mine fires and methane explosions. For the Siberian Federal district, a characteristic situation in 2017 was that «the share of the population living in unfavorable conditions from air pollution throughout the Federal district was 42%» [7]. Coal mining is associated with the transformation of vast areas of land, the destruction of the geological structure of the soil and leads to the complete disappearance of the fertile layer of the earth. Coal mining enterprises use hundreds of hectares of land for their work, and therefore as a result of their functioning, the relief and ecosystem of large areas of the region changes, forests are cut down, many natural landscapes are flooded or dried up. All this requires careful study of the regulatory framework in the field of environmental protection, adaptation of environmental standards to modern technological conditions of production development. First of all, it concerns the formation of ecological consciousness of the population and the creation of conditions for improving the "quality of life", contributing to the comprehensive development of human potential and stimulating people in their daily lives (management, labor, everyday, etc.) [8] to comply with the requirements of environmental norms and standards.

Adopted at the legislative level, environmental standards, first of all, have technological features of their application, which, in turn, implies a certain culture of their application. Depending on the subjects of application (public authorities, local governments, employers, employees, etc.), their General culture, the implementation of «cultural capital» in terms of critical conditions R. Solow. It is worth noting that the action of the «socio-cultural law» (A. S. Akhiezer) allows to predict and overcome social conflicts, because «the observance of the unity of social relations and culture is a vital imperative, which is the law of reproductive activity» [9, p. 482]. Effective overcoming of social contradictions allows to reduce the level of social tension in the society, which has arisen as a result of concrete practice of implementation of existing environmental standards. The General level of ecological culture as a whole contributes to the harmonization of the system «nature-society» in the framework of coevolution [10]. In our view, the reflection of the culture of application of environmental standards can be seen at the regional level.

One of the mechanisms for the removal of socio-cultural and social tensions is the adoption and implementation of the state program «Ecology, subsoil use and rational water use» for 2017-2021 [11]. Its objectives, in particular, include sustainable provision of the economy with mineral resources, and the task is «to reduce the overall anthropogenic impact on the environment by improving the environmental efficiency of the economy» and «conservation and restoration of biological diversity of the Kemerovo region». Attention is paid to the effective use and reproduction of mineral resources, as well as the effectiveness of public administration in the field of ecology and subsoil use.

From the point of view of changes in the regulatory framework,which contributes to changes in the environmental situation, it should be noted the amendments made in 2017 to the Federal law of 30.03.1999 № 52-FZ «On sanitary and epidemiological welfare of the population», establishing standards for the calculation and assessment of risks to human health. This allows the gradual introduction of a risk-based approach in the environmental safety management system. Medical institutions controlled by state structures are involved in this process. The result of their joint activities is the resolution «On the use of risk assessment methodology for quality environmental and public health management in the Russian Federation», etc. [7]

The situation around single-industry towns, their development and functioning deserves special attention, since most coal enterprises (mines, sections, etc.) are concentrated in them. They are the city-forming, and historical accumulation of social problems in the field of underfunding, and in some cases of «optimization» of medicine, education, culture, etc. leads to escalate the level of social tension. The negative role, in this case, belongs not only to representatives of local business structures, but also to municipalities as carriers of the state model of management. It is a question of implementation of strategy of minimization of expenses on reproduction of the human capital, the social transformations [12] occurring in the conditions of implementation of projects economically significant for the state, but exploiting for this purpose the available potential within the local labor market.

The objectives of the priority program «Integrated development of single-industry towns» aimed at the strategic development of single-industry towns in Russia until 2025 have not been achieved. The reason for this is not only its early completion (January 1, 2019), but, as noted by the Accounts Chamber of the Russian Federation, and the lack of «strategic national priorities» [13]. These include: the lack of a «diversified approach to state support of single-industry towns depending on their socio-economic and geographical characteristics», the loss of the most qualified personnel who are forced to leave due to the lack of prospects for changes in the field of social infrastructure policy, as these workers were the basis and prospects for the development of city-forming enterprises. This aspect is a significant brake on the implementation of the diversification program as one of the innovative mechanisms [14] of social-economic and socio-cultural development of single-profile territories at the regional level.

The current environmental situation requires the adoption of operational decisions that would be a compromise for all stakeholders (public authorities and management, business leaders, employees, the public, etc.). An important aspect, in this case, is the solution of the problem of adaptation of workers of industrial (coal) enterprises to meet the requirements of modern environmental standards, because it depends on the individual worker the final implementation of all regulations. Successful adaptation can be considered if the person is satisfied with the quality of the new social space and acquires a higher social status in an environment with changed conditions (claims).

Features of ecological consciousness are determined not only by the system of value orientations, but also differentiated depending on belonging to a certain social group. The content base of ecological consciousness depends on many factors, among which there are professional activities and functional orientation of consciousness. Transformation of consciousness of workers in the field of environmental consciousness takes place within the framework of the system of transforming training. It is aimed at the "perspective transformation" of the personality, which has three dimensions: psychological (changes in the understanding of oneself); value (revision of the belief system) and behavioral (changes in lifestyle) [15].

Transformative learning is the expansion of consciousness through the transformation of the basic picture of the world and the characteristics of their individuality, while transforming learning is facilitated by controlled processes, such as gaining access to the symbolic content of the subconscious and critical analysis of the hidden prerequisites [16]. An important component of transformative learning for the individual is to change his or her frame of mind by critically rethinking beliefs, and by consciously creating and implementing plans that will lead to new ways of defining worldview paradigms. This process is fundamentally rational and analytical [17]. All this, ultimately, contributes to the change in the criteria of economic behavior aimed at the effective use of the principles of sustainable development [18] and adaptation of environmental and economic indicators of industrial enterprises [19] in mining regions, as well as the formation of innovative socio-cultural practices in the field of ecology and environmental consciousness.

4 Conclusion

The cyclical development of the Kuzbass economy has a direct impact on the environment. Due to the decline in production in the mid-and late 90-ies of XX century in the region has improved the environmental situation. But the relative industrial and economic growth in the early 2000s again raised the issue of improving the conditions for the creation of environmental safety of the industrial region. Currently, the region is carrying out activities to improve the environmental condition, which indicates the desire for the realization and protection of human rights to a favorable environment and partial realization of the need for environmental safety [20, p. 8-9], which will ultimately contribute to the formation of environmental awareness and the introduction of various programs in enterprises for environmental training aimed at improving the overall environmental culture of the population engaged in specific environmentally hazardous areas of the economy.

References

1. V. Zolotukhin, A, Bel'kov, E. Stepantsova, M. Kozyreva, A. Tarasenko, E3S Web of Conf., **15**, 040115 (2017)
2. T. V. Kiseleva, V. G. Mikhailov, G. S. Mikhailov, IOP Conference Series: Earth and Environmental Science, **84**, 012044 (2017)
3. T. Gvozdkova, M. Tyulenev, S. Zhironkin, V. A. Trifonov, Yu. M. Osipov, IOP Conference Series: Earth and Environmental Science Current Problems and Solutions, **50:1,** 012010 (2017)
4. T. V. Galanina, M. I. Baumgarten, V. G. Mikhailov, T. G. Koroleva, G. S. Mikhailov, IOP Conference Series: Earth and Environmental Science, **50:1**, 012030 (2017)
5. D. Marasova, M. Ligatto, D. Cassiti, V. Zolotukhin, E3S Web of Conf, **105,** 04025 (2018)
6. *The Environmental Performance Index – a combined indicator of Yale Center for Environmental Law and Policy at Yale University* (Yale University, Yale, 2018)
7. *Report on the state and environmental protection of the Kemerovo region in 2017* (AKO, Kemerovo, 2017)
8. V. Zolotukhin, N. Zolotukhina, M. Yazevich, A. Rodionov, M. Kozyreva, E3S Web of Conferences, **21**, 04008 (2017)
9. A. S. Akhiezer, Socio-cultural dynamics of Russia, **1**, 804 (1997)
10. Gritskevich, E. F. Kazakov, A. V. Konovalov, International Conference on Research Paradigms Transformation in Social Sciences, **1,** 2357 (2017)
11. *About the approval of the state program of the Kemerovo region «Ecology, subsoil use and rational water use» for 2017 - 2021* (AKO, Kemerovo, 2017)
12. V. Zolotukhin, O. Zhukova, International Conference on Research Paradigms Transformation in Social Sciences, **1,** 1085 (2017)
13. *The problems of development of single-industry towns still need to be solved, but at a new level* (AKO, Kemerovo, 2019)
14. N. Egorova, N. Zaruba, T. Jurzina, V. Tumin, E3S Web of Conf, **105**, 04021 (2018)
15. Boyd, D. Robert, and J. Gordon Myers, International Journal of Lifelong Education, **7:4**, 261 (1988)
16. D. Elias, It's time to change our minds: An introduction to transformative learning. ReVision, **20:1** (1997)
17. Mezirow, J. Transformative Learning: Theory to Practice. New Directions for Adult and Continuing Education, **74**, 5 (1997)
18. S. Zhironkin, D. Khloptsov, N. Skrylnikova, I. Petinenko, O. Zhironkina, E3S Web Conf., **41**, 04010 (2018)
19. V. G. Mikhailov, T. V. Kiseleva, IOP Conference Series: Materials Science and Engineering, **11**, 012012 (2017)
20. N. A. Volkov, Special *report of the Commissioner for human rights in the Kemerovo region* (AKO, Kemerovo, 2017)

Expert Evaluation of the Risks in Coal-Mining Enterprises Based on Fuzzy Logic

Elena Raevskaya[1,*], *Alexander* Pimonov[1,2], and *Vladimir* Mihailov[1]

[1]T.F. Gorbachev Kuzbass State Technical University, 650000, Kemerovo, Russian Federation
[2]Institute of Economics and Industrial Engineering, Siberian Branch of the Russian Academy of Sciences, 630090, Novosibirsk, Russian Federation

Abstract: The article is about a complex approach based on the method of analyzing hierarchies and using of fuzzy logics for assessment of risks deals with innovations in coal mining industry. This approach allows evaluating both quantitative and qualitative risk indicators of compared alternative innovative projects, does not depend on a field of expertise, thus it makes possible to attract specialists with competencies in various fields of knowledge. Such kind of approach allows making a qualitative assessment of the situation on the basis of formalized logical conclusions, making decision making comfortable and accessible to any specialist. The proposed methodology is implemented as part of a decision support system.

1 Introduction

Currently, coal industry enterprises occupy one of the leading places in terms of the number of negative factors associated with production activities and affecting the environment. In 2017 coal-mining enterprises of the Kemerovo region emitted pollutants into the air with a total weight of over 915 thousand tons. It is 61.5% of all stationary sources in the region. The total volume of wastewater discharged by these enterprises amounts to almost 288 million m^3. The total amount of waste generated as a result of mining of all types of minerals for the period under review is 3.12 billion tons [1].

However, an analysis of the state of environmental protection measures in the regions of coal mining and processing shows [2] that at present most technical solutions and measures are aimed to neutralizing and eliminating the consequences of production activities, rather than eliminating the root causes and their sources. Reducing the negative impact of coal industry enterprises on the environment is carried out through partial restoration of the disturbed natural state, which is achieved through land reclamation, clarification and purification of wastewater, dust and gas recovery by heat and power plants and others. These measures are ineffective, because they do not solve the problems that cause negative phenomena, but only partially eliminate the consequences, which led to a sharp deterioration of the ecological situation in the regions and large economic costs.

Thus the question of the need for a preliminary assessment of the probability of the

* Corresponding author: pag_vt@kuzstu.ru

occurrence of a negative factor with the aim of its further exclusion becomes urgent. This, in turn, compels business leaders and authorities to pay special attention to the development and implementation of innovative processes in enterprises. However, innovation processes are accompanied by a high degree of risk associated with the implementation of the introduced innovation [3]. So, in modern conditions of tightening national environmental legislation and increasing the sensitivity of the main technical and economic indicators of an enterprise to its environmental and economic characteristics, the process of managing environmental and economic risks is particularly relevant. The infrastructure of coal-mining enterprises is based on the integrated functioning of structural units in various fields of activity.

Managers constantly face multidirectional tasks, decisions on which cannot be made using the same factors and models (comparison of enterprises by the level of investment attractiveness, evaluation of the effectiveness of mines, open-pit mines, coal processing plants, comparison of the competitiveness of coal grades and products of processing, comparison of the ecological state of the areas of mining operations, comparison of the technical level of mining machines and equipment, selection and evaluation of engineering and technical workers in the coal industry) [4].

Persons making decision on these tasks periodically face the need for an integrated assessment of the potential of innovative projects. When carrying out such an examination, a large number of functional indicators [5] (mining-geological, technological, technical, economic, ecological, social) must be taken into account, which may have a different nature - quantitative and qualitative. Thus, an important task is the formation of methods and algorithms for evaluating of ecological and economic risks of mining enterprises which allow taking into account the maximum possible number of factors associated with the implementation of innovation with minimal time, labor costs and a low proportion of subjectivity.

2 Management of ecological and economic risks

Many foreign and domestic scientists are engaged in research and management of ecological and economic risks. Thus, a common risk management algorithm was formed, consisting of the following mandatory steps:

1) identification of ecological and economic risks;
2) assessment of ecological and economic risks;
3) management of ecological and economic risks, including the development of directions for their reduction and elimination.

At the first stage, the main factors influencing the formation of environmental and economic risks and potential directions of possible losses from their implementation are determined.

The main task of the second stage is to determine the economic damage and the probability of occurrence of a risk event.

The complexity of this stage is the using of adequate methods of assessing economic damage and the likelihood of the implementation of environmental and economic risk, ensuring maximum reliability of the result.

Some authors detail the main stages of managing environmental and economic risks, highlighting in independent sections the definition of the structure of possible damage and graphical representation of the law of its distribution. In this case risk management is the formation of a prototype decision base, taking into account the control of risk exposure measures [5].

One possible option for investigating environmental and economic risks is the using of an approach consisting of four stages, where the first involves identifying the hazard, for assessing directly the environmental risk [6, 7].

Makhutov [8] differentiates strategic economic risks in the following way: local; object; local; regional; national; global; planetary. In the same paper, the authors calculate these risks in a classical way as the product of damage to the probability determined by the expert method. This approach should also be used for environmental and economic risks, which in many cases are transformed into traditional financial and economic risks.

An important problem of an adequate calculation of environmental and economic risks is the choice of a method for assessing the likelihood of an adverse event. In practice, the most common are three main methods [9]:

1) statistical - is based on the analysis of statistical data on similar events that occurred at similar sites in a given territory;
2) analytical - investigation of cause-effect relationships in the territorial production system, allowing to estimate the likelihood of risk occurrence as a complex phenomenon;
3) expert - is based on an assessment of the likelihood of adverse events by analyzing the results of a survey of experts.

For the most qualitative and accurate evaluation of the probability of occurrence of adverse events, it is necessary to use all methods simultaneously with checking the convergence of the results.

A number of specialists believe that in addition to the above methods, math simulation ensures greater reliability of the result obtained [9].

Some experts [10] interpret zones of environmental and economic risk as a normal distribution, analogous to depicting speculative risks in the financial sector, when any deviation from the "normal" situation is viewed as a risk event (profit, loss, gain, loss, etc.).

In accordance with this approach, the following differentiation of environmental and economic risks is presented with the possible probability of their occurrence:

1) is critical with probability 0,00 – 0,19;
2) dangerous with probability 0,20 – 0,36;
3) allowable with probability 0,37 – 0,62;
4) acceptable with probability 0,63 – 0,79;
5) background with probability 0,80 – 1,00.

Practical application of the theory of ecological and economic risks is reflected in the analysis of the company's environmental and economic indicators [11–12], when one of the simple options for assessing environmental and economic risks is to determine the relative weight of fees for excess environmental impact in the overall value of the fee. Analysis of the data of large chemical enterprises has shown that this value can reach 90%, characterizing the ineffectiveness of the conducted environmental policy and adversely affecting the company's technical, economic and financial indicators, reducing its profit. Another aspect of this problem is the inadequate rationing of the potential negative load on the environment, where compliance with standards is technologically impossible.

Estimation of environmental and economic risk can also be simplified by means of a risk factor (*KR*):

$$KR=ED\,/\,P \tag{1}$$

ED – economic damage from negative impact on the environment, ruble;

P – payment for negative impact on the environment, ruble. Areas of risk factor values are presented taking into account the considered risk zones:

1) neglectful – до 0,2;
2) acceptable 0,2 – 0,5;
3) critical 0,5 – 1,0;
4) unacceptable – upper 1,0.

3 Methodology for risk evaluation based on fuzzy logic

The analysis of the main approaches to the study and assessment of ecological and economic risks has shown that, despite the variety of scientific papers on this issue, the management algorithm must necessarily include elements such as identification, assessment and measures to reduce or eliminate risks. Another classic problem in the management of ecological and economic risks is the search for adequate methods for determining the likelihood of occurrence of a risk event, ensuring maximum reliability of the result obtained.

Thus, the analysis of the problem has shown that there is no universal formalized methodology which would allow a qualitative effective evaluation of environmental and economic risks.

So there is a need to find a method which would allow an unlimited number of indicators to be used in order to assess the likelihood of risk occurrence, and in addition would make it possible to make estimates on the basis of indicators having a different nature - quantitative and qualitative.

It Should be noted that for the evaluation of risk it is advisable to use not only numerical methods of data processing [13], but also to carry out a purely qualitative assessment of the situation on the basis of logical conclusions, presenting the obtained quantitative values of variables as some linguistic parameters [14], in combination with the method of analyzing the hierarchies of T. Saati [15], which allows you to rank the compared alternatives and conduct their pairwise comparison for each possible type of risk.

The advantages of using this approach include the possibility of using the expert's experience. As one of the tools for supporting decision-making in a condition of uncertainty, one can use the apparatus of the theory of fuzzy sets. The basis of the theory of fuzzy sets is the production of fuzzy logical inferences, that is, the derivation of conclusions in the form of a fuzzy set, which corresponds to the current input values using a fuzzy knowledge base and fuzzy operations [13–15].

The proposed methodology for a comparative expertise of risks and determining degree of importance in comparison with each other is based on the definition of linguistic variables and their values for each type of risk, the probability of which arises in the compared alternatives.

The technique using the hierarchy analysis method and the fuzzy logic apparatus consists of four stages.

Stage1. Preliminary examination of the compared alternatives for all possible types of risk. The purpose of the preliminary examination is to exclude the obviously unlikely risks on the selection criteria, so that at the next, estimated stage, the experts do not have to assess almost impossible risks.

Stage 2. Using of the method of analysis of hierarchies (MAI) T. Saati [14].

Stage 3.Combination the criteria as fuzzy sets.

Stage 4. Ranking of variants on the basis of fuzzy set-criteria, which correspond to Bellman-Zadeh scheme. When assessing indicators, experts set the lower estimates - pessimistic and the upper estimates – optimistic. Further processing of Fuzzy formulated opinions of experts is proposed to be carried out by dephasing of the obtained fuzzy partial estimates and their further processing in a dephased form [14]

Thus, the task of decision making under the given technique is as follows.

There is $\mathbf{V} = \{v1, v2, \ldots, vn\}$ – a set of alternatives that are subject to multicriteria analysis;

$\mathbf{C} = \{c1, c2, \ldots, cm\}$ – a set of quantitative and qualitative types of risks by which alternatives are evaluated;

$\mathbf{B} = \{b1, b2, \ldots, bk\}$ – competence of estimates k of experts conducting experts' assessment.

The problem is to arrange the elements of the set **V** by the criteria from the set **C**, taking into account the competence **B** of the expert estimates. For example, in the role of qualitative criteria for comparing alternatives, we will use the possibility of a mountain impact (K_1), the release of gas into the atmosphere (K_2), exogenous and endogenous fire hazard (K_3), pollution of the landscape (K_4); the formation of a depression funnel (K_5); traumatic hazard of personnel (K_6). In accordance with the above list of criteria, one can distinguish linguistic variables X_{ij}, where i is the number of the alternative; j is the type of risk.

We define the range for all linguistic variables as $U_i = [0, 1]$ for $i = 1, 2, 3, 4$; $j = 1, 2, 3, 4, 5, 6$.

The sets of values of linguistic variables (term sets) for the third criteria are formulated as follows:

T (X_{i3}) = increased fire hazard + fire and explosion hazard + typical fire hazard + typical moderate fire hazard + reduced fire hazard.

Further, the following stages of the examination are expected:

1) fuzzy variables of each term set are used as qualitative estimates of alternatives according to one of the criteria;
2) to obtain evaluation of the alternatives under review, the method of expert evaluation is applied and survey of the expert group is conducted;
3) the compatibility functions of linguistic variables are formed as fuzzy subsets $U_i = [0, 1]$ of the following form, for example, for type of risk K_3:

$$M(\text{increased fire hazard}) = \int_{0}^{0,25} (1-4u)/u \quad (2)$$

$$M(\text{fire and explosion hazard}) = \int_{0}^{0,25} (4u)/u + \int_{0,25}^{0,5} (-4u+2)/u \quad (3)$$

$$M(\text{typical fire hazard}) = \int_{0,25}^{0,5} (4u-1)/u + \int_{0,5}^{0,75} (-4u+4)/u \quad (4)$$

$$M(\text{typical moderate fire hazard}) = \int_{0,5}^{0,75} (4u-2)/u + \int_{0,75}^{1} (-4u+4)/u \quad (5)$$

$$M(\text{reduced fire hazard}) = \int_{0,75}^{1} (4u-3)/u \quad (6)$$

4) using average estimates of alternatives for each of the criteria, a matrix of average estimates is formed;

5) in view of the fact that the criteria have different degrees of importance, their coefficients of relative importance are determined using the hierarchy analysis method, then

the Eigen values of the matrix are calculated and the coefficients of the relative importance of the criteria are calculated;

6) a fuzzy set of estimates of the criteria given by experts is modified by raising to a power corresponding to the coefficient of relative importance of the criterion;

7) calculation is made of set D needed for selection of the preferred alternative based on the Savage criterion (max (min)). To do this, select the maximum value of the indicator from the minimum for each alternative. The alternative thus obtained is given preference in selecting a project for implementation [15].

Thus, the decision-making become formalized and eliminate subjectivity of experts.

4 Conclusion

Using of proposed methodology forms the decision-making procedure for assessing the possibility of occurrence of ecological and economic risks as a clear and formalized, excluding possible subjectivity, taking into account the views of various experts.

The developed methodology is as part of an decision support system which allows taking into account criteria of risk which can have a different nature - quantitative and qualitative, using the method of comparisons of T. Saati.

The developed software automates the decision-making process in tasks which are difficult to quantify to any numerical analysis, such as competitive selection of innovations for their subsequent implementation, makes possible a preliminary assessment of the possible benefits and risks associated with the implementation of the evaluated project, allows minimize the influence of the human factor, thereby increasing the objectivity of the decision. The practical importance of this paper is to increase the efficiency of ecological and economic analysis of the enterprise, aimed at supporting the adoption of environmentally-safe management decisions.

The proposed methodology is focused on a wide range of people, due to the professional specifics of the selection-related tasks, for example: managers, sociologists, politicians, researchers, jury members, etc. Due to a mechanism for taking into account the opinion of several experts, it is possible to use experience of specialists with competences in various fields knowledge, which will make the result of decision making more objective and qualitative.

The use of the approach can reduce intellectual and time costs during the deciding making on the implementation of innovations.

References

1. G.S. Trushina. Ugol', **10**, 98 (2018)
2. Y.A. Fridman, G.N. Rechko, E.Y. Loginova, Journal of Mining Science, **51:5**, 924 (2015)
3. A. Pimonov, E. Raevskaya, T.Sarapulova, E3S Web Of Conferences, **15**, 01021 (2017)
4. Y.A. Friedman, G.N. Rechko, A.G. Pimonov., Regional Research of Russia, **2:3**, 206 (2012)
5. S.G. Tyaglov, T.V. Kushnarenko, A.A. Khokhlov, M.A. Qeropyan, European Research Studies Journal, **20:1**, 198 (2017)
6. J.R. Olsen, J.H. Lambert, Y.Y. Haimes, Risk Analysis, **18:4**, 497 (1998)
7. T.M. Tikhomirova, A.G. Sukiasyan, Economy of Region, **4**, 164 (2014)
8. N.A. Makhutov, M.M. Gadenin, Ecology and Industry of Russia, **20:1**, 47 (2016)

9. N. Tikhomirov, T. Tikhomirova, E. Khamitov, V. Ponomarev, European Research Studies Journal, **20:2**, 321 (2017)
10. A.D. Murzin, T.Y. Anopchenko, Asian Social Science, **10:15**, 249(2014)
11. I.A. Ageev, V.N.Burkov, V.I.Zinchenko, T.V. Kiseleva, Automation and Remote Control, **66:6**, 995 (2005)
12. V.G. Mikhailov, N.N. Golofastova, T.V. Galanina, T.G. Koroleva, Ya.S. Mikhailova, IOP Conf. Ser.: Earth Environ. Sci., **50:1**, 012038 (2017)
13. L.A. Zadeh, Information and Control, Fuzzy sets, **8:3**, 338 (1965)
14. L.A. Zadeh, Information Sciences, Similarity relations and fuzzy orderings, **3:2**, 177 (1971)
15. W. Yoram, T.L. Saaty, Management Science, **26:7**, 641 (1980)

The Unity of the Trajectory of Sustainable Development of the Mining Region and Ensuring its Environmental Safety

Oksana Sheveleva[1,*], *Ekaterina* Slesarenko[1], *Natalia* Kudrevatykh[1] and *Tatiana* Mamzina[1]

[1]T.F. Gorbachev Kuzbass State Technical University, 650000 28 Vesennya st., Kemerovo, Russian Federation

Abstract. Environmental safety can be considered as an integral and one of the most important parts of the concept of sustainable development of a coal-mining region, since the environmental risks occupy one of the leading places among the risks of backbone enterprises of such regions due to the particularities of the coal industry and its impact on the environment. In the article, environmental safety is assessed through indicators of the quality of atmospheric air and water, as well as investments in the restoration and protection of the environment; discrepancies were revealed between the amount of financial investments allocated to nature restoration measures and the results of the anthropogenic impact of coal mining enterprises on the biosphere; emphasis was placed on the need to improve the efficiency of funds allocated for environmental protection and rational use of natural resources. Considered priority areas of environmental activities implemented in the framework of the environmental policy of JSC SUEK-Kuzbass. The effectiveness of the system of environmental remediation measures developed by the Company, their adequate funding, makes it possible to place the Company among the industry leaders in environmental and social responsibility for its activities.

1 Introduction

Currently, the main factor in the sustainable development of mining regions is the change in the quality parameters of production processes based on the introduction of advanced, environmentally friendly technologies designed to minimize the harm done to the environment, man and the animal world.

The Kuznetsk coal basin, located in Kemerovo region, is one of the largest coal deposits in the world, so the problem of the environmental safety state is extremely important for Kuzbass. The region possesses significant mineral deposits and a variety of natural resources. Several industries are developed on the territory of Kemerovo region - coal mining, metallurgy, mechanical engineering and power engineering. However, Kuzbass specializes primarily in coal mining - there are more than 380 mines and open cuts in the region.

* Corresponding author: shob.fk@kuzstu.ru

The most important centers of the coal industry are the cities of Prokopyevsk, Mezhdurechensk, Belovo, Berezovsky, Kemerovo, Novokuznetsk, Leninsk-Kuznetsky, Novokuznetsky and Prokopevsky districts. There are more than two hundred mines and open cuts on the territory of the above cities and districts. There are two large coal basins - Kuznetsk coal and Kansk-Achinsk lignite - in the region. The share of Kemerovo region accounts for mining more than 75% of coking coal.

2 Research methodology

The iinformation about the state of the natural environment (changes in weather, climate, the state of the animal and plant world around, etc.) has been actively used in human life since ancient times. Recently, changes in the environment are increasingly occurring under the influence of the human factor. The study of anthropogenic changes is the basis of environmental monitoring, which is an information system for analyzing and predicting the state of the environment and the management actions developed on this basis that are necessary to resolve issues in the field of environmental safety.

In Russia, considerable attention is paid to the environmental safety at the federal level, a number of legislative and regulatory acts have been developed, including: Federal Law "On Environmental Protection" No. 7-FZ dated January 10, 2002 (Chapter X. State Environmental Monitoring).

The regional level deserves special attention, since it is here where we can take into account the factors characteristic of the economies of specific territorial entities within the country [1-3], and therefore at this level specific problems can be considered and special programs for studying the environment state and the impact of anthropogenic factors on it can be developed.

3 Results and discussion

In order to identify the level of severity of environmental problems in Kuzbass, the comparative analysis of indicators of environmental monitoring of the quality of atmospheric air and water resources of Kemerovo region and the Siberian Federal District was conducted (Table 1). The indicators “Emissions of pollutants into the atmospheric air from stationary sources” and “Proportion of captured and treated atmospheric pollutants in the total amount of pollutants from stationary sources” are included in the State Program of the Russian Federation “Environmental Protection” for 2012-2020.

Table 1. Indicators of environmental monitoring of the quality of atmospheric air and water.

Indicators	Years							
	2010	2011	2012	2013	2014	2015	2016	2017
1. Emissions of pollutants into the air from stationary sources, kilotons								
Siberian Federal District	5868	5919	6017	5816	5570	5688	5605	5764
Kemerovo region	1411	1390	1360	1356	1332	1344	1349	1488
Share of Kemerovo region in Siberian Federal District, %	24.1	23.5	22.6	23.3	23.9	23.6	24.1	25.8
2. Captured air pollutants from stationary sources, kilotons								
Siberian Federal District	21998	23420	20896	19459	19469	18497	17401	18012
Kemerovo region	4823	6321	5334	4512	4636	4683	4234	4422

3. The proportion of captured and treated air polluting substances in the total amount of pollutants from stationary sources, %								
Siberian Federal District	78.9	79.8	77.6	77.0	77.8	76.5	75.6	75.8
Kemerovo region	77.4	82.0	79.7	76.9	77.7	77.7	75.8	74.8
4. Discharge of polluted wastewater to surface water bodies, million cubic meters								
Siberian Federal District	2218	2142	2077	1941	1750	1696	1654	1640
Kemerovo region	700	661	572	598	478	462	444	433
Share of Kemerovo region in Siberian Federal District, %	31.6	30.9	27.5	30.8	27.3	27.2	26.8	26.4
5. The volume of recycling and successively used water, million cubic meters								
Siberian Federal District	15739	16093	17052	16288	15997	16504	15978	1587 0
Kemerovo region	4524	4759	5114	5043	4766	4895	4891	4895
Share of Kemerovo region in Siberian Federal District, %	28,7	29,6	30,0	30,1	29,8	29,7	30,6	30,8

The amount of emissions of pollutants into the air from stationary sources in the Siberian Federal District in 2010–2017 varied from 5.5 to 6 thousand tons, and in the Kemerovo Region - from 1.3 to 1.5 thousand tons. Thus, the share of emissions of pollutants into the atmospheric air in Kuzbass accounted for 22.6-25.8%, i.e. almost a quarter of all pollutant emissions in the district. Moreover, the highest concentration of emissions was noted in 2017 (1,488 thousand tons), which led to an increase in the share of Kuzbass in the overall structure of emissions of the Siberian Federal District to 25.8%. According to this indicator, the Kemerovo region hold the second place (behind only the Krasnoyarsk Territory) among the 12 regions of the Siberian Federal District, which indicates the severity of this problem for our region. In 2017 compared to 2010, the amount of pollutant emissions into the atmospheric air in Kuzbass increased by 5.5% (from 1,411 thousand tons to 1,488 thousand tons), while the similar indicator in the District decreased by 1.8%. At the same time, the share of coal production in Kuzbass accounts for 61.5% of the total amount of pollutant emissions into the atmospheric air (915 thousand tons), i.e. the coal mining industry is the main air pollutant in the region.

According to the indicator "The proportion of captured and treated air polluting substances in the total amount of pollutants from stationary sources," the Kemerovo Region is in 7th place among 12 regions of the Siberian Federal District. The percentage of captured and treated atmospheric pollutants in the total amount of pollutants in Kuzbass is comparable to that in the District. In 2017, there is a slight lag in the indicator for the Kuzbass compared to the District data - by 1%, which is not critical because the situation changes from year to year, with a slight change in the direction of increase and decrease.

However, the situation looks quite different if we consider it in terms of absolute indicators, namely: 25.2% of non-treated air pollutants in 2017 in Kuzbass amount to 1,488 thousand tons of emissions per one region - Kemerovo region, while 24.2% of non-disposed substances in the District is 5764 thousand tons, which is 480 thousand tons per one region of the District (taking into account that there are 12 regions in the Siberian Federal District), i.e. emissions in Kuzbass are 3 times higher than the average in other regions of the Siberian Federal District.

The situation of polluted wastewater discharge into surface water bodies is being improved in the both in the Siberian Federal Distrcit and Kuzbass. Thus, in the Siberian Federal District, discharges for 2010–2017 decreased by 26.1%, and in the Kemerovo region - by 38.2%. However, despite the rapid decline in discharges in the region compared to the District, the Kuzbass and in 2017 ranks second in this indicator among the 12 regions of the Siberian Federal District, and it accounts for more than a quarter of the total polluted wastewater discharge in the District.

Table 2. Composition and structure of investments in fixed assets, aimed at environmental protection and ration use of natural resources.

Indicators	*Years*								
	2010	*2011*	*2012*	*2013*	*2014*	*2015*	*2016*	*2017*	*2018*
Investments in fixed assets, including	714.9	1161.6	1713.1	1404.4	2833.1	2058.1	1668.9	3150.1	2954.2
1) on protection and rational use of water resources	251.2	556.0	1459.1	1143.2	2025.7	1107.0	1255.5	2651.7	1847.3
- the same as a percentage of investments in fixed assets	35.1	47.9	85.2	81.4	71.5	53.8	75.2	84.2	62.5
2) on protection of atmospheric air	292.5	506.4	105.7	144.6	67.6	337.4	220.1	385.3	584.5
- the same as a percentage of investments in fixed assets	40.9	43.6	6.2	10.3	2.4	16.4	13.2	12.2	19.8
3) to other activities (including protection and rational use of land)	171.2	99.2	148.3	116.6	739.8	613.7	193.3	113.1	522.4
- the same as a percentage of investments in fixed assets	24.0	8.5	8.6	8.3	26.1	29.8	11.6	3.6	17.7

Within 2010-2017, the amount of investment in fixed assets aimed at environmental protection and rational use of natural resources in the Kemerovo region increased by 4.1

times. The largest share of investments throughout the analyzed period (except for 2010) is accounted for by investments in the protection and rational use of water resources (in 2017 it is 62.5% of the total investment in fixed assets). The amount of investments in this environmental protection area over the eight analyzed years increased by 1,596.1 million rubles, or 7.4 times.

Return on such investments was expressed:

- in reducing the polluted wastewater discharge into surface water bodies by 38% or 267 million cubic meters;
- in the growth of savings in fresh water intake through the use of recycling and reverse water supply systems, which reflects the indicator of the volume of recycling and successively used water - there is a rapid growth in the Kuzbass (an increase of 8.2%) compared to the increase in this indicator for the SFD only 0.8% (Table 1).

Return on investments in fixed assets for air protection was less productive - with an increase in the amount of investments by 292 million rubles, or almost 2 times, emissions of pollutants into the air from stationary sources increased, and the proportion of captured and treated air pollutants decreased. It is quite possible to recognize such a situation as logical from the point of view of the formation of the structure of investments in fixed assets, since the share of investments in air protection in the total amount of investments decreased by 21.1%.

When analyzing the ratio of investments (table 2) and commissioning of facilities for environmental protection and rational use of natural resources (table 3), there is no obvious correlation between these indicators. So, the amount of investments in 2010 being 251.2 million rubles; the capacities of the commissioned wastewater treatment facilities were only 0.7 thousand cubic meters per day, and a year later, with the amount of investments exceeding only 2.2 times, the capacities of the commissioned facilities were 220 times more!

Table 3. Dynamics of commissioning of facilities for environmental protection and rational use of natural resources.

Indicators	Years								
	2010	**2011**	**2012**	**2013**	**2014**	**2015**	**2016**	**2017**	**2018**
Wastewater treatment facilities, thousand cubic meters per day	0.7	154.5	34.8	110	69.9	124.4	22.9	94.0	189.8
Plants for trapping and treatment of harmful substances from waste gases, thousand cubic meters per hour	600	1036	471.7	322	-	75.7	27.6	16.0	30.0

The situation of efficiency of investments in the protection of atmospheric air the looks even more ambiguous - the amount of investment is increasing annually, while reducing the commissioning of facilities for capture and treatment of harmful substances from waste gases. At the same time, according to statistics, the type of economic activity "Mining" is the main source of emissions of pollutants into the atmosphere. The emissions figures are comparable with two other activities - “polluting sources” (“Manufacturing industries” and “Providing electric energy, gas and steam”), but the emissions for these two types of activity even in total terms are 2 times lower. As for all other types of economic activity, the exceedance of emissions in mining is hundreds of times higher! In this regard, it is completely natural that the investment in environmental protection measures in mining should be maximal.

However, according to statistics, the efficiency of disposal of harmful substances in the mining industry of Kuzbass is quite low - the proportion of captured and treated harmful substances from all sources of pollution in mining ranges from 43% to 50% in 2013-2017 years, i.e. more than half of the harmful substances remain in the atmospheric air, adversely affecting the quality of life of people, an increase in morbidity and mortality, and a decline in the birth rate.

Table 4. Dynamics of current (operational) environmental costs.

Indicators	**Years**					
	2012	**2013**	**2014**	**2015**	**2016**	**2017**
Total, mln. Rub., including	5708	7579	10224	6858	7250	9688
1. air protection and climate change mitigation, mln. Rub.	1764	2097	3240	1920	2114	3614
- the same as a percentage of current environmental costs	30.9	27.7	31.7	28.0	29.2	37.3
2. wastewater treatment, mln. Rub.	2599	4129	5314	3543	4292	5051
- the same as a percentage of current environmental costs	45.5	54.5	52.0	51.7	59.2	52.1
3. other activities in the field of environmental protection, mln. Rub.	1345	1353	1670	1395	844	1023
- the same as a percentage of current environmental costs	23.6	17.8	16.3	20.3	11.6	10.6

In 2012-2017, in Kuzbass, an increase in the current (operational) costs of environmental protection was by 3,980 million rubles or by 69.7%, including rising costs for air protection and climate change mitigation, amounted to 1,850 million rubles (or 2 times), wastewater treatment - 2452 million rubles (94.3%), waste management - 251 million rubles (51.4%). The remaining costs decreased to a critical level, for example, over the past five years, the cost of ensuring the radiation safety of the environment decreased by 17 times, and in 2015–2017 the costs of research and development activities to reduce negative anthropogenic environmental impacts, and biodiversity conservation and protection of natural territories.

The largest share in the structure of current (operational) expenditures on environmental protection in Kemerovo region over the past five years was attributed to wastewater treatment costs. So, in 2017, the share of these costs amounted to 52% of the total amount.

Thus, for the regions with the predominant location of the coal mining industry (including Kuzbass), the following main environmental problems are characteristic:

- topsoil disturbance, as well as its contamination by waste generated by the operations of mines and open cuts;
- negative impact on water bodies as a result of discharges of pollutants into them, as well as washing off surface pollution from the territories of mines, open cuts and dumps;
- plant cover depletion due to open cut mining, often causing complete destruction of soil and vegetation;
- increasing emissions of pollutants into the air from stationary sources, and reducing the proportion of captured and treated air pollutants;
- exacerbation of the sanitary and epidemiological situation, causing a deterioration in the health of the population, an increase in morbidity and mortality.

All these problems are often conditioned by the lack of incentives for the environmental activities of coal enterprises. It is more profitable for the economic entities to pay for the negative impact on the environment than to carry out reconstruction and modernization of production in order to reduce the harmful effects.

The experience of coal industry enterprises that are leaders in social accountability and corporate sustainability and implement a ramified and ambitious environmental policy, deserves special attention in this regard.

One of these companies that have been mining coal in Kuzbass since 2001 is JSC "SUEK". Coal is mined in 17 mines and 12 mines, which are located in seven regions of Russia. From 2005 to 2017, coal production increased by 39%. The volume of "SUEK" coal reserves is 5.61 billion tons - according to this indicator, "SUEK" ranks fifth in the world. SUEK's strategy provides for sustainable long-term competitiveness based on an integrated chain of production, transportation and marketing of coal, and provides for strategically important areas, including a special emphasis on sustainable development, including investment in environmental, public and social projects to improve well-being of the regions in which it operates.

In the Kemerovo region, JSC "SUEK-Kuzbass" mines coal in 2 open cuts and 9 mines. In order to improve the quality of coal, it is processed in 4 concentration plants in the region. The activities of JSC "SUEK-Kuzbass" are related to various factors that have an impact on the environment (Table 5), and since exceeding the permissible levels of impact on natural resources is fraught with fees for pollution in excess of the established limits, administrative liability, or suspension of activities, then to reduce environmental risks an environmental policy has been developed in the enterprise to reduce the impact of these factors.

Table 5. Lines of environmental policy of JSC "SUEK-KUZBASS".

Factors affecting the environment	**Types of monitoring**	**Monitored Indicators**	**Priority areas of environmental activities**
air pollutant emissions	monitoring of air pollutant emission sources	air quality of the working area, atmospheric air at the industrial site and in the sanitary protection zone	gradual reduction of emissions of pollutants (above all, greenhouse gases and solids) into the atmosphere
discharge of pollutants into surface water bodies	monitoring of surface water bodies above and below the sewage discharge point	the quality of wastewater discharged from technological facilities and entering the treatment plant, treated wastewater, surface water	gradual reduction of discharges of polluted wastewater into water bodies
soil disturbance in the process of mining	waste disposal sites monitoring	waste disposal sites	arrangement of waste disposal sites to reduce the man-made impact on the environment

Following on the principles of the environmental policy, JSC "SUEK-Kuzbass" introduced an environmental management system certified in 2009 for compliance with the requirements of the international standard ISO 14001: 2004. Environmental indicators are monitored by accredited sanitary and hygienic and sanitary and industrial laboratories. The company has defined goals and objectives for meeting the standards for discharges and emissions of pollutants and mined-land reclamation; hazards were identified, environmental risks were estimated, measures aimed at minimizing these risks were developed, i.e.

comprehensive and diverse work is underway to reduce the negative impact on the environment in the region.

However, despite fairly close attention to the issues of ensuring environmental safety at the federal level, the problem of the anthropogenic impact of industrial enterprises on the ecosystem, as well as improving the efficiency of the implemented environmental protection and recovery measures, remains one of the most pressing. Thus, a number of scientists [4, 5] recognize the threat of low social accountability and corporate sustainability as one of the most acute in addressing issues of improving environmental safety. The problem of accumulated waste from industrial enterprises, as well as their storage and disposal is widely investigated in the works of such authors as M.A. Gasanov, K.A. Kolotov [6], A.G. Koryakov, M.V. Kulikov [7], M. Kwiatkowski [8]. The relationship between environmental protection and investment in innovation processes is presented in the works of C. Barra, G. Bimonte, L. Senatore [9]. The influence of coal mining enterprises on the level and quality of life is described in the works of K. Mustafina [10], V.A. Ermolaev, A.V. Selyukov [11] and many other scientists.

At the same time, the successful global and regional experience in improving the environmental safety of industrialized regions [12, 13] provides a rich set of techniques and methods for their implementation.

4 Conclusion

Despite the presence of coal mining companies such as SUEK, a rather impressive number of mining companies use dirty technologies, without considering the most efficient methods for the conservation of mined-out rock masses and land reclamation. Rent-oriented behavior does not allow their owners to spend on these goals more than is required by law. In such conditions, the presence of even 20% of coal mining enterprises in the industrialized region, focused on reducing the anthropogenic impact of their activities, is unlikely to be able to fundamentally resolve the situation. The following systematic and comprehensive works in this direction are required:

1) increase of standard requirements for already operating coal mining enterprises in terms of investments in environmental innovations;
2) environmental certification of production of all enterprises of the coal industry taking into account the requirements of international standards;
3) development of a system of benefits for enterprises implementing measures to ensure the environmental safety;
4) use of higher tax rates for the production of environmentally hazardous products;
5) extended application of environmental insurance;
6) tightening the rules for coal mining enterprises when issuing mining licenses, in particular, reducing the corruption component in this aspect. There are frequent situations when regulatory authorities warn coal companies in advance about inspections, measurements of harmful emissions are carried out within a specified time, so violations, as a rule, are not detected;
7) changing the attitude of owners to environmental problems, promoting their social accountability and corporate sustainability;
8) introduction as a mandatory condition for the operation of coal mining enterprises of the gradual transfer of their production facilities to resource-saving, waste-free technological solutions, which allow minimizing the release of harmful substances into the environment.

References

1. M. Cehlár, L. Mihok, Theoretical and Empirical Researches in Urban Management, **8:4,** 60-72 (2013)
2. S. Zhironkin, D. Khloptsov, N. Skrylnikova, I. Petinenko, O. Zhironkina, E3S Web Conf., **41**, 04010 (2018)
3. S. Zhironkin, S. Demchenko, G. Kayachev, M. Ryzhkova, O. Zhironkina, E3S Web Conf., **41**, 04011 (2018)
4. O.V. Zonova, E.V. Nekhoda, Coal in the 21st Century: Mining, Processing and Safety, **1,** 125 (2016)
5. E. Ganebnykh, T. Burtseva, A. Petuhova, A. Mottaeva, E3S Web Conf., **91**, 08035, (2019)
6. M.A. Gasanov, K.A. Kolotov, K.A. Demidenko, E.A. Podgornaya, O.V. Kadnikova, IOP Conf. Ser.: Earth Environ. Sci., **50:1**, 012025 (2017)
7. A.G. Koryakov, M.V. Kulikov, Economics and Innovation Management, **4**, 16-26 (2018) DOI: 10.26730/2587-5574-2018-4-16-26
8. M. Kwiatkowski, Archives of Waste Management and Environmental Protection, **12:4**, 40 (2010)
9. C. Barra, G. Bimonte, L. Senatore, *Innovation Processes and Environmental Safety* (IEPA, Rome, 2018)
10. K. Mustafina, Alma Tourism, **5**, 109 (2012)
11. V.A. Ermolaev, A.V. Selyukov, Journal of Mining and Geotechnical Engineering, 2(5), 42-55 (2019). DOI: 10.26730/2618-7434-2019-2-42-55
12. R. Chuturkova, A. Simeonova, J. Bekyarova, N. Ruseva, V. Yaneva, Journal of Environmental Protection and Ecology, **12:3**, 805 (2011)
13. Nurongguli Tuoheti, Maimaitiaili Abulaiti, Zafar Ahmed, Namaitijiang Maimaiti, Journal of Politics and Law, **6:3**, 160 (2013)

On the Issue of Food Security of Coal Mining Region

Piotr Kosinskiy[1], *Natalia* Zaruba[1], *Natalia* Egorova[1], *Aleksey* Kharitonov[1]

[1]T.F. Gorbachev Kuzbass State Technical University, 28 Vesennyaya, Kemerovo, 650000

Abstract. The problems of food security of Kemerovo region are studied in the article. It was found that agribusiness and its performance are significantly affected by the pronounced industry orientation of the region, in which coal mining is intensively developed, which is accompanied by the annual land retirement. Assessing the food security of the region in terms of balanced nutrition in terms of calories, protein, fats and carbohydrates in food, one could conclude that they are not consistent with the vital needs of people. The imbalance of nutrition of the population affects its health, which, according to epidemiologists, is commensurate with genetic, active chemical or infectious negative effects on humans. To address food security of the region, a forecast of food consumption by the population was developed taking into account scientifically-based nutrition standards, taking into account: the level of money income and purchasing power of the population, effective demand for agricultural products; production capacity of agricultural industry and production facilities. It is proposed to apply a synergistic approach based on the creation of an agri-food cluster.

1 Introduction

Kemerovo Region is a coal mining region dominated by coal industry, including numerous coal open-pits. In 2017, the region ranks third in the Siberian Federal District in terms of coal production, in the amount of 241.5 million tons, 60 percent of which is accounted for by open-pit mining. Coal mining enterprises occupy large areas of land and, along with a negative impact on the environment, polluting the air and water resources, contribute to land retirement, including rock dumping on them. High-intensity coal mining in the region, the consequences of a high urban saturation are accompanied by signs of land degradation, including its fertility, a decrease in the population of animals, and therefore a loss of renewable natural resources.

Coal mining in the region is increasing annually, which entails an increase in retirement, as a rule, of highly fertile land. According to expert estimates, an average of 10 hectares of land is required to be disturbed to produce one million tons of coal by open-pit mining. In conditions of unfavorable habitat, along with other measures to ensure the normal living conditions of the population, an important factor is its balanced diet.

However, in recent years, the region does not provide itself with food from its own sources. The emphasis in the nutrition of the population is shifting towards the consumption of bread and bread products, potatoes, which increases the content of carbohydrates in their diets and contributes to the imbalance of nutrition, which affects health.

We emphasize that the food supply and food security of the region are the main factors determining social peace in the country and its specific regions, and the guaranteed availability and affordability of food for the entire population of the country ensures an active and healthy life. Without a systematic scientific state approach to addressing the issues of food security of the population, without developing solid conceptual solutions in the field of food security, the absence of solution for this problem can lead to irreversible processes. Food security has become the most acute of many problems that have arisen during the transition of the country to the market economy, affecting every inhabitant.

The experience of foreign countries and companies shows that there is a change in the development paradigm aimed at approaching food security. In some countries, urban food strategies were developed to create a healthy and prosperous community through the reengineering of a regional food and agricultural system. For example, in Canada, urban food strategies are developed by interdisciplinary teams, local governments, nonprofit organizations, and universities to provide a wide range of food and rural planning. In such teams there are farm logistics expert, farmers, retailers, and even site creators [2, 12].

2 Materials and Methods

The level of food security is directly dependent on the development of the agricultural industry. In the conditions of budgetary shortfall, the development of the agricultural sector can be promoted by a steady attraction of investments in food production, the development of interregional cooperation of integration ties. Food self-sufficiency is the basis of food security, which determines the degree of dependence on external food supplies. The agricultural activities, its production and economic performance are influenced by exchange processes in the external environment, represented by participants in foreign and domestic markets and including material, energy, labor, information and other resources.

Trends in the development of agriculture in the region indicate high prices for resources and the disparity between prices for industrial and agricultural products. It is the prices of resources that currently determine the competitiveness of agriculture, and entry into WTO can disrupt metabolic processes and, as a consequence, lead to the loss of stability of agriculture functioning [6, 7, 10].

Food provision of the population is an economic relationship in society that arises when members of society are provided with food that meets scientifically-based standards. It is entrusted to the state, which is obliged to create conditions that guarantee the stable and efficient use of food stocks.

Food security in the region should be considered as a multidimensional phenomenon, characterized by a combination of economic, social and political aspects. Food provision is characterized by several forms of food consumption, which depend on the average daily intake of an individual: the extreme forms of food providing problem are *chronic hunger*; problems arising from drought, flooding and other unforeseen events cause *epidemic hunger*; problems caused by non-compliance of food consumed with scientifically-based nutritional standards, including the caloric content of food [3, 8, 11].

The problem of food supply of the population is especially acute in the regions with intensive development of coal mining and metallurgical industries, whose enterprises are located in rural areas. This circumstance entails not only land retirement, but their disturbance in the areas bordering on industrial enterprises, as well as the outflow of highly professional specialists from agricultural enterprises. At the same time, agriculture is the main source of food supply in the region.

Kemerovo region is a region of this type. Agricultural land as of 01.01.2018, occupies 27.3 percent in the land structure; industry, transport, communications and other special purpose industries account for 1.7 percent; forest land - 55.9 percent; settlements occupy 4.1 percent; protected areas and objects - 8.5 percent. The region is characterized by a high population density of 28.39 people per square kilometer. In the Siberian Federal District this rate is 3.75, in Russia - 8.56.

The agri-business and its performance is significantly affected by the pronounced industrial orientation of the region, in which coal mining is intensively developed, which is accompanied by the annual land retirement. As of 01.01.2018, the coal enterprises employed 79.1 thousand hectares, which are irretrievably retired

Agriculture in the region occupies a small share in GRP, which varies from 2.8 percent to 4.4 percent in recent years. This is due to the fact that the region's economy has an industrial orientation. In 2017, the volume of agricultural production, including agricultural companies, individual entrepreneurs and households increased by 2013 to 31.8 percent and amounted to 58.5 billion rubles in current prices. The technical equipment of agriculture in the region remains wanting. In 2013-2017, the tractor fleet decreased by 19.4 percent (by 386 units), combine - by 21.4 percent (by 111 units). Agricultural enterprises write off annually 5-20 percent of agricultural machinery.

As a result, the load on the tractor and combine fleet increases. For example, in 2017 there were 468 hectares of arable land per tractor, which is 75 hectares more than in 2013 per tractor; the load on the grain harvester increased by 24 hectares over the same period. This fact entails a violation of agrotechnical terms of cultivation of agricultural crops, which increases the already high risk for the agricultural sector.

We emphasize that the share of the economically active population of the region employed in agriculture in 2017 was 32.9 thousand people (2.7 percent). The forecast of food security through 2020, calculated on the basis of rational consumption standards, predicts the following volumes of production will be reached: bread and bread products - 277 thousand tons, meat and meat products - 198 thousand tons, milk and dairy products - 898 thousand tons, 687 million eggs, vegetables and potatoes are 370 and 264 thousand tons, respectively [4].

The basis of the forecast of consumption of food by the population is based on factors related to: the level of monetary incomes and purchasing power of the population, effective demand for agricultural products; production capacity of agricultural industry and processing capacity in the region; dynamics of prices for agricultural products, assortment and availability of substitute products in the regional market.

In 2017, the consumption of bread and bread products, potatoes by the population of the region was 9.3 and 10 percent more than rational consumption rates. The population, to a large extent, is under-consuming milk, vegetables, fruits and berries (46.4, 46.1, 42.1 percent, respectively), which indicates an imbalance in nutrition. The shortage of food is compensated by supplies from the regions of the Siberian Federal District and neighboring countries (Azerbaijan, Belarus, Kazakhstan).

Assessing the food security of the region in terms of balanced diet in terms of calories, protein, fats and carbohydrates in food, it can be noted that they do not correspond to the vital needs of people. Energy and food requirements depend on how hard the person performs, on his gender and age. It is required to consume 96-108 grams of proteins; 406-440 grams of carbohydrate per day for men at the mature age of 10-60 years. According to statistics for 2017, the protein content in the diet of the population in Kemerovo region was lower by 19.4-31.4, carbohydrates by 86.4-120.4 grams than acceptable daily intake.

The imbalance of nutrition of the population affects its health. According to international statistics, the dependence of birth rate and mortality of the population on health service is 10 percent, and 50 percent is the quality of food, housing conditions, employment. Epidemiologists claim that a negative effect of insufficient and unbalanced nutrition on a person is commensurate with genetic, active chemical or infectious effects.

Modern state policy towards the agricultural sector, which provides the population with food, is focused on its sustainable development, which is its most important task. This is determined by the following: a) the mission of providing the population with food is entrusted to the agricultural sector; b) as a result of institutional changes in recent decades, the development of agriculture had negative trends; d) the state policy towards agriculture and the peasantry is implemented according to the residual principle. This circumstance does not contribute to the stable development of the agricultural sector at the regional level.

Many agricultural enterprises operate in the conditions of limited access to the food market and outlets. Due to the low financial security of agricultural enterprises and their low investment attractiveness, an economic recession is observed [1; 2]. In addition, there are problems associated with management, personnel and financial support of the agricultural business entities, the inertia of development, affecting the competitive properties of agricultural products, which is reflected in poor economic performance and failure to quickly respond to the constantly changing market environment.

The unfavorable macroeconomic situation predetermined the disparity between the profitability of agriculture and the profitability of other sectors, which impede the agricultural reproduction on a simple scale. The minimum level of profitability must correspond to 25-27 percent in order that agriculture could develop as a system.

3 Results

Since agriculture is an unstable system, it cannot self-develop without external influence. This circumstance requires the search for such approaches that can contribute to take the agricultural sector to a qualitatively new level of management and the creation of conditions for mutually beneficial interconnections of all agribusiness enterprises that meet the conditions imposed by the market.

In our opinion, this task can be solved within the framework of the agri-food cluster, as an organizational and economic mechanism that can actually contribute to the growth of agricultural production, improve the living standards of rural residents and conserve the environment for the generations to come. The feasibility of creating agro-industrial integrated formations at the present stage is due to the increase in competitive advantages and the synergistic effect of uniting parastatal business entities and including scientific and technical institutions and technology parks based on cooperation.

The emergence of a synergistic effect is due to the coordinated interaction of heterogeneous processes, which can be the initiator of a specific synergistic effect in specific conditions, which will allow the system to make the transition to a whole new level of development.

It is appropriate to use a synergistic approach to agriculture, which have the attributes of open, artificial, unstable economic system, depending on the impact of internal and external factors on it in the form of import pressure and climatic conditions.

We emphasize that the agricultural sector, as an economic system, is inherent: integrity and a single socio-economic territorial structure, within which economic actors are interconnected and interact in the process of production, distribution, exchange and consumption.

A distinctive feature of the regional agri-food cluster is: combining the efforts of participants in the food market, which will improve the competitiveness of the structure, reduce risks in the process of redistribution of commodity flows, and more quickly adapt to changes in market conditions and policies for providing the population with food.

 https://doi.org/10.1051/e3sconf/201913403008

Here you can see the general properties characterizing the cluster and synergistic approaches. This conclusion is supported by A.S. Khukhrin, who proved the similar properties of cluster and synergetic approaches in his research. These similarities have system problems at the macroeconomic level, do not affect specific enterprises, and therefore the network formed by them and within which there are interactions and interconnections, is to be studied.

According to synergetic, open and, to a certain extent, non-equilibrium systems are characterized by development through increasing their complexity and orderliness. One development cycle of a system simultaneously combines two phases: a period of smooth and predictable linear change during evolutionary development, which can later lead a system to the unstable state; the system acquires a stable state as a result of a surge that instantly takes it out of the crisis.

The synergetic approach to the stable development of agriculture is based on the fact that the activities of cluster members are carried out in an interconnected and mutually beneficial manner. The synergistic effect was calculated using the data of state statistical reporting forms of Kemerovo region business entities. In relation to the cluster, the synergistic approach is expressed in gaining the ultimate profit by its participants, taking minor management actions, which is a rather attractive side of the formation of such self-organizing economic structure as an agri-food cluster.

Based on the existing factors, a methodical approach to the interaction of entities within the agri-food cluster, based on a synergistic effect, is proposed. Produced value added and financial synergy obtained by participants in the agri-food cluster form a synergistic effect.

Saving on the scale (cost reduction due to increased production volumes, manifested by a decrease in fixed costs in their total volume); savings in diversity (coverage) when existing fixed assets are used in the production, raw materials and materials, allowing expanding the range of produced goods, will form value added.

The criteria for the feasibility of creating an agri-food cluster are: synergistic effect, higher competitiveness of cluster entities, processing and sale of food products without intermediaries, expanding the market segment, which will provide the population with food according to scientifically-based nutrition standards [4, 5] (Table 1).

Table 1. Calculation of the synergistic effect of the region's agri-food cluster, million rubles

Indicator	**Years**					
	2015	2016	2017	2018	2019	2020
Profit	12319	14561	17211	20343	24046	28422
Current period production costs	14402	14114	13691	12869	12226	11981
Estimated risk	521,1	615,9	728	860,5	1017	1202
Synergistic effect	403.4	4380.09	4777.04	5362.57	6046.6	6824

Calculations show that by 2020, the estimated profit will reach 28.5 billion rubles, 2.3 times higher than in 2015 due to a 16.8 percent decrease in production costs of the current period. The synergistic effect will be 6824 million rubles.

Thus, the proposed methodological approach to the calculation of the synergistic effect gives grounds to consider the agri-food cluster as a promising structure in the region that will combine the interests of agricultural producers, processors and consumers of food products. The feasibility of applying a synergistic approach to the development of agriculture in the region and food security is that the agrifood cluster is a self-organizing integration form that will lead to the intensification of the formation and development of sectoral clusters. When forming a cluster, one should make choices in elaborating a development strategy, coordinating development goals and objectives adequately to industry performance, and in accordance with the strategic goal.

An important circumstance is the state food policy in support of the development of domestic agriculture. Food policy of the state should be implemented using benchmarking, as the most effective method of extrapolating the experience of foreign countries.

It is more important for the state to support the concentration of food production, which consists in concentrating all stages of production and sales cycles, which will have a significant impact on reducing production costs and the cost of food, and as a result, increase food affordability for the population. Such a concentration can be carried out only within the framework of the agri-food cluster.

4 Conclusion

Summarizing the above, we can conclude that the formation of food security at the regional level is a multifactorial system, including: affordability and physical access of the population to safe and nutritious food of its own production in accordance with scientifically-based consumption standards; the resistance of the economic system against the internal and external impacts of economic, political nature and the ability of the system to withstand and minimize the costs associated with the negative effects on the system that have occurred; taling agriculture, as the main supplier of food, to a qualitatively new level of management that will ensure food security in the region.

To ensure the food security of coal mining region, it is proposed to apply a synergistic approach based on creating an agri-food cluster, whose activities will ensure optimal mutually beneficial interconnections of all participants, obtaining a synergistic effect, affecting not only the economic activities of its participants, but also the functioning of other clusters.

References

1. A.D. Sheingate, *The rise of the agricultural welfare state. Institutions and interest group power in the United States, France, and Japan* (Princeton University Press, Princeton, 2001)
2. *Agricultural Policies in OECD Countries. Monitoring and evaluation 2007* (OECD, Paris, 2007)
3. D. John Shaw, *Global Food and Agricultural Institutions* (Taylor & Finance group, Routledge, 2009)
4. E.A. Fedulova, A.V. Medvedev, P.D. Kosinskiy, S.A. Kononova, P.N. Pobedash, Foods and Raw Materials, **4:2**,157 (2016)
5. E.A. Fedulova, A.V. Medvedev, P.D. Kosinskiy, S.A. Kononova, P.N. Pobedash, Foods and Raw Materials, **4:1** (2016)
6. O. Flaten, S. Hisano, Nougyou to Keizai Agriculture and Economy, **73:8**, 129 (2017)
7. D. Giovannucci, S. J. Scherr, D. Nierenberg, C. Hebebrand, J. Shapiro, J. Milder, K. Wheeler, *Food and Agriculture: The Future of Sustainability* (Taylor & Finance group, Routledge, 2012)
8. H. Joseph Hülse, *Science. Agriculture, and Food Security* (NRC Research Press, Ottawa, 1995)
9. T. Jurzina, N. Egorova, N. Zaruba, P. Kosinskiy, E3S Web Conf., **21**, 04010 (2017)

10. M. Cellura, F. Ardente, S. Longo, Journal of Environmental Management, **93:1**, 194 (2012)
11. L.J. Van der Meer Cornelis, *Japanese Agriculture. A comparative economic analysis* (Taylor & Finance group, Routledge, 1990)
12. Allen Patricia, *Torether at the table.Sustainability and Sustenance in the American Agrifood System* (The Pennsylvania State University Press, Providence, 2004)

Approaches to Evaluation of Environmental and Economic Damage to the Kuzbass Agglomeration Caused by Coal Mining Industry Development

Piotr Kosinskiy[1], *Vladimir* Merkuriev[1], and *Aleksey* Medvedev[2]

[1]T.F. Gorbachev Kuzbass State Technical University, Kemerovo, Russian Federation
[2]Kemerovo Institute (branch) of Plekhanov Russian University of Economics, Kemerovo, Russian Federation

Abstract. The article is devoted to the study of coal mining industry problems in a region and environmental and economic damage caused by its activities. It is found that the intensive development of coal mining industry is accompanied by negative environmental consequences associated with the impact on social and economic indicators of a region and the quality of life of the population. The methodology for analysis and evaluation of environmental and economic damage to a regional economy based on the loss of gross regional product (GRP) and associated with the loss of working time due to the incidence of the population, determined by the level of environmental pollution, was developed. The developed methodology solves the problem of evaluating the impact of environmental components not only on environmental and economic damage, characterized by GRP losses, but also on the quality of life of the population. In addition, the optimization nature of the presented model allows finding the economic potential of producers and the objective efficiency of enterprises, which will make it possible to compensate for damage caused by environmental pollution that is adequate to the negative consequences of their activities.

1 Introduction

The modern development of economy is characterized by high rates of coal mining industry development, contributing to environmental impairment in some regions of the country. The focus on traditional GRP growth indicators, income, welfare, and its negative environmental consequences are becoming the object of close attention of scientists and doers.

The present-day situation shows a certain inconsistency in the evaluation of the use of natural resources and their losses, the impact, in this regard, on the environment, environmental and economic damage to the economy of the country and its individual regions.

The underestimation of environmental and economic damage leads to a distortion of economic development indicators, which is fraught with the risk of an inefficient choice of social and economic development lines in general.

The lack of perfect methods for evaluating environmental and economic damage to the economy does not allow evaluating the consequences of coal mining industry development with a high degree of confidence and providing the rationale for introducing the environmental component into the context of social and economic development aimed at improving the quality of life and human well-being. This is especially important for the regions in which coal mining, metallurgical, chemical industries of the economy are prominent, which entails the presence of significant areas of disturbed land. These regions of the Siberian Federal District, which have a raw material orientation, include the Kemerovo region.

In the region, more than half of GRP is generated by coal mining industry, which is a leading branch of the regional economy. Therefore, the development of the methodology for evaluating environmental and economic damage caused by the intensive development of coal industry, which allows for timely and effective action by authorities, is a pressing issue.

2 Materials and Methods

Coal mining industry, which includes numerous open pits, occupies large areas of land in the region [1] and "... is a source of negative environmental impact through air pollution and water contamination, contributing to the arable land degradation, including the storage of rock on them [4]". Coal mining in the region is increasing annually, which entails an increase in the alienation of agricultural, as a rule, highly fertile land. According to experts, mining of one million tons of coal requires an average of 10 hectares of land to be disturbed [11].

It should be emphasized that about 60 percent of coal in the region is mined using open cut methods. During the analyzed period (2013-2017), the area of disturbed land increased by 7.77 percent, from 73.3 thousand hectares to 79.1 thousand hectares [1]. The standard cost of the reclamation of nonfarm soils in 2014 was 1020 thousand rubles per hectare [2].

Underground coal mining is accompanied by the arrangement of sites for storing coal and rock, ventilation technology for underground workings, degassing, and contributes to the entry of methane into the atmosphere, which represents the main sources of air pollution. The stored rock contains significant amount of coal, and, when interacting with the atmosphere ignites spontaneously, releasing sulfur dioxide, resinous substances in the form of sublimation products, carbon monoxide.

Open cut mining technologies include blasting operations accompanied by gas-dusting, which is a pollutant of atmospheric air. High-intensity coal mining in the region, the consequences of a high degree urbanization are accompanied by signs of land degradation, including its fertility, a decrease in the animal population, and hence the loss of renewable natural resources. At the same time, the intensive coal mining industry development increases emissions of volatile substances into the atmosphere and pollutes the air. In 2017, the region ranks third in the Siberian Federal District in coal production, producing 241.5 million tons. Increasing volumes of coal mining are mainly accompanied by open air pollution. In 2017, pollutant emissions amounted to 1,487.6 thousand tons or 552.0 kilograms per inhabitant. Coal mining enterprises of the region emitted 8.8 percent more emissions in 2017 than in 2013.

There are more than 200 types of various substances that pollute the atmosphere. "Among them there are highly toxic and carcinogenic substances in the form of soot, lead, carbon oxides and nitrogen. The structure of pollutants to the greatest extent contains elements that are produced by stationary sources and are solids with a volume of 146.8 thousand tons. In addition, the volumes of gaseous and liquid substances occupy 1340.8 thousand tons" [1, 4].

An equally serious problem is methane and its emissions. The most promising way to solve this problem is to use methane for fuel, which will significantly reduce emissions of pollutants into the atmosphere.

Modern regional economic policy is aimed at the use of natural resource reserves, allowing the dynamic development of coal mining and metallurgical industries. From the point of view of the survival of the regional economy and the stable social activities, such development is fully justified. However, this does not solve the problem of improving health as one of the most important components of the quality of life, which is the main indicator of social and economic environment of the country and its individual regions.

The habitat and its quality is characterized by the health of the population, and the dynamics of the incidence of the working age population disease and mortality show the well-being and negative phenomena of regional development. The above mentioned indicators are the main problems for the Kemerovo region. Replacement of the population has a negative trend (Table 1).

Table 1. Replacement of the Kemerovo region population in 2013-2017.

Indicators	Years				
	2013	**2014**	**2015**	**2016**	**2017**
Population as at the beginning of the year, thousand people	2742.4	2734.1	2717.6	2708.8	2694.8
Birth rate (per mille)	13.6	13.2	12.5	12.1	10.5
Death rate ((per mille)	14.5	14.6	14.5	14.3	14.1
Infant mortality (per mille)	7.3	7.2	6.7	7.3	6.9
Natural population decline (per mille)	- 0.9	-1.4	-2.0	-2.2	-3.6
Life expectancy at birth, years	67.7	67.85	68.31	68.72	69.33

The population of the region in 2013-2017 decreased by 1.7 percent from 2,742.4 thousand to 2,694.8 thousand people. In the same period, the birth rate per mille decreased by 3.1 (23.9 percent). And although there was a positive trend in mortality and infant mortality, the demographics remain rather disappointing. The Kemerovo region has one of the highest population density rates among the regions of the Siberian Federal District, there are 28.15 people per square kilometer. At the same time, in the Siberian Federal District this rate amounts to 3.75, in the Russian Federation - 8.58 [3, 7-11].

As in the most regions of the country, the demographics can be characterized by a decrease in the population due to low life expectancy at birth and a decrease in migration growth. In 2017, 1,515 people less than in 2016 arrived to the Kemerovo region. The consequences of the negative impact of the negative environment on the demographics in the region are respiratory diseases, neoplasms, infectious and parasitic diseases, injuries, accidents at work and in everyday life.

The data of State Budgetary Healthcare Institution "KOMIATS" indicate a rise in the overall incidence of the population by 1.5 percent. In 2017, there is an excess of mortality over the birth rate of 9.8 thousand people, or 25.7 percent. Infant mortality (children under 1 year per mille) tends to decrease.

However, it remains quite high due to the pregradient growth of diseases and a number of pathologies caused by the toxic effects of chemical compounds found in the air, water bodies and in general, the environment on the mother and fetus [3].

Summarizing the above, we can conclude that adverse environmental factors have a significant impact on health and, consequently, on the quality of life of the population, and there is no real improvement in the environmental profile.

3 Methodology for evaluating environmental and economic damage to the Kemerovo Region economy

The consequence of the negative impact of environmental factors on human health is underproduced GRP. Table 2 shows the cost of underproduction of GRP generated by the impact on water and air environment.

Table 2. Indicators for evaluating the damage resulting from the incidence of the population caused by environmental factors by groups of diseases, 2013-2017.

Indicators	Years				
	2013	2014	2015	2016	2017
Population, thousand people	2742.4	2734.1	2717.6	2708.8	2694.8
GRP, billion rubles	668.3	752.0	843.3	865.3	10581.1
GRP, thousand /person/year	243.7	275.5	309.9	318.9	391.6
Emission rate, thousand tons	1356.0	1332.0	1344.5	1349.5	1487.6
Dirty discharge, million cubic meters	598	478	462	444	433
The incidence of disease per mille	1620.0	1576.0	1586.0	1596.0	1645.4
Average annual number of man-days lost due to illness	15.0	14.8	12.6	12.7	12.7
Cost of underproduction of GRP, million rubles	29712.7	30899.1	31084.5	31280.0	32249.7
GRP losses, %	4.4	4.1	3.7	3.6	3.4

Calculated by the authors according to the data of Kemerovo Regional Medical Information and Analytical Center (State Budgetary Healthcare Institution "KOMIATS")

Table 2 reflects the cost of underproduced GRP generated by the incidence of the population, which suggests the rationale for developing automated tools (models, methods, algorithms) for analyzing this phenomenon.

4 Results

Currently, there are no advanced methodologies for evaluating environmental and economic damage to the economy caused by environmental factors, and the existing methods based on minimum and maximum levels of morbidity, mortality caused by their impact, the real state of affairs, in our opinion, do not fully reflect the real state [5].

In this study, the authors propose a method for evaluating environmental and economic damage to the regional economy (GRP loss) associated with the loss of working time due to morbidity, determined by the level of environmental pollution. This problem can be solved using the model of economic development of the region, the criterion of the efficiency of which is to maximize the cost of GRP, taking into account the interests of the producer, consumer and the tax center of the region, which include the environmental component.

Let's consider the following substantive statement of the problem of evaluating environmental and economic damage caused by the environmental component.

The following economic agents operate in the region: producers (P), consumers (C), regional manager - tax center (T). The producer (P) organizes the release of the n-th number of product types, including innovative ones. It acquires (rents) basic production assets, buildings, land, etc., for the n-th number of enterprises of individual industries.

At the same time, the consumer (C) is provided by the producer (P) with funds for living in the form of wages, insurance contributions to social funds and spends them on his/her consumer spending (consumer basket) (CB). By collecting tax payments, fees, fines and realizing the redistribution of financial resources in the form of subsidies, the regional manager (T) organizes the interaction between (P) and (C).

It is assumed that (P) can produce products using environmentally dirty technologies, for which, according to certain rules, it is fined by (T), increasing the financial resource at (T)'s disposal. As a result of exposure to pollutants, the (C) incidence rate increases, which leads to a loss of working time and, accordingly, an increase in production costs of (P) for labor and, therefore, the cost of underproduced GRP.

It is required to determine the level of GRP losses associated with an increase in the (C) incidence, as well as the optimal investment in ОПФ (purchased equipment) and production (revenue from sales), which characterizes a regional economic development project, in which income and expenses (discount balance) of (P) and (C) and discounted tax flows of (T) within the time bucket of T will take into account the interests and natural limitations of the activities of each of the economic agents to the maximum.

The setting of a mathematical problem includes the following notations: average cost of basic production assets – ck, k-th type, k=1,..., n; average design capacity of basic production assets of k-th type – Vk; average operation time of basic production assets of k-th type – Tk; costs per unit of k-th type - Pk; forecasted demand for k-th product type – qk; specific and maximum pollutant emissions per production unit of k-th type – ξk, Dk, respectively; amount of revenue (P) from the sale of manufactured products – R; Z – total current (depreciation, wages, taxes, raw materials, materials, fines, etc.) costs (P); Am – total depreciation charge; S0 – depreciable value of basic production assets; Wb – balance sheet profit (P); W – net profit (P); C0 – initial fund (C); I0 – maximum investment amount; value added tax (НДС); N1, N2, N3, N4, property tax (НИ), profits tax (НП), insurance contributions (СВ) of enterprises (P); total environmental fines of enterprises; corresponding taxes and fees (rates) - α1, α2, α3, α4; α5 – penalty rate (expertly determined share of proceeds (P)); Dot1, Dot2 – (T) subsidies for (P), (C); β (labour consumption) – the share of total revenue from the sale of products of (P) entering the wage fund; p (consumption of materials) – share of total costs of (P) used to purchase raw materials; r – discount rate, T – time bucket.

The balance equations of own funds of economic agents have the form:

(P): $$DP = W + Am + Dot_1; \quad (1)$$

(C): $$DC = C_0 + \beta R + Dot_2 - \sum_{k \in CB} q_k; \quad (2)$$

(T): $$DT = \sum_{i=1}^{5} N_i - \sum_{i=1}^{2} Dot_i. \quad (3)$$

4.1 Conditions restricting the activities of regional economic agents

The primary limitations are the solvency of producer and consumer: DP≥0, DC≥0.
The limitedness of investment and subsidies can be expressed as

$$(P): I \leq I_0,$$

$$(T): \sum_{i=1}^{2} Dot_i \leq Dot_0 . \tag{4}$$

Where, (P) is the first condition limiting the maximum investment volumes, and the maximum amounts of subsidies received by agents (P) and (C) from the agent (T) are the second limitation. The meaning of this task lies in the fact that the subsequent restrictions affecting the volume of production of each type of product in terms of value do not exceed the performance of ОПФ and the demand for it.
(P), (C) and (T) performance criteria are:

$$J_{(P)} = T \cdot \frac{W + Am}{1 + r_э}, \tag{5}$$

$$J_{(C)} = T \cdot \frac{\beta R}{1 + r_э}, \tag{6}$$

$$J_{(T)} = T \cdot \frac{N_1 + N_2 + N_3 + N_4 + N_5}{1 + r_э}. \tag{7}$$

It should be noted that the specified performance criteria reflect a strategic (long-term) character, which manifests itself throughout the regional development time bucket, $r_э = \dfrac{rT}{1-(1+r)^{-T}} - 1 -$ static analogue of the project discount rate recalculated for the entire time bucket T.

Let's introduce the following variables: average value of k-th type - x_k; x_{n+k} – k-th product type sales revenue (k=1,…,n); x_{2n+1} = Dot_1, x_{2n+2} = Dot_2, x_{2n+3}= I. Then $Am = \sum_{k=1}^{n} \frac{T}{T_k} x_k$, $R = \sum_{k=1}^{n} x_{n+k}$ – sales revenue for all products; $S^0 = \sum_{k=1}^{n} \left(1 - \frac{T}{T_k}\right) x_k$; $N_2 = \alpha_2 S^0$. Taking into account the introduced notations, the net profit (P) can be calculated by the formula $W = (1-\alpha_3)\left((1-p-\beta)R - Am - N_1 - N_2 - N_4 - N_5\right)/(1-p)$, and the mathematical model of the problem to be solved will take the following form:

$$NPV_{(P)} = T \cdot \sum_{k=1}^{n} \frac{(-\gamma_k - 1 - r_э) x_k + \eta_k x_{n+k}}{1 + r_э} \rightarrow max, \tag{8}$$

$$NPV_{(C)} = T \cdot \sum_{k=1}^{n} \frac{\beta(1+\alpha_4) x_{n+k}}{1 + r_э} \rightarrow max \tag{9}$$

$$NPV_{(T)} = T \cdot \sum_{k=1}^{n} \frac{\sigma_k x_k + \tau_k x_{n+k}}{1 + r_э} \rightarrow max, \tag{10}$$

$$\sum_{k=1}^{n} \gamma_k x_k - \sum_{k=1}^{n} \eta_k x_{n+k} + x_{2n+1} \leq 0, \tag{11}$$

$$\beta \sum_{k=1}^{n} x_{n+k} + x_{2n+2} \leq C_0 - \sum_{k \in CB} q_k , \tag{12}$$

$$\sum_{k=1}^{n} \sigma_k x_k + \sum_{k=1}^{n} \tau_k x_{n+k} + x_{2n+1} + x_{2n+2} \leq 0 , \tag{13}$$

$$x_{n+k} \leq q_k \ (k = 1, \ldots, n), \tag{14}$$

$$-\delta_k x_k + x_{n+k} \leq 0 \ (k = 1, \ldots, n), \tag{15}$$

$$\sum_{k=1}^{n} x_k \leq I_0 \text{ ,} \tag{16}$$

$$\sum_{k=1}^{2} x_{2n+k} \leq Dot_0 \text{ ,} \tag{17}$$

$$\alpha_5 \xi_k \frac{x_{n+k}}{P_k V_k} \leq D_k \tag{18}$$

$$x_k \geq 0 \ \ (k = 1,...,2n+3), \tag{19}$$

where $\delta_k=(P_k V_k)/c_k$ – return on basic production assets of k-th type;

$$\gamma_k = -\frac{T}{T_k} + 1 + \frac{1-\alpha_3}{1-p}\left(\frac{T}{T_k} - \alpha_1 + \alpha_2\left(1 - \frac{T}{T_k}\right)\right), \tag{20}$$

$$\eta_k = \left(1-\alpha_3\right)\left(1 - \frac{\beta + \alpha_1 + \alpha_4\beta + \alpha_5\xi_k}{1-p}\right), \tag{21}$$

$$\sigma_k = \alpha_1 - \alpha_2\left(1 - \frac{T}{T_k}\right) + \frac{\alpha_3}{1-p}\left(\frac{T}{T_k} - \alpha_1 + \alpha_2\left(1 - \frac{T}{T_k}\right)\right), \tag{22}$$

$$\tau_k = -\alpha_1 - \alpha_3 + \alpha_3 \frac{\beta + \alpha_1 + \alpha_4\beta + \alpha_5\xi_k}{1-p} - \alpha_4\beta \text{ .} \tag{23}$$

The developed model (1) - (12) is a three-criterion linear programming problem. Its multi-criteria analysis is performed by bringing it to an equivalent single-criterion problem, which has a convex linear combination of criteria [5].

The model contains the most important (P), (C) and (T) activity limitations that impede them to operate in a crisis mode, which implies: ensuring solvency (4) - (6) of the mentioned economic agents; the amount of production based on their demand for it is formed [5], not allowing the risks of its overproduction.

Inequality (6), at the same time, can be interpreted as the efficient use of ОПФ, or the degree of scientific and technical progress. Conditions (9) and (10) are aimed at limiting the volume of investments and subsidies, which are financial resources circulating in the financial system. (11) limit emissions of pollutants to their maximum permissible values.

A feature of the model is its optimization nature, which allows identifying the economic potential of (P) activity. This is an important argument in favor of its use due to the fact that the current practice of compensating the damage caused by environmental pollution in the form of an environmental fine is focused on indicators of current accounting reports of enterprises, which are often underestimated in the interest of minimizing tax and environmental costs; performance indicators of enterprises being underreported. Using the objective, automated tools for determining the economic potential of (P) can be the basis for the efficient management of the regional environmental policy in the interaction between (P), (C) and (T). The next feature of the proposed model is its linear and statistical nature. This predetermines the use of the model as an efficient algorithm in the numerical analysis of the automated rationale for the adopted and approved decisions developed by the regional authorities for various aspects of economic development.

For example, using the proposed model, as the authors see, if there is an appropriate statistical base, it is possible to solve the problem of evaluating the impact of the environmental component not only on environmental and economic damage, characterized by GRP losses, but also on the quality of life of the population. Preliminary calculations performed using the bench-mark data by means of the package [5] showed the existence of a solution to the problem described by model (1) - (12).

5 Conclusion

Summarizing the above, we can conclude that there is considerable environmental and economic damage to the economy of the region caused by the intensively developing coal mining industry. In addition, this circumstance entails a profile that is characterized by a high anthropogenic load and the level of the incidence of diseases, exceeding the average figure in Russia by more than 2 times. The cost estimate of damage caused by the development of coal mining industry in the region shows an increase in the form of GRP losses generated by the adverse effects of the environment on public health. As a result of the low life expectancy of the population and the state of health, the population of the region decreases, which is fraught with high economic risks not only at the present time, but also in the future.

Therefore, the primary task of the regional authorities, in our opinion, is the development and adoption of immediate decisions aimed at the use of automated, objective tools to evaluate and balance the costs of improving economic efficiency and preserving the environment of the region. The developed methodology solves the problem of evaluating the impact of environmental components not only on environmental and economic damage, characterized by GRP losses, but also on the quality of life of the population. In addition, the optimization nature of the model makes it possible to identify the economic potential of producers and the objective efficiency of enterprises, which will allow adequate compensation for the damage caused by environmental pollution.

The methodology for evaluating environmental and economic damage to a regional economy is universal, and can be used by state authorities in adjusting future programs for the region's social and economic development, and other regions having the similar economic development level.

References

1. S. Zhironkin, S. Demchenko, G. Kayachev, E. Taran, O. Zhironkina, E3S Web of Conferences, **105**, 03008 (2019)
2. V.G. Mikhailov, N.N. Golofastova, T.V. Galanina, T.G. Koroleva, Ya.S. Mikhailova, IOP Conf. Ser.: Earth Environ. Sci., **50:1,** 012038 (2017)
3. V.G. Mikhailov, T.V. Kiseleva, V.A. Karasev, G.S. Mikhailov, V.A. Skukin, IOP Conf. Ser.: Earth Environ. Sci., **66:1**, 012008 (2017)
4. S. Zhironkin, M. Gasanov, G. Barysheva, K. Kolotov, O. Zhironkina, E3S Web Conf., **15**, 03012 (2017)
5. A. Krechetov, A. Khoreshok, V. Blumenstein. E3S Web Conf., **21**, 00001 (2017)
6. N. Kudrevatykh, T. Snegireva, A. Tselischeva, E3S Web Conf., **15**, 04006 (2017)
7. P.D. Kosinskiy, V.V. Merkuryev, A.V. Medvedev, Coal in the 21st Century: Mining, Processing and Safety, **366** (2016)
8. N. Zaruba, N. Egorova, P. Kosinskij, E3S Web Conf., **15,** 04003 (2017)
9. S. Kostyuk, N. Bedarev, O. Lyubimov, A. Shaikhislamov, E3S Web Conf., **21,** 01002 (2017)
10. A. Krechetov, V. Blumenstein, E3S Web Conf., **15,** 00001 (2017)
11. V. Trifonov, O. Loyko, D. Nesteruk, S. Zhironkin, E. Strekovtsova, AIP Conference Proceedings, **1800**, 050009 (2017)

Digital ecosystem of the region and the competitiveness of Kuzbass coal enterprises

Lyudmila Samorodova[1] , *Larisa* Shut'ko[1] , *Yulia* Yakunina[1] , *Oleg* Lyubimov[1] , and *Péter* Kovacs[2]

[1]T.F. Gorbachev Kuzbass State Technical University, 650000 Kemerovo, 28 Vesennyaya St., the Russian Federation
[2]University of Miskolc, H-3515 Egyetemváros, Miskolc, Hungary

Abstract. The issues related to the formation of the digital ecosystem in the coal-mining region within the framework of the world project "industry 4.0" are investigated. It is proved that the prospects for the development of the coal-mining region are associated with its export orientation and diversification of the economy, its complication. Technological momentum based on the use of digitization and cyber-physical systems will lead to a significant reduction in the cost of time for the implementation of manufacturing operations, which enhances the market potential of manufactured production, its competitiveness by reducing production costs, the energy costs of its manufacture and, accordingly, offered prices. The most active processes are the formation of the digital ecosystem and the use of the "Internet of Things" in the mining industry, which indicates a new stage of restructuring in this industry. Russia has accumulated "positive" experience related to the implementation of projects "smart mine" and "intellectual open-cast mine" in Kuzbass mines and coal cuts. The increasing complexity of the regional ecosystem through digitalization and automation of production will contribute to the growth of competitive advantages and competitiveness of Kuzbass coal-mining enterprises.

1 Introduction

Competitive struggle for leadership between countries, regions and companies is taking place in the modern world in the field of digital economy on the basis of new forms of organization of production, providing a significant increase in productivity in the industrial sector of the economy and the spread of synergetic effect to other industries and fields of activity.

The strategy of industrial development of Russia within the framework of the program "Digital economy of the Russian Federation" developed in 2017-analogue of the German program "Industry 4.0" (2011) – was a response to changes in the ecosystem [1] in the fourth industrial revolution [2]. According to the experts, the industrial revolution of 2020-2035 and the implementation of the road maps "TechNet" and "Development of technologies in the field of the Internet of Things" of the program "Digital economy of the Russian Federation" by 2024 will lead to an increase in labor productivity in Russia by 30%. By 2035, the increase in labor productivity should be 95%, new sectors will produce about 10% of GNP. "TechNet" is the direction of the program of transition to the fourth technological order in Russia, in the industrial direction of the development of the "Internet of Things" connected with digital design and modeling, 3D printing and robotization of production facilities [2]. According to the Ministry of Industry and Trade, mainly the Russian industrial revolution will work on the domestic market. The share of machinery and equipment in Russia's exports will not grow too much – from 8% to 13% by 2024. The share of Russian engineering companies in the world market will increase to 1.5% by 2035 , which may well be ensured by the preservation of the current positions on the world market by "Rosatom" structures. Over the next three to four years, the EU will invest about 140 billion euros in Industry 4.0, and labor productivity should increase by 18 % [3].

Prospects for the development of the coal-mining region are traditionally associated with its export orientation and diversification of the economy [4-5]. In the context of the fourth industrial revolution, countries and regions with a high economic complexity index (ECI) or economic complexity index will compete in the markets. In accordance with the method of calculation of this indicator, the detection of the level of complexity of the economy of the country (region) is based on "two main criteria:

1. Economic diversity is a criterion reflecting the number of different goods a country (region) is capable of producing.

2. Economic universality is a criterion that shows the number of countries (regions) that can also produce these goods" [6-7].

Thus, competitive advantages of the territory (the company), is able to produce (and export) complex industrial goods that can not be repeated by others in the economy or the market participants. Moreover, diversification of products and technological progress in one sector generates synergetic development of other related industries [8].

2 Materials and Methods

It should be noted that the well-being of the population in countries with a high index of economic complexity is higher than in other countries with lower values of this indicator. According to the rating compiled in 2015, the level of complexity of the country's economy was as follows: The 1st place − Japan, the 2nd place-Switzerland, the 3rd place − Germany and the 4th place − South Korea, the 9th place − the USA, the 19th place – China (2014), the 33rd place – Canada, the 50th place – Russia (2014).

Russia is the 16th largest export economy in the world. In 2016, exports to Russia amounted to 282 billion dollars. Imports total $ 180 billion and imports $ 180 USD billion. This resulted in a trade surplus of us $ 102 billion US dollars, resulting in a trade surplus of us $ 1.25 USD billion. In 2016, Russia's GNP was $ 1.28 trillion US dollars. The United States, and its GNP per capita amounted to 23.2 thousand dollars. USA. In the ranking by the level of economic complexity in 2016, Russia took the 34th place [9-11]. The impetus for the complication of the Russian economy was the active development of import substitution, due to international sanctions.

The results of the research of Russian regions conducted by Russian economists correspond to the standard conclusions:

1. The greatest potential for export complication is found in regions with a high level of complexity of economies.

2. The growth of the economies of the regions is largely associated with the diversification and complication of exports, including commodity regions are less likely to export development.

This is due to the fact that "in these regions, there are only a small number of production factors, such as human capital, infrastructure, equipment or certification, with the help of which new export goods can be produced. Creating much of the spectrum of factors needed for diversification is expensive and risky. Great opportunity for diversification exists in those regions where a significant part of the ingredients for the production of new export commodities already exists. The regions with certain economies are mainly located in the Western part of Russia. The industrial diversification of the Eastern and other less industrialized regions is therefore more costly. To complicate their export baskets, significant investments are needed, not only of a productive nature. Large expenditures are available to the budgetary constraints of industrialized countries" [12]. Table 1 shows the values and place of a number of Russian regions by the level of the index of economic complexity.

Table 1. Ranking of regions in 2013 [13].

Region	The index of economic complexity (ECI)	
	Value	Rating
Kemerovo Region	0,104	52
Krasnoyarsk Region	0,057	56
Moscow Region	1,362	1
Novosibirsk Region	0,306	27
Tomsk Region	-0,005	60
Khanty-Mansiysk Autonomous Okrug – Yugra	-0,020	62

The development of the coal-mining region in the future is associated with an increase in the potential of its economic complexity, digitalization of the industry and access to new principles of business process management and data exchange at the enterprise level (Fig.1). The implementation of the program "Development of technologies in the field of the Internet of Things" allows to implement for the cross-cutting integration of business processes through the use of different sources of information from different enterprises and optimization of management of different resources through the entire value chain in automatic mode. Ultimately, an ecosystem of enterprise-level and region is formed, covering enterprise partners, and the benefits from cooperation through the development of control system for production, more than competition with each other. In accordance with the concept of ecosystem by J. Moore "for the implementation of any innovation partners-customers and provider partners are needed. And the more radical (and often more valuable) the innovation, the more, deeper and wider other players, especially consumers, should be involved. In modern conditions cooperation, cooperation with suppliers and consumers, exchange of experience with producers occupying close market niches become much more important factor of success, than irreconcilable competition" [15]. The system of interaction between the participants of the digital ecosystem allows them to use the common infrastructure and interface of the platform, to create new products that could never create one by one, as well as to create their own products, and to move the common solution.

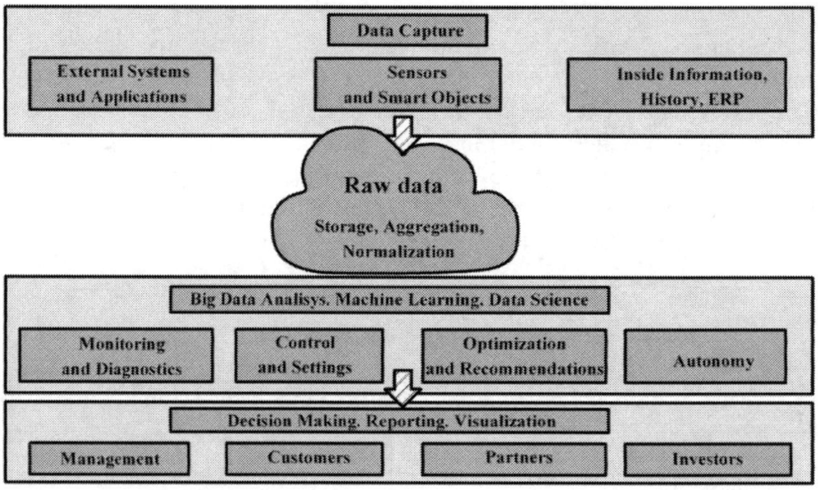

Fig.1. The scheme of control and data exchange.

The Central element of the Internet of Things ecosystem and complex projects with a high degree of automation, a large number of participants and connected devices are the Internet of Things (IoT) platform, which acts as an intermediary as a set of devices transmitting data in a wide range of formats using various communication protocols. The platform provides their collaboration, makes it possible to develop user applications and services. As a rule, developers support the most complete list of preinstalled sensors and devices, standards, protocols, analytical tools to ensure the rapid integration of solutions. According to optimistic forecasts of Gartner, the overall economic effect of the introduction of the Internet of Things in all sectors of the economy on a global scale will be 1.9 USD trillion by 2020.

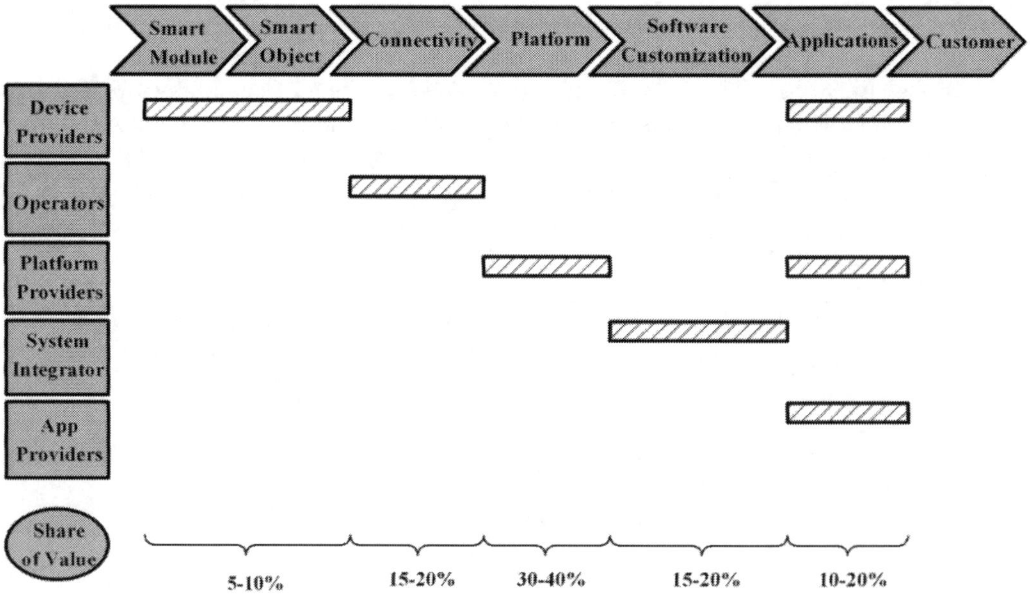

Fig. 2. The added value chain of the IoT project.

In a digital environment, the organization of work varies across the value chain (Fig.2). Integration of software, work in the cloud storage, data analysis have to be combined with the design, production and technical support of complex "physical" equipment. The classic control model breaks down with the advent of smart devices. There is a steady user interaction, monitoring equipment, recommendations, predictive maintenance, optimization, replacing, adding new features, and so on. New functional links appear in the structure of enterprise management:

– data management division;

– continuous product development unit-DevOps is responsible for the operation of "smart" devices and their optimization once the devices are already in use. It already employs programmers-developers (dev), IT specialists, production and service personnel (ops).

– new development methodologies are being implemented in organizations within separate development departments (Development/Software Engineering), IT operations department (Technology Operations), quality assurance department (Quality Assurance).

– customer experience management department, which oversees the use of equipment, creates recommendations for improvement, provides constant interaction with the client.

3 Results and Discussion

In a digital ecosystem of the organization IT division and R & D are closely cooperate for the speedy implementation of improvements and updates. The effect that the company receives as a result of organizational changes depends on the art of processing the obtained data. Thus, information becomes an asset that affects business performance. In addition, the more partners in a single network and the economy, the smarter it is, because machine algorithms allow to identify patterns, provide models and generate recommendations. Ultimately, the use of digital and cyber-physical systems will lead to increased productivity through a substantial reduction in production time, reduction of production costs and consumption of electricity for the manufacture of products, which increases its market potential and the possibility of reducing supply prices. Qualitative analysis of data according to the estimates of foreign experts allows to increase productivity from 3 to 15 %. The two main directions of the program "Digital economy of the Russian Federation", "Internet of Things", and "automation and robotization of production processes" – cover all processes of mining production [16].

In modern conditions, there is a transformation of coal mining companies into digital companies, in which many standard business processes are already automated, the hierarchy and relationships within companies are changing, the order of interaction with the entire industry ecosystem with competitors, suppliers and customers. Automation of operations continues in the area of operations management, and the information system makes the process cheaper, more reliable and more efficient. The interrelationship between industrial wireless infrastructure (IoX), traditional industrial enterprises and "Industry 4.0" within the digital ecosystem of the region is shown in figure .3.

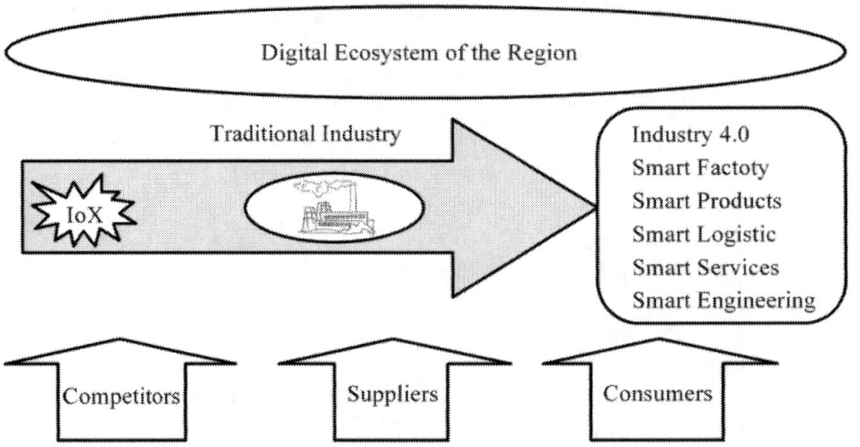

Fig. 3. Digital ecosystem of the region's economy.

Siberian coal energy company (SUEK) is already at the stage of digital transformation, and its main product is "digital coal". Coal as a" digital product" means that due to various information systems and automated systems (trade and logistics systems), closely integrated with each other, production systems and the ERP core, it is possible to control the movement and quality of each shipment of coal at minimum operating costs to the end user. The formation of a flexible production chain is of particular importance for coal sales, especially in the spot market, which brings the company's main margin [17].

Currently, there is already a positive experience of application in the mines of SUEK complex "Smart mine", developed by NPF "Granch" as a single information-but-management infrastructure designed to monitor and control any process equipment in the mine, provide communication and alarm, surveillance, warning and search for people caught by the accident. In addition, it includes a multi-functional measuring system lamp control. The system makes it possible to locate miners in the mine with an accuracy of up to 20 m, as well as to monitor methane concentration. "Smart mine" implements the so-called "triangle of miner's safety". A miner is always in touch; ground services always know its location; the atmosphere around it is always under the control of the controller. The decrease of the complex "Smart mine" in comparison with alternative systems (in the Russian market the nearest competitor firms from Australia) is that the system operates on the basis of the principle of determining the location of the object and its trajectory of movement throughout the entire length of the mine at any point; maintains continuous communication with each underground worker; all messages to workers are transmitted by a human voice, the system of "communication" with the workers is built into the individual lamp on the miner's helmet. In addition, there is also a sensor for gas analysis of the atmosphere.

In this respect, the miner also acts as a means of moving the sensor along the mine. According to the technical level, this warning device with the functions of a head rechargeable lamp is a modern communicator, in many respects exceeding the well – known brands, with the functions of a measuring device. Due to the stationary fixed sensor, which is equipped with a mine and moving (on the miners' helmets) in the entire shaft is continuous large-scale scanning gas analysis, the mine is under constant control. Moreover, information about methane or other problems in the mine is automatically sent to all mine managers and sent to the head office, which guarantees that appropriate measures are taken if necessary. In general, this Russian technology is at least 40% cheaper than imported analogues.

The mines "Uzhnaya" (holding company "Sibiskiy Delovoy Souz" SDS) and "Kotinskaya" ("SUEK-Kuzbass"), "Russkiy Ugol", "Raspadskaya" have already acquired the status of "smart" mines. Also, 20 coal mines of Kuzbass were supplied with various modifications of the "Smart mine" system [18]. In the near future, it is planned to use at coal mining enterprises equipment with sensors, drones, which will have to transmit an image, produce a 3D-scan of mine workings, measure the temperature, as well as the gas situation.

"VIST Group" (SKOLKOVO Foundation) has been working on the project "Intellectual open-cast mine" for several years [18]. The structure of the system "Intellectual open-cast mine" consists of 4 blocks:

- I block-– autonomous dump trucks, remotely controlled equipment (excavators, loaders, bulldozers, drilling rigs), autonomous railway transport;
– II unit – control center, including: software control of autonomous technology; automatic dispatching and optimization; work places remote control;
- III unit-– wireless data transmission system;
- IV block– -high-precision satellite and inertial navigation systems.

The stages of implementation of the project "Intellectual open-cast mine" are as follows:

- the robotic transportation of rock mass by dump trucks on a fixed route between stationary points of unloading and loading;

the robotic transportation by truck of the rock mass between the excavators and unloading point without equipment remote control of excavators and other equipment;

- the robotic transportation by dump trucks of rock mass using remote-controlled equipment (excavators, bulldozers, loaders, etc.).

Today, the automated control system mining complex named "Kar'er", which is the basis for the construction of robotic mining production, is implemented in large mining enterprises of "SUEK", "Holding company "SDS-Ugol", "Mechel-Mining", "Managing company "Kuzbassrazrezugol", "Managing company "Colmar", and others. In general, the creation and commissioning of the system "Intellectual open-cast mine" will allow to reduce the production costs of open-pit mining; ensure safe coal mining in hard-to-reach and difficult climatic regions; "soften" the problem of shortage of qualified personnel.

The technology "Intellectual open-cast mine" is being implemented today in the open-cast mine "Pervomayskiy" (on the site of Sokolovskoye field of the Kemerovo region) by "Holding company "SDS ". This section uses an electronic blasting system. The exposed rock is sent to the filling waste spaces, and the use of heavy equipment allows to reduce the impact on the environment. The dispatching system is based on the use of the GLONASS system. Land reclamation is applied during the whole time of the enterprise work, there is no noise or dust, as in other sections. In the short term, "SUEK" intends to form an "experimental fleet of unmanned dump trucks".

4 Conclusion

The fourth industrial revolution, which is based on the use of the "Internet of Things" in industry, digitalization of production characterizes the new stage in the development of coal-mining regions, increasing the complexity of their economies and the competitiveness of enterprises in the field of coal mining and enrichment. The article shows that the Kuzbass coal region is actively involved in the processes of formation of the digital ecosystem and the implementation of road maps "TechNet" and "Development of technologies in the field of the Internet of Things" in the framework of the Russian project "Digital economy of the Russian Federation". The transformation of the mining companies in the digital company, the introduction of technologies of "Smart mine", "Intellectual open-cast mine" allows you to create a new product such as "digital coal", to optimize the business processes and the structure of enterprises, reduce production costs and increase productivity, to build relationships with competitive rents, suppliers and consumers within the emerging regional digital ecosystems.

References

1. J. Moore, *The Death of Competition* (Collman, New York, 1996)
2. K. Schwab, *The Fourth Industrial Revolution* (Logan, New York, 2016)
3. *Approval of the program "Digital economy of the Russian Federation" by the Government of the Russian Federation from July 28, 2017 No. 1632-R// 31.07.2017.* URL: http://government.ru/
4. S. Zhironkin, S. Demchenko, G. Kayachev, E. Taran, O. Zhironkina, E3S Web of Conf., **105**, 03008 (2019)
5. C. Bülow, *Das Kusnezk-Becken in Sibirien: Entwicklungsstrategien zur Modernisierung einer altindustriell geprägten peripheren Region* (Springer, Berlin, 2017)
6. H. Klüter, Geograph. Rundsch., **49**, 723 (1997)
7. C.A. Hidalgo, R. Hausmann, PNAS, **30**, 123-131 (2009)
8. S.A. Zhironkin, K.A. Kolotov, A.E. Genin, F.V. Agafonov, S.A. Kovalevsky, IOP Conf. Ser.: Earth Environ. Sci., **50:1,** 012011 (2017)
9. V. Trifonov, O. Loyko, D. Nesteruk, S. Zhironkin, E. Strekovtsova, AIP Conference Proceedings, **1800**, 050009 (2017)
10. Economic Complexitiy Rankings (ECI). URL: https://atlas.media.mit.edu/en/rankings/countryeci/
11. M. Cehlár, K. Teplická, A. Seňová, 11th International Multidisciplinary Scientific Geoconference and EXPO - Modern Management of Mine Producing, Geology and Environmental Protection, SGEM 2011, **1**, 913-920 (2011)
12. R. Hausrnmann C., Hidalgo, S. Bustos, M. Coscia, *The Atlas of Economic Complexity: Mapping Pathsto Prosperity* (MIT Press, Boston, 2011)
13. I.L. Lyubimov, M.A. Gvozdev, M.V. Kazakova, V.K. Nesterov, Journal of New Economic Association, **2:34**, 94-122 (2017)
14. *The rating of the Russian Regions.* URL: https://iep.ru/files/text/nauchnie_jurnali/kazakova_NEA_2-2017.pdf
15. M. Prokudina, O. Zhironkina, O. Kalinina, M. Gasanov, F. Agafonov, E3S Web Conf., **21**, 04003 (2017)
16. L. Samorodova, L. Shut'ko, Yu. Yakunina, O. Lyubimov, E3S Web Of Conferences, **41**, 04011 (2017)
17. S. Zhironkin, O. Aleshina, V. Gorev, Y. Gunyakov, O. Zhironkina, E3S Web of Conferences, **105**, 04001 (2019)
18. Ju. Platkin, L. Platkina, Mining industry, 1:137, **22-28** (2018)

The Managerial Aspects of Integrating the Sustainable Development Principles into Practices of Mining Companies

Yuliia Lazarenko[1,*], and *Olga* Garafonova[1], *Vyktoriia* Marhasova[2], *Svetlana* Grigashkina[3], *Olena* Kozureva[4]

[1]Kyiv National Economic University named after Vadym Hetman, Management Department, 03057, 54/1 Pr. Peremogy, Kyiv, Ukraine
[2]Chernihiv National University of Technology, Accounting, Tax and Audit Department, 14035 Chernihiv, Ukraine
3 T.F. Gorbachev Kuzbass State Technical University, Russia
[4]National University of Pharmacy, Management and Administration Department, 61002, Kharkiv, Ukraine

Abstract. In order to successfully achieve sustainable development objectives, modern enterprises have to establish a framework and implement appropriate managerial tools for the effective integration of sustainability into business operations. The paper aims to clarify the concept of sustainable development in the mine management context, highlight the main principles of sustainable mining and devise appropriate tools and mechanisms from a managerial viewpoint to implement the strategic sustainability priorities into business practice. Based on a literature review of research on sustainable mine management issues, the main dimensions of the sustainability concept are highlighted and the specific goals corresponding to each of the priority focus areas are identified in the paper. As a result of the study, the key organizational aspects and firm-level mechanisms that should be primarily taken into account from a management decision-making viewpoint to embed the strategic sustainability principles into business practices and corporate activities of mining operators are pointed out. From a practical perspective, the findings can be used by mining operators for defining strategic sustainability priorities and setting related corporate objectives, as well as devising guidelines for the effective implementation of the integrated sustainable mine management system.

1 Introduction

There is no doubt that in the modern business environment it is crucial for organizations to change their entrepreneurial mindset, current management practices as well as established business models in order to keep pace with the economic trends and social needs. As the

* Corresponding author: yuliia.lazarenko@kneu.ua

business conditions change, it is becoming clear that companies should move away from the traditional management approach in which the primary focus is on maximizing profits, regardless of the cost to others. Extractive industries such as mining, logging, the extraction of minerals, oil or natural gas are nowadays confronted with serious challenges of sustainable development, in particular, with regard to environmental and social issues. Thus, there emerges an objective need for mining companies to implement and integrate the principles of sustainability into all aspects of their operational activities.

This research focuses on defining a theoretical basis for a deeper understanding of the specific characteristics of sustainable development in the mining sector and is intended to identify its significant dimensions as well as priority focus areas and corresponding objectives with regard to mining operators. Based on a conceptual review of the literature related to the sustainable mine management issues, key organizational aspects and firm-level mechanisms which should be primarily taken into account from a company's viewpoint to embed effectively and efficiently the strategic sustainability priorities and principles into the corporate activities of mining operators are pointed out in the paper.

2 Materials and Methods

In general, the sustainability concept in the mining sector represents the idea of how industry companies can contribute to sustainable development. Based on the overview of existing approaches and definitions, sustainable development can be presented as an integrative concept that embraces a range of interconnected aspects, namely, the economic dimension which focuses on the economic needs of society and economic development, the social dimension that refers to social and cultural needs, and the environmental dimension with regard to the management of natural resource base, and the maintenance of the ecosystem integrity [1]. In the context of the mining industry, sustainable development is aimed at striking a balance between achieving the benefits of the mining projects and in the same time the improving of the economic, social, and environmental sustainability, with a specific emphasis on the optimization of positive impacts. Sustainability as a complex and multidimensional concept presumes that it is necessary to operate taking into account the interconnections between economic, sociocultural and environmental aspects, as well as the need for the fair distribution of benefits and opportunities. In this vein, sustainable development can be presented through an integrated approach that is aimed at building a prosperous community by addressing economic, environmental, and social issues and avoiding in the same time the ecosystem degradation and overconsumption of the key natural resources [2]. With this regard, the challenges of the modern turbulent business environment and the evolution of the sustainable development at the firm level have heightened the need for devising and implementing balanced approaches to the management of mining operators, in order to ensure an effective integration of the economic, social and ecological norms, principles, values and priorities into corporate activities.

This paper presents a theoretical framework for investigating the main determinants of sustainable development in the mining sector and strives to provide a basis for defining the core organizational factors and focus areas at the firm level which are from a decision-making perspective of critical importance for effectively managing the process of embedding the sustainability values and principles into the business operations of mining companies. The paper is structured as follows: first, a descriptive overview of the theoretical aspects of sustainable development in the mining sector is provided. Then, based on a review of the existing sustainable mine management literature and thematic research, the specific characteristics of sustainable mining are considered from a management viewpoint. As a result of the study, through the integrated approach to sustainability, the main organizational factors and intra-firm mechanisms are summarized that are of particular importance for integrating long-term strategic priorities related to sustainable development into business practices of mining companies.

3 Results and Discussion

In recent decades the concept of sustainable development has become a significant economic factor. Accordingly, one of the overarching goals for the mining operators is today to find ways and devise effective implementation mechanisms by which the mining industry can address the sustainability-related challenges and contribute to sustainable development. Given the global relevance of the issue, the study of the most important barriers and possible opportunities for achieving sustainable goals in the mining industry has gained importance within the academic community during recent years. However, although it is an essential issue for mining enterprises to develop sustainability strategies, appropriate methods, organizational factors, and managerial mechanisms of integrating sustainability into business practice are still not sufficiently researched.

As the overview of the underlying theoretical concepts has shown, the sustainable development concept originates from the resource management and is applied particularly with regard to environmental protection issues and maintaining natural ecosystems. But sustainability as a broad and multifaceted concept cannot be considered exclusively from an environmental perspective [3]. A multidimensional approach to understanding sustainability in the mining sector assumes that three key focus areas should be highlighted: environmental dimension, economic dimension, and sociocultural dimension.

Sustainable mining can be characterized as one wherein the proceeds generated from the mining projects are on a continuous basis reinvested in the sustainable undertakings and in community support, including projects which are aimed at maximization of environmental, economic and social benefits, e.g. development of the health-care services, improving access to education, cultural promotion etc. Thus, from a management perspective, the practical implementation of the sustainable development paradigm is reflected in relevant business decisions designed not only to enhance the efficiency of the company's operations and multiply its assets, but also contribute to the well-being of society [4]. In order to fulfill their obligations relevant to the attainment of sustainable development, it is necessary for mining companies to be involved in strengthening the multi-stakeholder interactions, including public and private sector actors that enable to incorporate knowledge and values from different sources and to develop integrated strategies [5]. Accordingly, following the conceptualization proposed by A. Jabłoński (2013), at the company level sustainability decision-making is associated with such processes as improved communication, integration, cooperation and partnership with customers, suppliers and a wide range of stakeholders, quality, agility, as well as innovation activity and teamwork [4; 6].

In the light of the multi-stakeholder approach, the sustainability concept in the context of the mining industry can be considered as a result of dynamic interaction among key actors, such as mining companies, government, local communities and authorities, non-governmental and civil society organizations, as well as industrial unions and environmental organizations. The evidence shows, that there is a link between a company's financial performance and its social and environmental initiatives: business activities, which are carried out in accordance with the principles of open cooperation, mutual benefits, and common interests allow for sustainable development in social and environmental aspects as well as economic growth of the company [1; 7]. However, it needs to be pointed out that given the complexity of the task of harmonizing the contradictory interests and priorities of the above-mentioned stakeholders, there is an important issue for modern companies to accommodate divergent views and find a compromise to ensure stakeholders' needs and requirements can be achieved [8]. The balancing of contradictory interests and values is therefore a necessary prerequisite for providing constructive interactions and maintaining integrative ways of working with diverse stakeholder groups. In this vein, it is important to develop new forms of cooperation, collaboration, and coexistence among the relevant stakeholders, with tolerance to ambiguity, complexity, and uncertainty of such interrelationships [5].

Given the common sustainable development challenges, it is evident that current managerial practices in the mining sector need to be revised to contribute to the community development, maintain natural ecosystems' functionality and be compliant with the sustainable development objectives. According to the approach proposed by *J.A. Botin (2009)*, the sustainable management of mining operators can be described and explained by the integrated model that encompasses foundational elements, such as a business culture, in which sustainability is recognized as a priority (including corporate vision, mission, and corresponding goals), strong leadership and motivation (in particular, regarding the human resources management), and an appropriate organizational structure which seeks to ensure long-term economic gains through sustainability [1].

With reference to the existing literature related to the sustainability concept in the mining sector, the main focus areas that are identified as being particularly relevant for managing mining towards sustainable development, specific objectives corresponding to each of the priority dimensions as well as potential benefits and the intended outcomes of the implementation of sustainability programs from a company's viewpoint, were summarized and presented in Figure 1, which is developed by the authors adapting [1-6; 10-16].

As indicated in Figure 1, the environmental dimension of sustainability can be defined as physical sustainability, alludes to the importance of maintaining the natural environment. The economic aspect of sustainability emphasizes the need to ensure long-term economic development and provide high living standards as well as equality with regard to sharing financial benefits. The third major dimension of sustainability is a sociocultural aspect that encompasses social welfare, human well-being, and social responsibility to provide a more balanced distribution of wealth, an appropriate infrastructure, cultural development, and other related factors.

Evidence shows, that modern mining operators undertake a wide range of activities and programs that are conducive to the achievement of sustainable management objectives for each focus area. These activities can be broadly grouped into the following categories: the protection and maintenance of the natural environment, providing for the workers' needs in health care in safety, development of the local communities, and stakeholder engagement [1; 9]. With regard to the mining sector the concept of sustainable management envisages that the operational activity of the company should be carried out in accordance with principles that presume an implementation of economic as well as social development programs in an effective manner, pursuing corporate social and environmental responsibility initiatives, compensating the local communities the spillover costs, and fair distribution of the benefits and surpluses from mining operations.

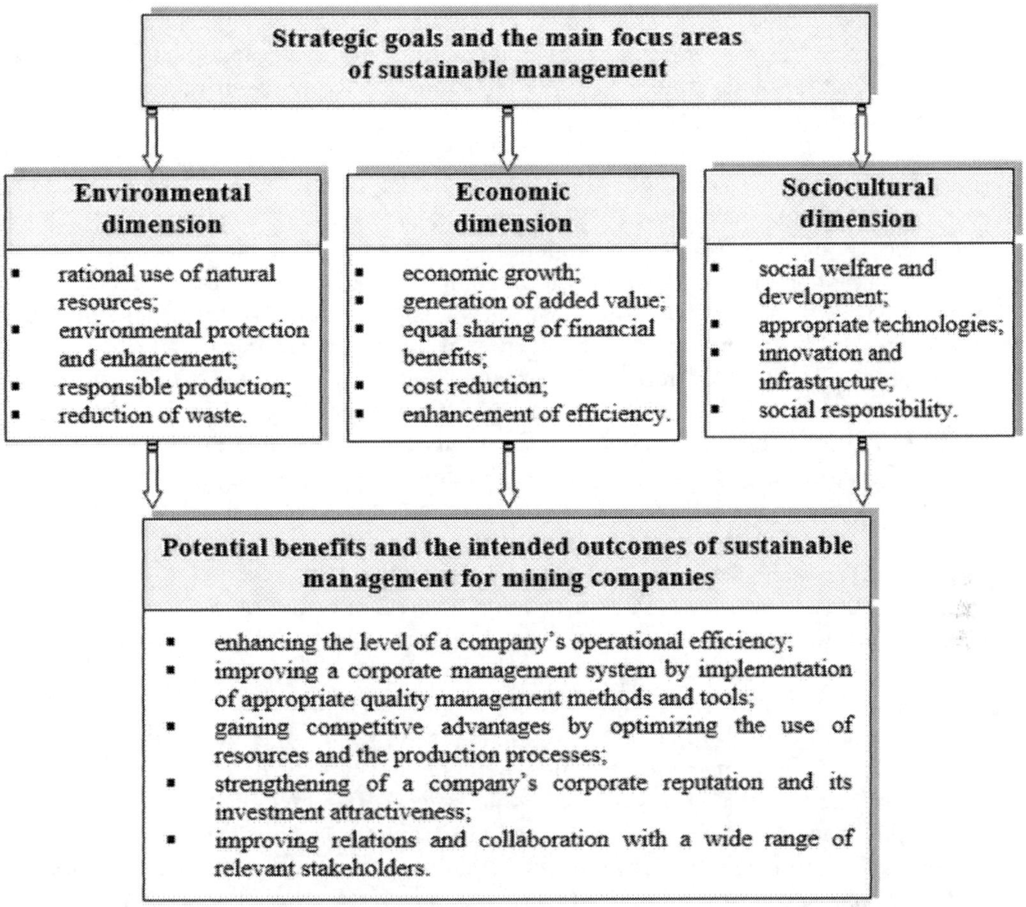

Fig. 1. Strategic goals and the potential benefits of implementing sustainability programs for the mining companies.

From an organizational viewpoint, it is important to integrate a concept of sustainable development into a company's corporate strategy and reconcile the main principles of sustainability with the operational activities and the current management system at all levels of the organization. Among the factors that promote the integration of sustainability values and principles into the organization, a special reference should be made to the corporate commitment to sustainable development and a well-developed sustainability-oriented organizational culture. In addition it requires adequate information management tools and techniques as well as specific intra-firm integration mechanisms that should be developed to reflect the organizational particularities and the specific nature of the mining industry, such as an organizational sustainability framework, appropriate policies and standards, managerial procedures and operations, as well as an integrated sustainability management system. To achieve the main sustainable development objectives, the values and principles of sustainability should be also accepted and implemented across all the levels of an organizational structure of a company. Corporate activities that aim at developing a common sustainability awareness, for example through various training programs and workshops can be also pointed out as an important step towards integrating sustainability values into the organization's strategy and business processes and deeper understanding of the social, environmental and economic context within which it operates. To summarize, successful implementation of the sustainability principles in the mining companies is influenced by multiple managerial and organizational factors, among which the following should be highlighted: a corporate culture, a management and leadership style, a current business strategy of a company, a type of organizational structure along with appropriate managerial tools and techniques that are contributing to sustainable-oriented activities.

According to *J.A. Botin (2009)*, the integration of the sustainable development principles into business practices of organizations that operate in the mining sector requires specific integration mechanisms, such as standards and systems, positions, roles and committees, public reporting, and independent assurance [1]. In this view, we share the opinion that integrating sustainability into the operational activities of a company should be considered as a managerial rather than as an engineering task and implemented at all functional levels of an organization. From a management perspective, to embed the sustainability principles into the corporate strategy and strengthen organizational performance purposeful actions need to be appropriately planned, coherently and specifically implemented and continuously monitored. Following the approach proposed by *N. Oertwig et al. (2017)*, the appropriate planning of operational actions of a company and the availability of evaluations for the systematic monitoring are identified as essential factors for a successful implementation of the sustainability strategy at the organizational level [10]. The adequate measurement tools, control mechanisms and communication with regards to the aspect of sustainable development require the interaction between various actors, evaluation methods and operational data in order to integrate sustainability into a comprehensive strategy of a company [10; 11].

In preparing a framework for business sustainability particular attention should be also paid to the development of engagement strategies for each of the key stakeholder groups as well as identification of main risks, uncertainties, technological, economic, environmental and other relevant conditions which could affect the implementation of a company's sustainable management system. With respect to mining activities, sustainable development presumes that mining companies interact with local communities and regional governments and take into account their interests in order to cover environmental as well as social aspects when it comes to making commercial decisions and developing technical issues [1]. Overall, the dynamics of the economic environment, the business model specificities, decision-making style as well as management attitude and behavior need to be recognized and taken into consideration to establish a comprehensive framework of sustainability integration within the organization. In this context, the achievement of a multi-stakeholder consensus around the principles and criteria of sustainable mining which requires extensive consultations and discussions with various actors could be challenging for the mine managers given the difficulty of reconciling divergent interests.

With regard to the implementation of new business models in the mining sector, attention should be drawn especially to the innovation which is considered to play an essential role in achieving sustainability by meeting the increasing social demand for the innovative products and services, without losing out on efficiency [11; 12]. In the case of mining sustainable development impose the need to promote business models that ensure financially viable and environmentally sound projects, which take into account the aspects of corporate social responsibility, bring lasting benefits not only to the investors, but also to the community, as well as contribute to economic development at the local and regional levels [13]. These key characteristics of sustainable mining should be taken into account by the companies that have as their objectives the effective poverty reduction and economic prosperity in the region. It needs to be emphasized that mining in the context of sustainable development concept can be also considered as a mechanism for economic growth and social development since it creates new wealth in terms of government income and employment opportunities.

4 Conclusions

Based on the findings of the study, it is possible to outline the main aspects which are of particular importance from a company's viewpoint for ensuring successful integration of the strategic sustainability priorities and principles into business practice and managing mining operators towards sustainable development:
- the effective implementation of sustainability aspects in the business practice requires appropriate changes in corporate strategy that are interconnected in an organization at all levels of decision-making;
- in order to achieve the sustainability objectives, corresponding to the environmental, economic and social dimensions, specific intra-firm mechanisms such as an organizational sustainability framework, appropriate policies and standards, roles and committees, managerial procedures and operations, as well as an integrated sustainability management system should be developed and applied;
- adopting a multi-stakeholder approach to sustainability it needs to be emphasized that it is critical for the companies to develop organizational skills and capabilities that facilitate interaction among various actors and ensure effective working relationship in collaborative networks.

To conclude, sustainable management in the context of the mining sector should be considered as a complex and multifaceted concept which strives to maintain a fine balance between the technical, financial, social and environmental aspects of mining activities. Appropriate managerial solutions for mining enterprises' sustainability development, the monitoring and evaluation of the integration process, and a pragmatic implementation strategy are necessary to provide the optimized decision-making and put established theoretical frameworks of sustainability into business practices.

References

1. J.A. Botin, *Sustainable Management of Mining Operations* (Colorado, US: Society for Mining, Metallurgy, and Exploration, 2009)
2. A. Ibiyemi, Y.M. Adnan, M.N. Daud, International Journal of Environment and Sustainable Development, **14 (2)**, 170 (2015)
3. M.A. Rosen, Glocalism. Journal of Culture, Politics and Innovation, **3**, 1 (2018)
4. S. Lorenc, O. Sorokina, Mining Science, **22 (2)**, 67 (2015)
5. J.-A. Everingham, Greener Management International, **(57)**, 91 (2012)
6. A. Jabłoński, *Models of Sustainable Business* (Publishing house Difin, Warsaw, 2013)
7. M. Grieg-Gran, *Financial Incentives for Improved Sustainability Performance: The Business Case and the Sustainability Dividend* (MMSD Project, London, 2001)
8. A. Lulewicz-Sas, *Corporate Social Responsibility versus Social Irresponsibility. Theory Approach* (Agencja Wydawnicza EkoPress, Białystok, 2013)
9. N. Yakovleva, *Corporate Social Responsibility in the Mining Industries* (Ashgate, Hampshire, 2005)
10. N. Oertwig, *Integration of Sustainability into the Corporate Strategy* (Springer International Publishing AG, Cham, 2017)
11. K. Maas, S. Schaltegger, N. Crutzen, Journal of Cleaner Production, **(136)**, 237 (2016)

12. R. Pfriem, R. Antes, K. Fichter, M. Müller, N. Paech, S. Seuring, B. Siebenhüner, *Innovationen für eine nachhaltige Entwicklung* (GWV Fachverlage GmbH, Wiesbaden, 2006)
13. M.I. Agienko, E.P. Bondareva, G.V. Chistyakova, O.V. Zhironkina, O.I. Kalinina, IOP Conference Series: Earth and Environmental Science, **50:1**, 012022 (2017)
14. A. Gürtürk. *Integrated Reporting and Sustainability-Related Assurance: Effects, Current Practice and Future Directions* (Kassel University Press GmbH, Kassel, 2017)
15. C. Marolla. *Information and Communication Technology for Sustainable Development* (Taylor & Francis Group, Boca Raton, 2018)
16. M. Koho, S. Torvinen, A. Torres, *Objectives, Enablers and Challenges of Sustainable Development and Sustainable Manufacturing: Views and Opinions of Spanish Companies* (IEEE International, Chicago, 2011)

Diversification Strategy and Sustainable Development of Mining Companies in the Eurasian Mining Region

Dmytro Voronets[1], *Olga* Garafonova[1,2,*] *Irina* Verezomska[3], and *Oksana* Oliinyk[4]

[1]Management Department, Economics and Management Faculty, Kyiv National Economic University, Kyiv, 03057, Ukraine
[2] Management Department, Economics and Management Faculty, Kyiv National Economic University, Kyiv, 03057, Ukraine
[3]Kyiv National University of Culture and Arts, Kyiv, Ukraine, 01601, Kyiv, Evgena Konovalca, 36
[4]Private Higher Educational Institution "Kyiv University of Culture", Kyiv, Ukraine, 01601, Kyiv, Evgena Konovalca, 36

Abstract. The aim of this study is to examine the relationship between diversification strategies implementation and sustainable development goals of mining companies of Eurasian mining region in modern conditions. Today Eurasia is very rich region with minerals-driven sustainable development goals and challenges. In this case it is important to improve strategies of mining companies of Eurasia and to listen to the biggest companies` experience. Diversification strategy is one of the most progressive way to develop mining companies wherein Global Sustainable Development Goals must be considered. Diversification of the enterprise could be realized in such ways: as the extension of its business activities to new areas or as the capital diversification — diversification of assets in company's investment portfolio. Over the past two years, from 2017 to the beginning of 2019, the number of diversified companies in the TOP 30 Mining companies of the world, according to the PWC consulting company and the Mining[dot]com resource, increased from 8 to 13. Thus, it is important to study the possibilities and features of implementing diversification strategies for mining enterprises of Eurasian mining region. In research it is presented the algorithm of the model, compiled on the basis of the development as applied to solving the problem of diversification of a mining enterprise.

1 Introduction

In spite of the strong operating performance, both investors and consumers seem to be down on the brand of mining. They question whether the industry can responsibly create sustainable value for all stakeholders. Discrete events, such as safety or environmental incidents, have contributed to these challenges.

* Corresponding author: ogarafonova@ukr.net

While they appreciate the efforts to improve operations and engineer superior results, it is clear that investors and other stakeholders are concerned that the industry is lagging when it comes to several factors that have not been a traditional focus of the mining industry. These include dealing with emissions, investing in differentiating technology and digitalization, engaging more proactively with consumers and building brand. Looking ahead to the rest of 2019 and beyond, we see a continuation of the strong operating performances, and pockets of progress in these contemporary factors. But we don't see any signs of a quantum shift in priorities that will allow the industry as a collective to keep pace with changes delivered in other sectors. Without such a shift, we expect the growing awareness gap between the brand of mining and the benefits of mining to continue to widen.

To restore faith in 'brand mining', leading miners need to prove they are keeping up with the pace of change. As an industry, this means transforming their reputation as efficient 'converters of dirt' to prominent builders of both economic and societal capital. Prioritizing green and customer-centric strategies, enabled by technology, will help earn the trust of stakeholders and enable miners to create sustainable value into the future. [1]

The mining industry impacts all 17 of the Sustainable Development Goals to varying degrees through mining itself, social investments, taxes and investment of public revenues.

Mining has the most direct impact on six goals:

- SDG15 (Ecosystem and Biodiversity Protection)
- SDG13 (Climate Action)
- SDG9 (Infrastructure, Innovation, and Industrialization)
- SDG8 (Employment and Economic Growth)
- SDG7 (Energy Access and Sustainability)
- SDG6 (Clean Water and Sanitation). [2]

2 Results

Eurasia has recovered from the recession of the 1990s and is integrating into the world economy—primarily through its abundant natural resources. The resource-rich countries of Eurasia have benefited from global economic growth. After all, Eurasia has more than one-third of the world's reserves of oil, gas, bauxite, and gold, and prices for these commodities have surged since 2000, boosting resource-related revenues. The other countries of Eurasia have also benefitted from the resource abundance of their neighbors through trade, capital flows, and remittances. Governments would do well to review strategies that rely on interventions to stimulate specific sectors or activities. Instead, it would be far more effective if Eurasian countries focused more on diversifying their national asset portfolios—that is, to ensure better balance between natural resources, physical and human capital, and economic institutions. [3,4]

Considered as a growth strategy the rationale of diversification, is for a company to explore new business areas that promise greater profitability. For a company to diversify, it needs to enter/expand in new markets or product lines which are related or/and unrelated to its existing businesses. [5] Diversification generally requires new skills, new techniques, and new facilities. As a result, it almost invariably leads to physical and organizational changes in the structure of the business which represent a distinct break with past business experience.[6]

The crux of the issue around diversification is one of minerals-driven sustainable development. The principal premise here is that while minerals are in themselves a depleting, non-renewable resource, the mistake is often made of considering minerals industries to be unsustainable. Without question, in themselves mines, and hence their industries, have defined lives. However, mining is a robust pioneering primary industry in difficult arenas such as deep rural areas and zones of high conflict. Typically, mining provides for the development of infrastructure and economic activity where few other sectors could or would venture. In this sense it has the propensity to spawn or catalyze the development of direct and indirect secondary and tertiary sectors that could, properly planned and managed, be sustainable.

Sustainability within the minerals sector depends entirely on the ability of governments' to:

a) plan their minerals economies in such a way as to maximize the development of mining-related secondary and tertiary sectors during the currency of mining operations;

b) encourage the development of non-mining related activities around mining service industries and mining infrastructure

c) in so doing, to diversify the use of mining infrastructure while it is being subsidized by mining sectors to the extent that the closure of the mines for which the infrastructure was developed does not cause its collapse;

d) actively encourage non-mining investment in mining regions to reduce the dependency of local, regional and indeed national economies on minerals and to simultaneously reduce the dependence on local infrastructures on mines.

The reality of the matter is however that most governments in developing economies are under-resourced for various reasons ranging from the lack of a firm tax base as a result of an under-performing or under-developed economy to the lack of economic planning experience and a dearth of well qualified economic planners. In this sense responsibility for achieving these objectives lies not only with the governments themselves and those intergovernmental and development organizations that assist them in their planning efforts, but also strongly with a responsible private sector. [7]

Economic diversification simply will not happen without an active planning process involving government, labor and private sector interests. In this sense, the planning for diversification goes far beyond the commonly accepted notions of:

1. Diversifying the use of mining infrastructure;
2. Promoting downstream and vertically-integrated minerals-based industries;
3. Planning for post-mining sustainable use of this infrastructure and well-being of the community;
4. The provision of investment incentives.

Diversification of the economy can take many indirect forms founded on the direct use of mining rents to:

1. Maximize the rents from mineral industries through constructive equity participation rather than relying purely on returns from royalties and taxation revenues;
2. Actively diversify the minerals sector itself;
3. Build a large and available mining skills base in order to facilitate the rapid development of mineral-based economic alternatives through investment in and support of appropriate training and educational facilities and programs;
4. Invest in infrastructure that will enable the development of otherwise marginal mineral deposits;
5. Invest in existing mining-related infrastructure to diversify the economic use of this infrastructure during the currency of mining operations;
6. Invest in non-mining related infrastructural and social development programs which will in turn generate immediate and sustainable economic activities not necessarily directly related to or dependent on mining;
7. Plan for and stabilize government spending though the judicious use of foreign exchange-based investment instruments to act as a buffer to volatile commodities markets;

8. Exert strict financial discipline to ensure adherence to these government spending programs;

9. Employ all available fiscal instruments to attract investment rather than resorting to grant based incentives; and

10. Constantly review mining and investment codes in close consultation with potential investors in order to encourage investment by demonstrating a willingness to co-operate with investors. [7]

The Table 1 [8] shows the ranking of mining companies as of April 2017 with division into mono-enterprises and diversified.

Table 1. Top 30 Mining Companies Market Capitalization USD Billions April 2017.

No	Company	Operations	Market cap – USD billions
1	BHP Billiton	**Diversified**	90.67
2	Rio Tinto	**Diversified**	74.04
3	Glencore	**Diversified**	55.98
4	China Shenhua Energy	Coal	54.96
5	Vale	Steel Raw Materials	48.31
6	Coal India	Coal	28.18
7	Southern Copper	Base Metals	28.04
8	Norilsk Nickel	Base Metals	25.30
9	Barrick Gold	Precious Metals	22.67
10	Anglo American	**Diversified**	21.28
11	Freeport-McMoRan	Base Metals	19.21
12	Hindustan Zinc	Base Metals	18.95
13	Newmont Mining	Precious Metals	18.03
14	Polyus	Precious Metals	17.34
15	Fortescue Metals Group	Iron Ore	14.68
16	Potash Corp	Potash	14.19
17	Fresnillo	Precious Metals	14.06
18	Newcrest Mining	Precious Metals	13.21
19	Teck Resources	**Diversified**	12.95
20	Ma'aden	**Diversified**	12.71
21	Goldcorp	Precious Metals	12.58
22	Vedanta	**Diversified**	12.08
23	Alrosa	Diamond	12.08
24	Franco Nevada	Precious Metals	11.69
25	South32	**Diversified**	11.30
26	China Coal Energy	Coal	10.40
27	Antofagasta	Base Metals	10.28
28	Zijin Mining	Precious Metals	10.22
29	PeĐoles	Precious Metals	10.21
30	Mosaic	Potash	10.07

Over the past two years, from 2017 to the beginning of 2019, the number of diversified companies in the TOP 30 Mining companies of the world, according to the PWC consulting company [1] and the Mining[dot]com resource [9,10], increased from 8 to 13. Based on the data in the Table 1 and data from PwC's 16th annual review of global trends in the mining industry [8] we can conclude that diversified industries are much more efficient in their core business than enterprises focused on one type of minerals.

One of the important and successful levers for managing mining in a market economy is its diversification. In a more general case, two procedures are distinguished:

• diversification of the enterprise, as the extension of its business activities to new areas (expanding the range of products or types of high services);

• capital diversification - allocation of money invested in the economy (for example, part of the profit from the main activity) in order to reduce risk or generate additional income.

Within the designated framework, the following types of diversification of mining enterprises can be distinguished:

1) the use (utilization) of the mineral product mined or obtained during the primary processing of ore mass (concentration);

2) integrated development of deposits, other than those identified as minerals for the extraction of minerals, minerals and mining;

3) the integrated development of mineral resources in the fields, with the exception of the two previous technologies, ensuring the production of both typical and non-traditional for the mining production of additional functions that allow you to create whole or part of the infrastructure and technological tools created to extract the main or additional minerals.

Common to the selected options is that in all cases, diversification is carried out on the basis of resources (minerals, mining, main production waste, etc.) that the enterprise has. The peculiarity of the first of these varieties of diversification is that the company invests insignificant, sometimes negligible capital on the formation of a line for the production of new (in addition to the main) products. The thing is that when extracting the main mineral resource, an enterprise inevitably extracts large volumes of various rocks from the subsoil, for example, during the excavation of underground mine workings or during overburden operations in quarries. It is known that rocks in their original or processed form are always in demand in the market. If the volumes of incidentally mined rocks are small compared with the volumes of extraction of the main mineral (ore), then the market value and selling price of them are formed only from the direct costs of their transportation and processing. In the case when the volumes of simultaneously mined and of interest on the market rocks are comparable with the volumes of extracted main minerals, the reserves of these rocks within the mine or quarry fields are calculated and approved as minerals in accordance with existing requirements. It also helps to reduce the cost of the main products of the enterprise and increase its competitiveness in the market. With diversification, the viability of the enterprise increases significantly, and the degree of its increase depends on the amount of additional income received by the company from the sale of "non-core" products. The enterprise as a result of the implementation of technological mining processes for the main project receives raw or salable ore and simultaneously various so-called "empty" rocks. However, the market price of products is greater, the higher the degree of processing of raw materials. Following this well-known truth, mining enterprises subject the mined rocks to various processing, the depth of which depends on market demand, technical capabilities and economic feasibility.

The second type of diversification of mining enterprises differs from the first in that, along with the disposal of incidentally mined rocks, they carry out specialized mining of certain types of mineral raw materials, which are absent in the bulk extracted from the subsoil, but are located directly in the mine field of the enterprise or in the zone of its economic influence. This is achieved on the basis of a more complete use of the enterprise infrastructure and, if required, with a certain qualitative and quantitative development. This kind of diversification of a mining enterprise is based on a deeper study of the market and is more responsive to demand and prices for various mineral raw materials, but requires the enterprise to have significant financial and production capabilities, and most importantly - the presence in the bowels of the relevant reserves of previously noted raw materials.

The first two varieties of diversification of mining enterprises considered in all aspects are closely related to the initial project for the extraction and processing of the main ore-mineral raw materials. True, at the final stages, the technological lines for the processing of primary and secondary raw materials can vary significantly.

3 Discussion

The task of diversification includes many aspects of the enterprise that change over time and have varying degrees of uncertainty. The solution to this problem is preceded by the development of an economic-mathematical model of the rational distribution of capital over time with the diversification of mining. This model allows to formulate a diversification strategy for the enterprise and choose the most effective of them from an economic point of view. The quality assessment of various directions of diversification of production in this case can be performed using the indicator of internal rate of return (GNI). However, taking into account the drawback of this indicator, which consists in the possibility of the existence of multiple roots for solving the equation, the model also uses additional criteria: net present value (NPV) and profitability index (IPR). The simultaneous use of these criteria for evaluating the effectiveness of the selected option allows you to take into account the disadvantages of each of them. The algorithm of the model, compiled on the basis of the development as applied to solving the problem of diversification of a mining enterprise:

1. At the first stage, the characteristics of various projects are compiled. They also introduce restrictions on investments, marginal rate of income, and the life of production;
2. For each project, possible cash flows are evaluated: positive, negative and net;
3. Establish the number of projects for further consideration;
4. For each project, the number of alternatives is determined. If they are, then conduct a marginal analysis;
5. Select the best projects from among alternative ones;
6. Form the number of projects for economic evaluation (non-alternative projects + the best of the alternative according to the results of marginal analysis);
7. The remaining projects are evaluated according to the GNI criterion (internal rate of return). At this stage, it is assumed that the start dates of the analyzed industries coincide;
8. According to the GNI indicator, the number of projects included in the enterprise development strategy is estimated. In this case, take into account the limitation of capital available to the enterprise for the implementation of projects;
9. Form an enterprise development strategy. Take into account the effectiveness of each of the projects, equity, as well as the possibility of reinvestment of income. There are two main development strategies;
10. According to the first strategy, all projects are ranked according to the GNI indicator. The sequence of input projects is determined, starting from the one that has the maximum value of GNI;
11. Allocate the second development strategy. Here, all projects are ranked by the deadline for the start of construction. On this basis, projects are ranked, starting with one that is characterized by a minimum deadline for the start of construction, taking into account the total possible amount of capital;
12. Assess the totality of all attracted projects according to the criterion of net present value (NPV). The assessment is carried out sequentially as subsequent projects are involved

in the strategy. Attracting a project is considered justified if it does not lead to a decrease in the NPV of the totality of previous projects;

13. Choose the most effective strategy for the development of the enterprise on the basis of comparing them according to the criterion of profitability index (IPR), determined by the ratio of positive and negative cash flows taking into account the time factor.

The adopted strategy determines both the most appropriate timing for the commissioning of each project and the appropriate distribution of investments over time. The two forms of diversification considered are ubiquitous in mining enterprises. And the matter is not only in the unequivocal profitability of this event, but also in the fact that enterprises radically and scale solve another problem - the environmental one. Mining and processing (especially in open pit mining) have a very negative impact on the environment. Moreover, the factor that plays the largest role in this process is the overburden rock stored in dumps.

4 Conclusion

The larger volumes of waste rock will be used as minerals, i.e. as raw materials for the preparation of products, the ecology of the mining enterprise will improve to a greater extent, and, as was shown earlier, the more profit it will receive. At the same time, it should be remembered that any waste is products for which application has not yet been found. It seems that most mining and processing enterprises remember this, expand the assortment of products from overburden rocks and successfully market it, competing with enterprises that are traditional suppliers of these products. There is hope that in the future the entire volume of waste in mining enterprises will be disposed of. In this regard, the preconditions for the restructuring of the raw material base of such a powerful industry as the building materials industry are being formed in the regions. The main idea is to completely abandon countless small quarries and quarries for the extraction of raw materials for enterprises producing construction materials in the entire country in large mining regions, and to receive them only from mining and processing enterprises, which will be much cheaper, better for quality and dramatically improve the environmental situation.

References

1. *Mine 2019. Resourcing the future*. URL: https://www.pwc.com/gx/en/energy-utilities-mining/publications/pdf/mine-report-2019.pdf
2. M.A. Klimovich, Economics and Innovation Management, **1**, 18-25 (2019) DOI: 10.26730/2587-5574-2019-1-18-25
3. *Diversified Development: Making the Most of Natural Resources in Eurasia* (International Bank for Reconstruction and Development, 2014)
4. N.V. Kudrevatykh, A.A. Galler, Economics and Innovation Management, **1**, 45-54 (2019) DOI: 10.26730/2587-5574-2019-1-45-54
5. İbrahim Anıla, IhsanYiğitb. Procedia - Social and Behavioral Sciences, **24**, 21494-1509 (2011)
6. S.A. Zhironkin, Ugol', 4, 29-31 (2001)
7. M. Cehlár, P. Rybár, J. Mihók, J. Engel, Journal of Mining and Geotechnical Engineering, **1:4,** 4-23 (2019). DOI: 10.26730/2618-7434-2019-1-04-23
8. S.M. Miliy, M.A. Rozhenko, K.A. Prokopenko, G.I. Korotkova, Journal of Mining and Geotechnical Engineering, **2:5**, 4-18 (2019). DOI: 10.26730/2618-7434-2019-2-04-18

9. *TOP 50 Mining Companies.* URL: https://datawrapper.dwcdn.net/geN1p/4/
10. *Value of top 50 mining companies surge $140 billion in 2017.* URL: https://www.mining.com/value-top-50-mining-companies-surge-140-billion-2017/

The Convergent Structural Base of Sustainable Development in the 21st Century

Sergey Zhironkin[1,*], *Michal* Cehlar[2], and *Vitaly* Zhironkin[1]

[1]Siberian Federal University, 660075 Lida Prushinskaya St. 2, Krasnoyarsk, Russia
[2]Technical University of Kosice, Faculty of Mining, Ecology, Process Control and Geotechnologies, 04200 9 Letná, Kosice, Slovak Republic
[3]National Research Tomsk Polytechnic University, 634050, 30 Lenina st., Tomsk, Russia

Abstract. Sustainable development is a symbiosis of technological and social progress, as well as responsible environmental management, should not rely on old industries that actively consume non-renewable natural resources, but on new, convergent-technological ones. In this paradigm, the environmental safety of future generations is formed by structural shifts in the economies of advanced countries caused by the expansion of technological convergence. Technological convergence – the result of accelerating scientific and technological progress in the 21st century – has a transformative effect on the economies of developed countries, forming a new type of sectorial genesis. It is not related to the deepening of sectorial specialization or inter-industry cooperation within the framework of the evolution of basic technologies, but to the emergence of technologies of a fundamentally new type, formed as a result of the merger of know-how from different industries. The structural shifts arising from technological convergence do not have the historical analogues and forms of regulation developed by industrial policies of developed countries. The article outlines the key statements of the the impact of convergent technologies on the structural base of sustainable development.

1 Introduction

At present, a new type of extended reproduction, which is essentially over-intensive and connected with a new integration of factors of production in new combinations of technology, cooperation and division of labor are being developed the countries of advanced market economies [1]. Technological breakthroughs of the 21st century, new forms of capital attraction and institutions of innovation investment create prerequisites for modernization of the system of productive forces. In this process, the diffusion of convergent technologies is connected with modification of economic relations in investment, innovation, credit spheres, in industry production, and transformation of ownership relations. The combination of such processes of modification of economic relations in the reproduction sphere creates conditions for structural shift of a new type, in which the structural-forming factor is not creation and commercialization of new industry technologies, but their interaction.

* Corresponding author: zhironkin@inbox.ru

2 Materials and Methods

Today, there is a prevailing technocratic approach to exploring the role of convergent technologies in the development of modern economy and disregard to the analysis of reproductive relations, through which technological convergence creates structural-transformative productive forces and modifies economic relations. For example, technological convergence was initially the object of sociological research in the 1990s (M. Castells [2]). Later, M.C. Roco, W.S. Bainbridge [3] identified convergent technologies as a synergistic result of interaction between representatives of four spheres of industry science: nano- and computational technologies, genetic engineering, neuroscience. Since 2001 under the auspices of the National Science Foundation of the United States, the so-called NBIC-initiative (nanotechnologies-biotechnology-IT-cognitive sciences) has been put forward, which denotes a convergent whole. The connection of technological convergence with the development of the national innovation system is seen as a connection of isolated trajectories of sectorial sustainable development [4], and as integration of innovative processes in basic and high-tech industries based on digitization (structural-and-technological convergence) [5].

The implementation of the conditions for the development of the reproductive process in technologically advanced countries leads to the fact that the structure of the economy begins to directly depend on the growth rate of production of intellectual products. These processes take it to a new level of capitalization and, consequently, research, production and market opportunities (Figure 1).

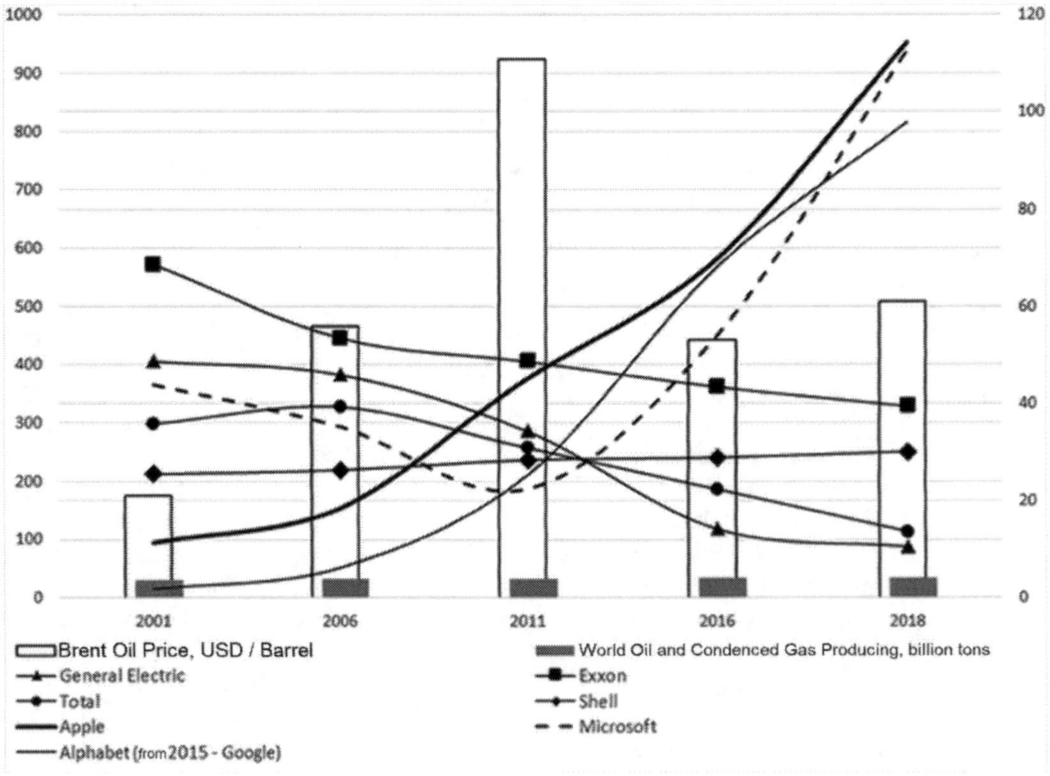

Fig. 1. The dynamics of the market capitalization of the largest international companies in the basic industries and the IT sector (annual average, billion dollars, left scale), oil prices and liquid hydrocarbon production volumes (right scale) [7] [8].

On the one hand, this is radically changing the structure of corporate property, employment and investment, and on the other, it accelerates and stabilizes macroeconomic dynamics. On the example of the United States, the trend of intellectualization and digitalization of the activities of the largest representatives of international business is clearly visible.

In terms of the development of productive forces, technological convergence means a new relocation of productive forces – both within production chains and spatially (including extraterritorial, in network form). As a result of this relocation, clusters of a new type are being formed, in which producers of the benefits of modern cycle of scientific and technological progress (such as modern medicines, global information access facilities, clean energy, etc.) create a "centric" network form of organization of the movement of resources, information and know-how, labour, and coordinate their activities on the basis of common economic incentives. Such relocation takes place in the form of the development of global network clusters of convergent technologies, using virtually unlimited sources of investment in the global capital market and a digital market for highly intelligent labour, creating products that exist mainly in intangible form, extraterritorially made and consumed.

3 Results and Discussion

Based on the above approaches to assessing the impact of convergent technologies on the development of national and global economy, we have highlighted the following principles of influence of convergent technologies on structural shifts:

1. The principle of accelerating the reproduction process during the expansion of converged technologies, according to which the connection of individual know-how and technological chains of different industries is accompanied by the connection of their investment flows.
2. The principle of the neo-industrial impact of technological convergence on the structural shift, according to which the separation of high-tech production from the innovative development of basic industries is unacceptable.
3. The principle of modification of economic relations in the system of public reproduction, connected with investment of innovations and their introduction, with creation of tax, financial and credit incentives of technological convergence, with acceleration of fixed capital renewal, with formation of new social relations and patterns of economic behavior.
4. The principle of targeted programming of convergent technologies development and investment in their mass production, diffusion in the basic sector of economy.
5. The principle of socializing technological convergence and smoothing out the negative consequences of the resulting structural shift.

The driving forces of technological convergence, such as modification of investment and change of nature of work in the process of development of fundamental research in the fields of NBIC, connection of the market of modern means of production and intellectual capital, transformation of property relations within the framework of collective investments, innovative entrepreneurship, public-private partnership in innovation define a set of conditions of influence of convergent technologies on structural shifts:

• reproduction and cluster conditions, which consist in accelerating the reproduction of fixed capital in modern manufacturing industries (mechanical engineering, radio electronics, production of synthetic materials and modern alloys, medicine, etc.).

• innovation and investment conditions of effect of convergent technologies on structural shifts are connected with neo-industrial development and modernization of industry, which should guarantee demand for NBIC-convergence products in the process of formation of post-oil energy (super-powerful power sources, bio-generators, etc.), technologies of superconductivity, recycling and energy-minimized industry, biochemical materials, nano-computers.

• institutional conditions include purposeful development of state and business institutions capable of reorienting investment, labor, and intellectual resources from extractive and manufacturing industries to industries in which convergent technologies are created and find demand. This means that innovations generated in the fields of information, bio-, nano-, cognitive technologies should be demanded in the traditional industries - biochemistry, radio electronics, light, food industry, polymer production, etc.

• the social conditions of the impact of technological convergence on structural shifts imply the purposeful formation of a sustainable social group of convergent technology developers, innovative entrepreneurs and start-up managers.

Thus, the technological determinants of structural changes in the modern economy are represented not only by new technologies. They also include fundamentally new forms of reproductive relations, methods of production and consumption, as well as consumer preferences, individualizing as digitalization unfolds. It is important that with a personalized approach to production there is no rise in price of the product, as one would expect. On the contrary, the sustainable development approach allows optimizing costs as much as possible provided that the production processes are deeply automated, oriented towards the maximum rational use of technical resources, efficient energy conservation and environmental management.

4 Conclusion

Technological determinants of the structural shift (radical irreversible changes in macro- and mesoeconomic proportions) can be considered those phenomena in the reproductive system that mediate the penetration of technologies into the processes of creation, distribution and consumption of goods, changing the picture of the involvement of production factors.

Firstly, inter-industry transfers and the international diffusion of innovations that generate technological convergence and create a synergistic effect of global scientific and industrial cooperation. The difference between the latter and the formation of global chains of movement of primary resources in the twentieth century. is the gain of all countries participating in innovative and technological cooperation, in contrast to the nonequivalent use of the raw materials potential of developing countries.

Secondly, the decentralization of production processes, their maximum automation and integration, focus on minimizing the share of material and natural resources.

Thirdly, the individualization of the manufactured product and the connection to the network of as many devices as possible (Internet of things, industrial Internet).

Fourth, the convergence of cognitive and information technologies, as well as nanotechnologies with their characteristic principle of suprasectoral functioning and a high level of integration potential.

Fifth, the shift in emphasis in basic research from the analysis of substances and processes to the synthesis of fundamentally new materials with new properties and characteristics. Their further construction at the atomic level with a reduction in the time gap between laboratory studies, the commercialization of innovations from 8-10 years in 1970-80 up to 2-3 years with access to a lower level of investment risk of innovation.

For the Russian economy, a positive technologically induced structural shift is the only way to reduce the technological gap. This in turn requires a review of the role of the state in regulating national reproduction. It is necessary to expand his partnership with high-tech business, creating maximum incentives for the inter-industry transfer of innovation and investment in convergent-technological production.

References

1. S.A. Zhironkin, K.A. Kolotov, A.E. Genin, F.V. Agafonov, S.A. Kovalevsky, IOP Conf. Ser.: Earth Environ. Sci., **50:1**, 012011 (2017)
2. M. Castells, P. Himanen, *The Information Society and the Welfare State: The Finnish Model* (Oxford University Press, Oxford, 2002)
3. M.C. Roco, W.S. Bainbridge, *Converging technologies for improving human performance: nanotechnology, biotechnology, information technology and cognitive science* (Kluwer Academic Publishers, New York, 2003)
4. A. Gate, University of Toronto Faculty of Law Review, **58:2**, 83–117 (2000)
5. T. Strader, *Digital Product Management, Technology and Practice: Interdisciplinary Perspectives: Interdisciplinary Perspectives* (Business Science Reference, Hershey, 2011)
6. *Top 10 most expensive companies in the world in 2019. Official website of the FXSSI news agency.* URL: https://ru.fxssi.com/top-10-samyx-dorogix-kompanij-mira
7. *Brent oil prices. The official website of the information and consulting agency RBC. Quotes section.* URL: https://quote.rbc.ru/ticker/181206
8. L. Dube, A. Faber, J. Struben, K.J. Srivardhini, Annals of the New York Academy of Sciences, **13311:1**, 144-159 (2014)

Human Potential in the System of Sustainable Development

Tatyana Snegireva[1,*], *Gennady* Kayachev[2], *Albert* Falaleev[2], and *Sergey* Kurgansky[3]

[1]Russian State Library, International Education Cooperation Sector, 119019 Vozdvizhenka st., 3/5, Moscow, Russia
[2]Siberian Federal University, 660075 Lida Prushinskaya St. 2, Krasnoyarsk, Russia
[3]Irkutsk National Research Technical University, 83, Lermontov street, 664074 Irkutsk, Russia

Abstract. The essence of sustainable development is that the progress of the current generation does not go against the interests of future generations. In the classical definition, the balance between generations is understood as the balance of needs. However, the word "needs" can be interpreted differently. First, we can talk about the ability to meet the needs in the context of the availability of natural resources. This means that we must leave to future generations the same amount of resources that we have, so that they can meet their needs with these resources. Secondly, we can talk about the ability to meet the needs as a standard of living. This means that for future generations it should be at least the same as the standard of living of our generation. At the same time, it is absolutely not necessary that this be achieved due to the availability of natural resources. For example, we can now invest some resources in the development of human capital and knowledge, and future generations will provide a higher standard of living through the growth of knowledge. Consequently, the factor of human capital, determining the pace of introduction of innovative technologies and moving away from resource-intensive industries, will determine sustainable development in the process of transition to a post-industrial economy.

1 Introduction

The founders of the concept of sustainable development drew attention to the fact that human activity has already gone beyond the balanced functioning of the Earth's ecosystem. The special importance of the problem of sustainable development is intrinsic to the mining sector of national economy which nowadays is a main source of the environment pollution [1-5]

The conclusion about reaching the limits of growth was based on a number of estimates. Thus, despite successful economic growth in some regions, since 1974, the decline in world production per capita began. At this turn the peak of the growth of living standards in the world (average per capita) was reached. From the mid-twentieth century, a fifth of arable land has been withdrawn from agricultural use, and recently every year 8.5 million hectares of the most productive regions of the planet go out of use. Improving the efficiency of the

Corresponding author: snegireva_tv@mail.ru

global agro-industrial complex no longer compensates for the absolute reduction of its resource base. Between the 1950s and the 1990s, the population doubled and the global economy grew almost 5 times, as a result of which the equilibrium in nature was disturbed and the preceding growth trends occurred. World economic growth reached a historic peak of 5.2% in the 60s, slowed to 3.4% in the 70s and 2.9% in the 1980s. In the 1990s, the growth rate of the global economy was about 0.5% per year, and per capita living standards continued to decline [6].

Thus, according to the estimates of the concept of sustainable development authors, the critical threshold for the sustainability of society has already been surpassed, since humankind consumes significantly more resources than the laws of stable functioning of ecosystems allow. Confirming the conclusion about the achievement of the planetary limits of economic growth and the state of "zero growth" (and even "dumping" of capacities in a number of regions), some authors drew attention to the factor of underdevelopment of traditional societies [7]. He assesses poverty as a major global problem, since it is both the cause and effect of excessive demographic growth.

The specified attention of the concept of sustainable development paid to the underdeveloped countries highlights the main contradiction of the global system as the contradiction between the North and the South. This contradiction is associated with more intensive use of human capital by the technologically advanced countries of the North, and with underdeveloped human potential in the resource-producing countries of the South.

The principle of maximizing the use of human potential as the basis for sustainable development was taken up by European Social Democracy, in particular the "Independent Commission on International Development Issues" by German Chancellor W. Brandt. In the 1980s, the Independent Commission on Disarmament and Security and the International Commission on Environment and Development were formed under the chairmanship of the leaders of the Social Democratic parties and the Prime Ministers of the Scandinavian governments. The result of their work were various memorandums, declaring the increasing importance of civil society, innovative and entrepreneurial initiatives in the modernization of the economy and society in accordance with modern environmental standards.

2 Materials and Methods

The Sustainable Development Strategy, approved by the UN General Assembly [8], provides for:

• meeting the needs of modern generations without threatening the ability of future generations to meet their own needs;

• respect for available global resources and environmental potential;

• early prevention of environmental problems as the most effective and economical means of achieving environmentally sound development;

• the major objectives of the environment and development policy should include the preservation of peace, the revitalization of economic growth and the alteration of its quality, the solution of the problem of poverty, the satisfaction of the immediate needs of people, the demographic problem, the preservation and strengthening of the natural resource base, the reorientation of technology and an integrated approach to environmental and economic issues in the decision-making process;

• equitable distribution of environmental protection costs, equitable distribution of economic development benefits between and within countries, as well as between current and future generations.

As has been repeatedly noted recently, the decisions of the UN Conference in Rio de Janeiro are practically not being implemented. These decisions were previously rated as unrealistic, although they were significant enough to mobilize public opinion. Thus, for example, G. Daily, an economist at the World Bank's Environmental Protection Department, said regarding "Agenda 21": "This discouraging document is nothing more than the joint snoring of idle economists in the deepest dogmatic slumber." [9]

At the same time, the discourse on discussing the problem of sustainable development is well-established. Consequently, discussions around sustainable development can be viewed as a subjective reflection of a latent, ill-understood, objective mental form. Consequently, explication is not a concept of sustainable development, but the underlying objective mental form and an analysis of strategies that are permissible within this form.

Evaluating the structure of the implicit objective mental form of the concept of sustainable development, let us fix its main aspects, on the basis of which the configuration of the use of human potential arises in a post-industrial environment-oriented economy.

The first cornerstone is "generation centrism". Of all the definitions of the concept of sustainable development, the most frequently used definition of the Brundtland Commission is: "Sustainable development is development that meets the needs of the present generation and does not jeopardize the ability of future generations to meet their own needs" [10]. The balance of intergenerational and intragenerational interests plays an important role in the concept of sustainable development. Intergenerational equilibrium is understood as an equitable distribution of resources between present and future generations, and the intergenerational aspect can be understood as equality in distribution within the same country and between different countries within the same generation.

The second cornerstone is "economocentrism". Extreme underutilization of human potential poverty – is assessed as a key global problem, the solution of which is a prerequisite and condition for solving other global problems (first of all, demographic and environmental problems). The Plan of Implementation of the World Summit on Sustainable Development (Johannesburg, 2–4 September 2002) also puts poverty overcoming at the forefront: "The eradication of poverty is the greatest global challenge facing the world today and is a prerequisite for sustainable development especially developing countries." [11] Thus, the concept of sustainable development focuses on the reorganization of economic activity.

The third cornerstone is "natural centrism". Attention is paid to the protection of the environment, the preservation of ecosystems, reproducible natural resources and biodiversity, the balance of demographic growth and biospheric restrictions.

On the basis of the three "whales" of the concept of sustainable development, an image of sustainable development economy emerges, which, firstly, uses natural resources rationally and, secondly, optimizes the intergenerational transmission of natural resources.

The question arises: in the framework of which economic system (model) is a possible economy of sustainable development? "Agenda 21" is permeated with the pathos of an effective market. At the same time, the very fact of adopting such a concept, as well as many reservations regarding the use of market instruments, testify, in our opinion, about the "failure of the market". Taking into account that "Agenda 21" contains a steady turnover of "economic and market", we can assume that sustainable development economics is not identical to a market economy, although it uses its tools. Therefore, a sustainable development economy can exist and develop in a non-market format.

In general, the economic life of a society includes a limited set of mutually complementary spheres of economic life, namely:

• team economy – the market economy;
• life supporting economy – the military economy,
• moral economy – the traditional economy,
• prestigious economy – the shadow economy.

Each of the spheres of economic life is characterized by a specific form of production and exchange, in particular: the market economy – commodity exchange, the moral economy – "help exchange", the prestigious economy – gift exchange, the traditional economy – service exchange, etc.

The focus of economic science is still the market economy due to the high uncertainty in the behavior of its agents. Recently, economists have begun to more actively study the institutions of the command economy (especially in the political economy of socialism) and the shadow economy. The economic anthropology (economic ethnology), the life-supporting economy, the prestigious economy, and the moral economy are examined. On the periphery of economic thought are the military economy and the traditional economy (studied mainly by historians).

Thus, when deciding on the systemic identity of a sustainable development economy, one has to make a choice from a limited set of ways to organize economic activity. By a combination of two criteria: the rational use of natural resources and the use of human potential in the translation of natural resources – the economy of sustainable development, should be attributed to the system of traditional economics.

3 Results and Discussion

The starting point for a precise exit of the Russian economy from the structural crisis should be the understanding that the initiation of new industrialization will lead to the need to develop a multilateral mechanism for involving human capital in it. However, the current methodology for studying structural changes in the economy is technocratic, and the postulates and theoretical conclusions based on research models of liberal methodology have failed to resolve the growing contradictions in the field of structural and technological renewal and reproduction of highly intelligent, productive, renewable driving forces for sustainable socio-economic development because the externalities for engaging human capital are: creation and distribution the use of new products and technologies, the rapid development of new technologies, their introduction into production and life, the ability for more complex and productive work, the generation of innovations, etc.

The content, structure and forms of labor and human capital, its capitalization are changing dramatically. A person's creative abilities, his creative competences expand the boundaries of production possibilities, from the production factor, human capital, namely, the use of human abilities, becomes the dominant of economic relations, and their development and growth in the ultimate goal of the functioning of the national economic and social subsystems. Under these conditions, there is an increasing demand for the concept of institutionalizing the involvement and integration of various forms of human capital based on the principle of social structuralism, permeating the entire structure of economic relations in the process of unfolding the new industrialization of the Russian economy.

The key prerequisites for the implementation of the new industrialization process in sustainable development are closely related to the expanding of human capital, and can be defined as follows:

1. Creation of viable conditions for the realization of human, as an active subject of the economic system, which, in the context of the development of environmental views on the process of the functioning of the economy, consists in creating effective and nature-oriented institutions ("rules of the game").

2. The subordination of the created mechanism is primarily to the interests of the person in order to ensure the possibility for his self-sufficiency, in conditions when the nature and functioning of new technologies require the employee to transition from the state of an external regulator to the state of the organic part of modern technologies.

The main principle of defining structural approach as a methodology of sustainable development is human-centeredness or human centrism, thanks to which the analysis of a person, or rather his (her) ability to generate new knowledge, as a key factor of the modernization process, manifests the principle of the primacy of the human role in economical and ecological relations through the prism of dialectics (understanding as "homoeconomicus", but as the basis of the structural development of the economy in the long term). This principle is that a person is a key factor in realization of positive structural changes and, at the same time, the main goal of the policy of sustainable development, since the socialization of convergent structure-forming technologies stimulate the expansion of the recycling as the industrial breakthrough of Russian economy.

Based on the new model of sustainable development, proposed by the authors, the directions of state policy formation were formulated considering human capital involvement, as follows:

• the state should give a person the starting point for investments to human capital due to the effective institutionalization of conditions for the growth of living standards, which will serve as a motivating factor for the realization of productive resource-saving technologies in the context of sustainable development;

• development of the country's production forces based on the latest technological structures in order to maintain the ability of individuals to use goods and services with the highest utility without damage growth for environment;

• the formation of a social basis oriented towards active self-realization, therefore the members of special social groups will be assured of the correctness of the development path chosen by the state and will become the basis for sustainable development. Under the conditions of sustainable development, these groups should be formed from scientists and engineers.

• creating conditions for improved communication of government, society and goods producers, which can be an effective tool for public control on nature preservation.

It was noted that the most effective and efficient transformation of the economy will be in the case of its implementation on the basis of a sustainable development initiated "from above".

Today, the state is required to purposefully form the institutional environment, to broader involvement of economy institutions, in order to develop industries with a multiplicative impact that meet the needs of the population, in particular in environment-friendly productions.

In the conditions of up-to-date efficiency of human capital involvement in the industrial sphere, it is necessary to form institutions that promote human development, the realization of its productive and social capital through the improvement of the system of material remuneration for productive work in the conditions of sustainable development.

Improvement of the material remuneration system should be focused on the reproduction of the labor force, which presupposes maintaining and restoring a person's ability to work, improving the skills of the workforce and implementing the function of forming future suppliers of labor, as well as motivating the realization of productive capital in industrial production.

4 Conclusion

Thus, the concept of sustainable development revives the values of the innovative economy, which explains the attention of its authors to developing countries. In the context of the importance of global natural resource constraints, the institutions of innovative economy are relevant for a sustainable development economy, and resource-saving technologies are the main form of human wealth.

References

1. S. Markov, J. Janočko, M. Tyulenev, Y. Litvin, E3S Web of Conferences, **105**, 01021 (2019)
2. M. Tyulenev, S. Markov, E. Makridin, Y. Lesin, V. Gogolin, E3S Web Conf., **105**, 02022 (2019)
3. V.V. Panov, O.S. Misnikov, Gornyi Zhurnal, **2015:7**, 108-112 (2015)
4. I.A. Mironenko, S.I. Protasov, Journal of Mining and Geotechnical Engineering, **1:4**, 24-34 (2019). DOI: 10.26730/2618-7434-2019-1-24-34
5. O.S. Misnikov, Journal of Mining and Geotechnical Engineering, **2:5**, 19-32 (2019). DOI: 10.26730/2618-7434-2019-2-19-32
6. S. Zhironkin, S. Demchenko, G. Kayachev, M. Ryzhkova, O. Zhironkina, E3S Web Conf., **41**, 04011 (2018)
7. S. Zhironkin, D. Khloptsov, N. Skrylnikova, I. Petinenko, O. Zhironkina, E3S Web Conf., **41**, 04010 (2018)
8. UN GA, *The US Sustainable Development Strategy* (United Nations General Assembly, Washington, 1990)
9. G. Daily, *On the critics of "Agenda 21"* (EPD WB, Washington, 2000)
10. L. Oliver, Modern Sustainability, **6:16,** 103-121 (2007)
11. J.R. Milligan, South Africa Journal of Economics, **123:13,** 65-79 (2004)

Effective cost allocation method in an industrial enterprise environment in mining industry

Kamila Janovská[1*], *Šárka* Vilamová[1], *Marian* Piecha[2], *Josef* Kutáč[1], *Roman* Kozel[3], and *Jana* Čitbajová[4]

[1] VSB-Technical University of Ostrava, Faculty of Materials Science and Technology, 17. listopadu 2172/15, 708 00 Ostrava – Poruba, Czech Republic
[2] Ministry of Industry and Trade Czech Republic, Na Františku 32, 110 15 Praha 1, Czech Republic
[3] VSB-Technical University of Ostrava, Faculty of Mining and Geology, 17. listopadu 2172/15, 708 00 Ostrava – Poruba, Czech Republic
[4] Technical University of Košice, Faculty of Mining, Ecology, Process Control and Geotechnologies, Institute of Earth Resources, Park Komenského 19, 042 00 Košice, Slovakia

Abstract. The ability to optimize costs based on knowledge of business processes during the period of change represented by the fourth industrial revolution is absolutely crucial. This paper is dedicated to the issue of cost management and allocation of overhead costs using the Activity Based Costing (ABC) method under the condition of an industrial enterprise. The aim of the authors was to propose a model that focuses on the allocation of overhead costs and calculation of assignment rates of overhead activities, which mutually cooperate. The proposed model enables objective stipulation of assignment rates of mutually cooperating overhead activities using exact economic-mathematical methods - the Leontief structural model. It was proven that use of the structural model, even in the case of multiple overhead and primary activities, leads to the considerable simplification of calculations when determining the assignment rates of overhead activities.

1 Introduction

In order to maintain the competitiveness of industrial enterprises, especially small and medium-sized enterprises, in the period of change represented by the fourth industrial revolution, the ability to optimise costs conditioned by a knowledge of corporate processes is a crucial skill at present. Without a detailed knowledge of costs, both those for individual processes and those for the final products of the industrial enterprise as a whole, industrial enterprises will be unable to define and adopt the right decisions to ensure sufficient profit and cash flow for the functioning and growth of the enterprise.

Current changes in the business environment can be summarised into the following points:

1. Significant growth of the structure of executed outputs,
2. Rising demands on the quality of provided outputs,

* Corresponding author: kamila.janovska@vsb.cz

3. Need to supply a wide range of outputs at shorter time intervals,
4. Shortening of the lifetime of offered products.

The outcome of these factors leads to changes in the cost structure, particularly by reducing the share of primary (variable) costs in relation to overhead (fixed) costs, which is given mainly by the increasing degree of manufacturing automation. The share of overhead costs needed to ensure support, servicing, information and controlling activities is rising. These conclusions logically result in the need for the better orientation of management measures not only with respect to the costs for final outputs, but also to the overhead costs for assurance processes [1]. Management measures do not refer merely to the need for managing the effectiveness of continuously rising overhead costs.

It is equally essential to address the issue of appraising the allocation of overhead (servicing) activities (e.g. transport, logistics, maintenance, sales), which are objects of overhead costs, to the primary activity (production). Subsequently, it is necessary to address the issue of distributing overhead costs in the primary (production) activities among outputs (products, activities) which are the final objects of overhead costs (see Fig. 1.).

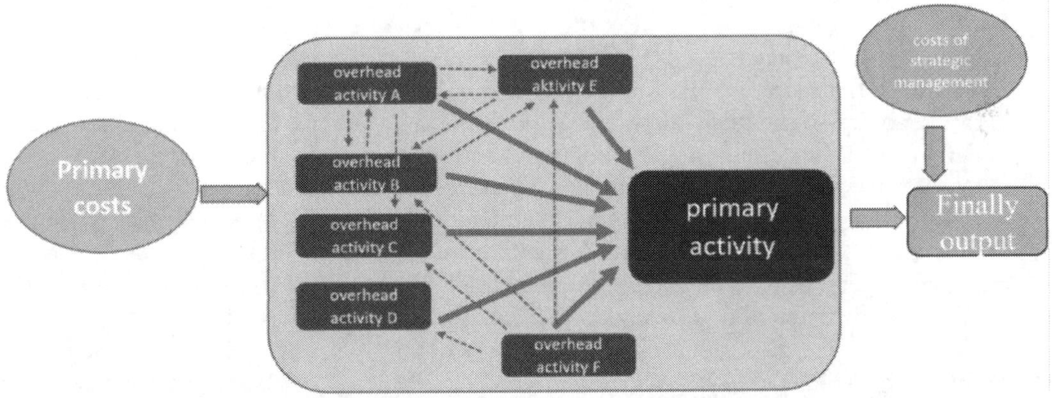

Fig. 1. Allocation of overhead costs of overhead activities to primary activities [actual processing]

The submitted article is dedicated to the issue of cost management and allocation of overhead costs using the Activity Based Costing (ABC) method under the condition of an industrial enterprise. Extensive literary research was conducted, which proved that the process of implementing the ABC method at domestic industrial enterprises is not very widespread and that the allocation of overhead costs for overhead activities to primary activities is particularly problematic.

2 Current state of scientific discovery of the examined issue - literary research

Most enterprises perceive the fact that in the period of change brought about by the fourth industrial revolution, traditional cost management methods will no longer suffice; it is necessary to start using management methods that can keep up with the dynamically evolving economic environment [2]. Hence, a process-based view of the enterprise, based on reviewing corporate processes and activities which create value for customers and generate costs, is being increasingly applied at industrial enterprises. This view of the corporate transformation process is employed in increasingly more new management instruments and methods, which include Activity-Based Costing. The principle of the ABC method differs from the traditional approach of calculation methods, because it divides consumed resources into activities with the aim of achieving the distribution of overhead costs based on the actual cause of their accrual. The ABC calculation returns to the cause and effect relation, because it strives not to divide resources into centres, but into activities, with the aim of distributing overhead costs according to the actual cause of their accrual.

As for application of the ABC method, it may be claimed that the level of acceptance into corporate practice is unfortunately low at domestic industrial enterprises.

The ABC method was created in the USA and its founders are deemed to be the American authors Cooper and Kaplan [3], who discuss it in the article "Measure Costs Right: Make the Right Decisions" of 1988. Corporate practice attempted to use the ABC method many years earlier; for instance Bouquin [4] mentions a certain degree of application of the ABC method in corporate management as far back as in the 1860s by the American company General Electric, which stood at the birth of many other commonly used management methods and approaches (e.g. the GE matrix, etc.). Since tine 1990s, the ABC method has been implemented and used by companies such as Coca-Cola, Apple Computer, IBM, HP, AT&T, P&G, Siemens and many others [5].

The processing industry with heterogeneous production and assembly technologies currently have the most experience in applying the ABC method. However, the method is fairly widely applied also in non-production sectors - e.g. in commercial companies, banking, insurance and transport [6]. Many authors (e.g. [7, 8]) agree that application of the ABC method is more beneficial at medium-sized and large enterprises. The reason is the high variability of products and services, the greater quantity of products and services, the wider range of customers, higher degree of automation of corporate processes and more extensive management structure of the enterprise, resulting in a relatively high share of overhead costs in total costs. As stated by one of many studies [9] concerning the degree of acceptance of the ABC method among enterprises, the lower degree of use among small enterprises may be caused by the relative high initial costs for implementation. Some studies state a degree of implementation exceeding 50% in the USA [10].

The oldest and most often cited studies concerning the ABC method include the study by Innes and Mitchel of 1991, which was conducted among the largest British enterprises at the time. The study was repeated twice with the same questionnaire, first in 1994 and then again in 1999. The study of 1999 offered interesting findings compared to the first survey in 1994. Between the follow-up studies, the number of enterprises using the ABC method did not increase, but the number of enterprises that refused to implement it did. Whereas in 1994 its distribution among production enterprises and service enterprises was comparable, in 1999 the ABC method was far more common in the financial sector (40% of enterprises) than in the production sector (14% of enterprises). A stronger inclination towards the ABC method in Great Britain was not confirmed even in another survey in 2007, the result of which include the finding of only 15% of distribution of the method amount major British enterprises in the reviewed sample. Another English-speaking country, where interesting research on the level of distribution of the ABC method was conducted, was Australia. Here, the identified distribution of the method was incomparably higher than in Great Britain. The survey conducted in Australia in 1998 indicated a degree of acceptance of the ABC method of 52%, while the subsequent survey in 2004), which focussed on medium-sized and large Australian enterprises, reported a 78% distribution of this method. The conducted scientific investigations indicate that the distribution of the ABC method in non-English-speaking European countries is lower. For instance, in his article "Activity-Based costing: how far have we come internationally?", it states that the lower distribution of the method in continental Europe is due to the use of different cost systems with a longer tradition and higher degree of elaboration than was the case in the United States before the introduction of the ABC method.

Only very few surveys of the use of the ABC method in corporate practice in the Czech Republic have been conducted. An overview of the status of practical application of this method in the Czech Republic is offered primarily by the survey conducted in 2004, which states that of a sample of 140 Czech industrial enterprises, 51% of enterprises have never examined the ABC system, a full 18% of enterprises have looked at the ABC system but then rejected its application, 23% of enterprises realistically considered the application of the method, 3% of enterprises where in the implementation phase and only 5% of enterprises were actively using it. Awareness of the method increases with the size of the enterprise, and consultancy agency often help with its implementation. A survey conducted in 2007 determined that there was a slight growth in use of the ABC calculation within this method, specifically this referred to use at 8% of Czech enterprises in the contacted selection sample.

The related issue of allocating overhead costs is addressed in his thesis by Novák - he maps the issue of overhead cost allocation and the use of modern cost management and calculation methods in the conditions of domestic manufacturing enterprises, the analysis of overhead cost structure and development and the creation of a simple guide on how to effectively approach the overhead cost management and allocation based on key rules and principles, while using various cost instruments. Of the total 8,070 enterprises address (enterprises in various sectors, based in the Czech Republic with more than 20 employees), 291 responded. Among other, they examined the distribution of the ABC method as an instrument for managing and measuring performance. Of the responds who replied, 22% used the method. ABC is among the above-average used management instruments in the paper and chemical production industry, and of the tertiary sector in the branches of transport and connections and activities concerning real estate and lease. On the contrary, low interest is apparent in the sectors of electrical instruments and construction. It may be assumed that of the companies that replied, there is a greater share of those interest in new management methods than in the total sample. For this reason, the distribution of ABC among Czech enterprises is probably lower than suggested.

3 Application of the ABC calculation method - allocation of costs for overhead activities to primary activities - proposed model

The fundamental step in application of the ABC calculation model is an activity analysis, consisting of several partial, consecutive steps which aim to calculate the individual costs for primary activities. Within the activity analysis, the specific partial steps are as follows:

1. Stipulate the activity benchmarks.
2. Stipulate the degree of activity output.
3. Calculate the individual (variable) activity costs.
4. Allocate the costs of support activity to primary activities.

The problematic part of allocating the costs of overhead activities to primary activities in corporate practice is the fact that overhead activity output is not consumed only by the primary activities, but also mutually by the other overhead activities. The problem then becomes the objective stipulation of allocation rates of outputs of mutually cooperating overhead activities, because their allocation links create mutual cycles.

Based on a discussion with experts from industrial enterprises, it may be stated that if they do address this issue, they do so in gradual steps (iterations), until the predefined accuracy is achieved, meaning a minimal difference in allocation rates in consecutive steps. The aim of all these steps is to calculate such cost rates (prices) for overhead activity output allocated to primary activities, which will ensure that the value of the product of these calculated rates and the quantity of overhead activity outputs (in the form of benchmarks) allocated to primary activities is equal to the sum of the values of primary costs of all overhead activities. This ensures that even if the overhead activities variously mutually transfer their outputs, in the end all of their primary costs will be allocated to the primary activity in the form of secondary costs, and these will then be distribute with the help of other activities in the form of benchmarks to the calculated outputs (products) of primary activities.

If we take into account that, in practice, there tends to be a high number of overhead activities, then the iteration calculation is usually very lengthy and ineffective, because if any changes occur in the course of calculation, e.g. in primary costs, the quantity or allocated outputs in the form of overhead activity benchmarks, all of the calculations must be conducted again and repeatedly. The number of overhead activities in the real conditions of industrial enterprises also determines the number of iterations leading to the calculation of final rates of the assigned outputs. In practice, this means that such iterations number in the tens. Such calculation is complicated, lengthy and every further intervention in the defined task is difficult (change of plan, calculation of variant solutions, divergence from the planned course).

Proposed model for allocation of overhead activity costs.

The proposed model focusses on the allocation of overhead costs and calculation of assignment rates of overhead activities, which mutually cooperate. The model enables the objective stipulation of assignment rates of mutually cooperating overhead activities using exact economic-mathematical methods - the Leontief structural model.

Calculation structure:

I. Composition of a chessboard table of the structural model - natural expression.

Based on the input data (management accounting and operative production records), a chessboard table of the structural model in natural expression is composed, which describes the conditions for the existence of a balance between resources and needs in a production - consumption system (industrial enterprise).

II. Composition of a chessboard table of the structural model - natural expression.

The benchmarks of the individual overhead activities need not be substantive identical (e.g. hours, t, CZK, unit...), which is why the chessboard table in natural expression must be modified into value expression.

III. Composition of Matrix A - matrix of direct consumption coefficients.

Matrix **A** is a matrix of technological coefficients, which expresses which products were directly consumed to produce the required output (final consumption vector) in the production - consumption system.

Calculation of Matrix A - matrix of direct consumption coefficients

$$x_{ij} = a_{ij}\, x_j \qquad i,j = 1,\, 2,\, \dots,\, n \tag{1}$$

where:

aij	the coefficient of direct specific production consumption of sector i in sector j (quantity of production of sector i needed to produce a production unit in sector j) (i, j=1,2,…,n);
xi	total value of production of sector i (i =1,2,…,n)
xij	total consumption of production of sector i in sector j of the given production - consumption system (i, j=1,2,…,n)

IV. Calculation of Matrix B - matrix of complex consumption coefficients

Matrix B is the matrix of complex consumption coefficients, which expresses the total quantity of production in sector i needed to produce a production unit in sector j designated for final consumption - sale. Indirect consumption is mediate consumption, given by the fact that the respective sector j needs, in addition to the respective sector i, the production of other system sectors, which also consume production from the respective sector i. Matrix *B* is an inversion matrix of matrix $(E - A)^{-1}$.

Calculation of Matrix B - matrix of complex consumption coefficients

$$B = (E - A)^{-1} \qquad (2)$$

where:

E matrix unit of the order n

B matrix of complex (direct and indirect) consumption coefficients of the production fo sector *i* in sector *j* b_{ij}; $b_{ij} \geq a_{ij}$; $a_{ij} \geq 0$

V. Calculation of the final value of assignment rates of outputs of mutually cooperating overhead activities.

4 Application of the proposed model at a specific industrial enterprise - discussion of achieved results

The structural analysis method was used in the application of exact decision-making methods when solving the issue of effective calculation of assignment cost rates for overhead activity performance. The text below simulates the calculation within 5 overhead activities based on their primary costs and the quantity of created and assigned output in the form of benchmarks. In the presented case, the overhead activities transfer their outputs among them in hours appraised at the respective cost rate. Table 1. contains the simplified input data for the calculation of assignment cost rates for outputs mutually assigned between overhead activities within the selected enterprise.

Table 1. Input data for calculating the assignment cost rates of outputs mutually transferred between overhead activities in the form of benchmarks using the ABC method at a selected metallurgical enterprise [own processing].

Input data for the calculation of assignment cost rates for outputs mutually assigned between overhead activities within the selected enterprise in the form of benchmarks								
Assigning activities						**Receiving activities**		
Overhead activities	**Primary costs for overhead activities [CZK]**	**Total quantity of assigned benchmarks per [hour]**	**Number of assigned benchmarks only to primary activities, [hour]**	**Number of assigned benchmarks only to overhead activities , [hour]**	**Number of received benchmarks from overhead activities, [hour]**	**Overhead activities**	**Number of assigned benchmarks only to overhead activities [hour]**	**Value of output assignment = Secondary costs of received activities [CZK]**
A	999 970	9 500	4 500	5 000	9 000	B	2000	210 520
						C	3000	315 780
B	2 000	23	13 000	10 000	5 000	A	3000	260 880

 https://doi.org/10.1051/e3sconf/201913403015

	080	000				C	5000	434 800
						D	2000	173 920
C	3 000	30	15 000	15 000	27	D	9000	900 000
	000	000			000	E	6000	600 000
D	3 999	39	19 000	20 000	11	B	3 000	307 680
	840	000			000	E	17000	1 743 520
E	4 999	47	22 000	25 000	23	C	19000	2 021 220
	860	000			000	A	6000	638 280

Based on the input data, a chessboard table of the structural model was compiled (see Table 2., 3.).

Table 2. Chessboard table of the structural model [own processing].

	Value of assigned outputs in the form of benchmarks (BM) for overhead activities [hour]					**Assigned outputs (BM) to primary activity [hour]**	**Assigned outputs (BM) total [hour]**
Activity	A	B	C	D	E	H	
A		2000	3000			4 500	9 500
B	3000		5000	2000		13 000	23 000
C				9000	6000	15 000	30 000
D		3000			17000	19 000	39 000
E	6000		19000			22 000	47 000

Table 3. Chessboard table of the structural model [own processing].

	Value of assigned outputs, [CZK]					**Assigned outputs to primary activity, [CZK]**	**Total primary costs for overhead activities, [CZK]**
Activity	A	B	C	D	E		
A		210 520	315 780			473 670	999 970
B	260 880		434 800	173 920		1 130 480	2 000 080
C				900 000	600 000	1 500 000	3 000 000
D		307680			1743520	1 948 640	3 999 840
E	638 280		2 021 220				4 999 860

The data from Table 3. was used to compile matrix A - matrix of direct consumption coefficients and subsequently to calculate matrix B - matrix of complex consumption coefficients (Table 4 and 5).

Table 4. Matrix A - matrix of direct consumption coefficients.

	A	B	C	D	E
A	0	0.105	0.105	0	0
B	0.261	0	0.145	0.043	0
C	0	0	0	0.225	0.26
D	0	0.154	0	0	0.349
E	0.638	0	0.674	0	0

Table 5. Matrix B - matrix of complex consumption coefficients, which respect and include the volume of mutual assignments between overhead activities

	A	B	C	D	E
A	1.050	0.116	0.148	0.038	0.031
B	0.312	1.048	0.225	0.096	0.060
C	0.166	0.059	1.185	0.269	0.236
D	0.321	0.201	0.346	1.087	0.421
E	0.782	0.114	0.893	0.206	1.179
	2.631	**1.537**	**2.798**	**1.696**	**1.928**

Calculated figures of the final value of assignment cost rates for overhead activities per benchmark unit for all the activities (overhead and primary) using the structural analysis method:

A: 2.631 * 105.26 = 276,994 CZK / benchmark unit
B: 1.537 * 86.96 = 133,742 CZK / benchmark unit
C: 2.798 * 100.00 = 279,849 CZK / benchmark unit
D: 1.696 * 102.56 = 173,999 CZK / benchmark unit
E: 1.928 * 106.38 = 205,041 CZK / benchmark unit

When using the iteration distribution of overhead activity costs to primary activities based on benchmarks - the ABC (Activity Based Costing) method on a chosen industrial enterprise, 193 calculation steps were needed using the iteration method to determine a result with zero divergence (the number of steps depends on the predefined degree of divergence). The achieved results indicate that the value of assignment rates to the benchmark unit calculated using the structural analysis method and repeated reading method are identical.

The proposed procedure uses the characteristics of complex consumption coefficients of the structural model, resulting from the inversion of the direct consumption coefficient matrix. These complex coefficients already include all the rates (assignments) between the matrix elements and thus allow, in our case, the stipulation of the size of assignment rates of mutually cooperating activities in a single step. The calculation if fast, accurate and enables change implementation and the processing of variant solutions in real time. It was proven that use of the structural model, even in the case of multiple overhead and primary activities, leads to the considerable simplification of calculations when determining the assignment rates of overhead activities. The final results are achieved in a single step, by inversion of the direct consumption coefficient matrix. The result is accurate, unlimited by a predefined difference between the rates of the last two steps.

The model is very clear and enables flexible reaction to changes in the input values, enables the fast calculation of variant solutions and evaluation of the impact intensity of changes in the input factors on the resulting assignment price of activities, and thereby on the cost level of individual calculated products.

Conclusions

The trend of progressive worldwide globalisation means that industrial enterprises are under rising competitive pressure and must build up their competitive edge by innovating transformation processes, developing human capital by high qualification which is characteristics for industrial companies or investing into new technologies. It is entirely obvious that enterprises should react to this situation also with the thorough management of their costs. Management decision-making may be assisted by SW products designed for specific purposes. Most industrial enterprises perceive this need and understand that in the period of change brought about by the fourth industrial revolution, traditional cost management methods will no longer suffice; it is necessary to start using management methods that can keep up with the dynamically evolving economic environment.

Calculation as one of the basic instruments of management accounting depicts both basic poles of the entrepreneurial process - naturally expressed output and its value characteristic. For this reason, it is a crucial tool which allows managers to identify the relations and behaviour of costs depending on the volume and structure of outputs, and stipulate the costs of corporate outputs as such. At present, there is a whole range of expert theses engaged in assessing the cost of enterprise outputs. The issue of calculating activity cost units with a focus on mutual consumption and self-consumption of the outputs of auxiliary activities is addressed by the thesis. Theses address the issue of effective cost management and optimisation of process and activity costs in metallurgical enterprises. Current competitive pressure obliges iron producers to increasingly include the costs for production and logistics processes as an important criterion in the decision-making process when comparing available raw materials and evaluating the efficiency of their use. The results of a case study concerning cost management at foundries in the Czech Republic are presented in the thesis.

A process-based view of the enterprise, based on reviewing corporate processes and activities which create value for customers and generate costs, is being increasingly applied at industrial enterprises. This process-based view of the corporate transformation process is employed in increasingly more new management instruments and methods, which include Activity-Based Costing. This method is primarily designated for industrial enterprises as an instrument for stipulating the assignment rates of mutually cooperating overhead activities, and is particularly suitable for small and middle-sized enterprises with diverse technological procedures, which do not have SW products from companies engaged in the consultancy or production and implementation of SW products.

The proposed method enables the stipulation of assignment rates for mutually cooperating activities using exact economic-mathematic methods - the Leontief structural model, even in cases when the individual activities cooperate and transfer their output (activity) between each other. The proposed procedure uses the characteristics of complex consumption coefficients of the structural model, resulting from the inversion of the direct consumption coefficient matrix. These complex coefficients already include all the rates (assignments) between the matrix elements and thus allow the stipulation of the size of assignment rates of mutually cooperating activities in a single step.

By applying this method, the lowest possible level of assignment costs for overhead activities to primary activities is calculated, thus ensuring the allocation of overhead activity costs to primary activities at a value that is equal to the value of the primary costs of these overhead activities.

The proposed model:

- Fully respects the mutual relations of assignments between activities.
- The final results - assignment cost rates of overhead activities to primary activities - are stipulated in a single step.
- Use of the structural model leads to the substantial simplification of calculations even in the case of multiple activities.
- The calculation enables change implementation and the processing of variant solutions in real time.

Costs and cost calculations hold a significant position in the data base of cost controlling. Controlling obtains the information needed to perform all of its functions by applying various methods. Exact methods of decision-making based on logical mathematical models may become an important source of quality information, as yet undervalued by corporate (controlling) practice. Yet, the mathematical apparatus and software support for their application in industrial corporate practice are sufficiently ready.

Acknowledgement

The work was supported by the specific university research of Ministry of Education, Youth and Sports of the Czech Republic No. SP2019/62.

References

1. M. Landa, *Účetnictví podniku: informační zdroj podnikatelských rozhodnutí* (Eurolex, Bohemia, 2006)
2. R. Kozel, Š. Vilamová, P. Baránek, V. Friedrich, Z. Hajduová, and M. Behún, Acta Montanistica Slovaca, **22**;4, 439-447 (2017)
3. R. Cooper, R.S. Kaplan, The CPA Journal, **60**:2, 38-48 (1990)
4. H. Bouquin, *Comptabilité de Gestion* (Economica, Paris, 2006)
5. J. Basl, P. Majer, M. Šmíra, *Teorie omezení v podnikové praxi: Zvyšování výkonnosti podniku nástroji TOC* (Grada Publishing, Prague, 2003)
6. B. Král, *Manažerské účetnictví* (Management Press, Prague, 2006)
7. D. Cagwin, M. Bouwman, Management Accounting Research, **13:**1, 1-40 (2002)
8. K. Krumwiede, Management Accounting, **79:10**, 32-36 (1998)
9. S. Jänkälä, H. Silvola, Journal of Small Business Management, **50**:3, 498-523 (2012) DOI: 10.1111/j.1540-627X.2012.00364.x
10. J. Majerová, The international spread of Activity-based Costing (Oeconomica Publishing House, Praha, 2010)

Use of modern computer technologies and 3D modelling in the study of historical montanistics

Henrieta Pavolová[1], *Zuzana* Šimková[1*], and *Miloš* Petráš [1]

[1] Technical University of Košice, Faculty of Mining, Ecology, Process Control and Geotechnologies, Institute of Earth Resources, Park Komenského 19, 042 00 Košice, Slovakia

Abstract. This paper points out the integration of using of modern computer technologies and 3D modeling of montaneous objects, which form an integral part of montaneous tourism. It describes in detail the chronological development of the term " montaneous ", which currently refers to all areas with interaction ties to primary, secondary and tertiary spheres of mining. At the same time, it points to an attractive form of making available objects, which no longer exists, resp. inaccessible historical montaneous objects using 3D visualization for the general public, which requires not only wider teams of high-quality programmers or graphic designers, but also experts who create historically relevant descriptive texts, diagrams or models, according which is possible their computer processing. Finally, it presents concrete 3D visualizations of historical montaneous objects in the territory of Eastern Slovakia.

1 Introduction

At presnet science, the concept of *Montaneous* is understood very inconsistently, diversity and often incorrectly, not only in connection with mining activities, therefore its methodological clarification is necessary at the beginning of the study. In Central European, or rather more precisely in the Czech-Slovak geographic space, the correct explanation of the term montaneous (Montanism) in the intentions of history, historical geography, environmental history or mining history must be understood in a broader conceptual context based on older scientific works. The concept of monaneous must be understood much broader than the mining itself (including mining of raw materials) and include also metallurgy, accompanying crafts and so on. Here it is necessary to point out at the beginning that we do not work with the term montaneous in the meaning of mountain, which is used as a terminus technicus eg. in geomorphology, landscape ecology, botany, zoology and the like.

Nowadays commonly used terms *montaneous, montanistics, montanist* are derived from the Latin word *Mons* in the meaning of mountain (m. No. *Montis*) [1], which we can translate in the primary sense as a mountain, mountain or mountain, respectively in the meaning of stone, rock, or boulder [2]. This term has been used to refer to historical mining (ie including metallurgy) since the 19th century [3]. Gradually, it began to settle down in

[*] Corresponding author: zuzana.simkova@tuke.sk

Austrian German [4], eg. *Montanuniversität Leoben* in Leoben, Austria, or the older term *K. by K. Montanärar* (Imperial Royal Montana Administration) and similarly.

The term montanistics comes from the German word *Montanistik (Montanwissenschaften, Montanwesen)*. It is an older term referring to the montan industry, ie. j. mining and metallurgy, e.g. *das Bergbau- und Hütenwessen betrefend* [1] *montanistic* t. c. *mining, upper* [5], *montanisticum, mining and metallurgy* [6] and similarly. It also denotes montane sciences, i. j. mining and metallurgical sciences (science of mining and metallurgy, *Bergbaukunde*), e.g. *Montanistics, mining science* [7].

Nowadays, the term "montanistics" cannot be simplified only to the synonym for mining, because the term "montaneous" is broader and much more complex than current (modern) mining and also refers to the metallurgical science segment and other related branches and employment. It is very closely related to economics, water management (drainage, use of hydropower, water transport), forestry (as a fuel base for smeltery or material for mines), transport, education, traditions and so on. It also concerns ironmongery, non-ferrous metallurgy, processing, oil and gas extraction, geodesy, power engineering, petrochemistry, geology, rock mechanics, petrography, engineering, prospection, geological and deposit prospecting, drilling and similarly. Nowadays, the field of mining sciences is more often referred to as *mining engineering* according to the English model. Nowadays (since the 20th century) montaneous is also one of the mining issues.

At present, scientific research uses computer technologies to a great extension. History in the Slovak scientific environment lags significantly behind. The only exception is a narrow team grouped around prof. P. Chrastina and Dr. L. Župčán at Trnava University. Leaders in historical research using modern technology are archaeologists at the Archaeological Institute of the Slovak Academy of Sciences in Nitra. These scientists hardly ever (except P. Chrastin) devote themselves to computer reconstructions in Montanistic.

2 Development of the research of historical montaneous using modern computer methods at the Technical University of Košice

From the above information we could conclude that the research of historical montanistics represents a multilateral issue of interdisciplinary nature, which requires a team of researchers assembled from montaneous experts. In recent years, researchers at the Technical University of Košice from the Institute of Earth Resources have been intensively dealing with the research of historical montanistics phenomena, who as part of interdisciplinary research, are gradually examining historical phenomena and processes in montanistics with the application of 3D models. They focus on areas with absent, respectively insufficient archival documents. It is the historical areas, where only minimal written and cartographic sources have been preserved, that represent the basal platform of research, where computer modeling is the basic source of (new) historical information. The first research team began to form at the end of the first decade of the new millennium under the leadership of a major montanistics and professor of mining history prof. Pavol Rybár.

It was this team that began to integrate modern computer technologies and 3D modeling, which they used in the field geotourism (*Information technologies in geotourism)* [8], in the reconstruction of historical montaneous territories and historical mining objects, to support the development of tourism (*Reconstruction of historical mining teritory by information technology* [9], *Aplication of GIS to the model of the old mining objects and their exploitation in tourism* [10]). They also used GIS to categorize and visualize historical data (Information technologies in montaneous tourism [11]), which were later applied in Central Europe (Mediaval salt routes leading through the territory of

Slovakia [12], Medieval salt routes in Slovakia [13], Salt routes in Slovakia in the middle ages [14]).

Since 2015, a team of experts in montanistics and mining history has been involved in the research of environmental history of the montaneous country in several partial research studies Computer modelling as a basic research and visualisation tool to research defunct historical mining technologies, using the example of cementation water mining in Smolník (Slovakia) [15], Relics of manual rock disintegration in historical underground spaces and their presentation in mining tourism [16], Historical blast furnace in Peklo valley of Ľubietová (Slovakia) and its reconstruction using 3D modelling [17]).

Currently, the research team of experts focus on using a complex of research methods to solve geotechnical problems such as reconstruction of historical water reservoirs, activation of old landslides, remediation of landslide slopes in a mining country with still active mining activity. He used methods of geodetic monitoring, 3D modeling of montaneous objects and geological environment, which he presented in his work Multidisciplinary research of landslide at UNESCO site of Lower Hodruša mining water reservoir [18].

He also focused on mapping of montaneous objects, modeling, virtual reconstructions and 3D visualization of extinct mining workings and montaneous facilities, utilization of results in support of montaneous tourism in former mining localities in the works Geographical-montanistics research in mining locality Smolník and virtual reconstruction of historical mining workings (Fig. 1.) [19], 3D visualization of old mining workings in Central Slovakia and their presentation on the internet [20], Field research and digital modeling of relicts of old mining shafts [21]. He uses non-standard methods and stereophotogrammetry for virtual reconstruction and derivation of positional-dimensional properties of objects of original montaneous landscape such as in work Using historic postcards and photographs for the research of historic landscape in geography and the possibilities of their digital processing [22].

Fig. 1. 3D model of mines in Smolník under the Rotenberg Hill (the second half of the 19th century)

Source: [19]

In solving and researching various (usually specific) pending historical problems in the field of montanistics, based on archival and field research, in the last year, in addition to 3D modeling, they also used methods of lidar scanning of montaneous landscapes presented in the works: 3D model of selected cultural monument based on contemporary historical and graphical sources [23], Display of the landscape and settlement at the confluence of Veľká Michalová and Rohozná in old maps [24], Reconstruction of the lost genius loci of the medieval landscape at the confluence of Kysuce and Radola, where the reconstruction of the defunct Gothic church from the 14th century was realized in 3D space, which was allocated in real country (Fig. 2.) [25].

Fig. 2. The reconstruction of the Gothic church in Radola

Source: [25]

3 Research of historical montaneous phenomena on the example of mining towns in Eastern Slovakia

The montaneous research was realized by a team from TU in Košice in many locations in eastern Slovakia. As an example we can mention Spišská Nová Ves, or its mining settlement Novoveská Huta, Gelnica and Smolník. We will present a different montanistics phenomenon in each area by using 3D models. In Novoveská Huta we will focus on the inaccessible mining underground of historical mines of copper-ore, from Gelnica we will present the reconstruction of Gelnice Castle as a historical point within the forming medieval mining town of Gelnica. In Smolník we will present through the 3D models of inaccessible montaneous underground, which were excavated in order to obtain copper by cementation.

 https://doi.org/10.1051/e3sconf/201913403016

The first reports of copper-ore mines in the Strážny Hill (898MASL) are from the 16th century. The most important area was Leithaus, later associated with the mining activities by Johanns. It was at the area of the Johani shaft where at the end of the 19th century was established the Novoveské Spa, whose relicts overlap the original montaneous heritage in the country. Based on the 3D model the underground inaccessible spaces of the copper-ore mines (some extinct) have been processed, as they were recorded by historical sources at the end of the 19th century (Fig. 3.).

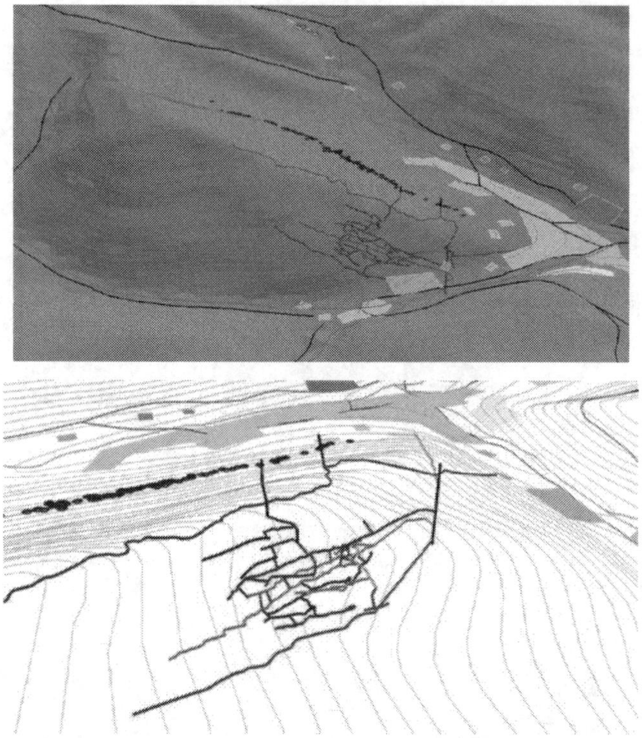

Fig. 3. 3D model of the historical montaneous copper-ore underground

Source: [15]

The research in Gelnica was focused not only on the visualization of the mining underground and the creation of 3D models of old mining spaces, but also on the visualization of ore veins in the Gelnice deposit area. For the purposes of the study, however, we mention as an example reconstruction 3D model of defunct Gelnica Castle, which was directly related to the occurrence of mining in the region. Gelnice Castle was built in the 13th century on the outskirts of the city in a location called Zámčisko in order to protect the city and its mines. Since the first half of the 16th century the castle was in ruins and served as a quarry for the construction of the city Gelnica. At present, only a few architectural elements have been preserved from the castle. Processed 3D reconstruction brings appearance of the castle before its demise. The gradual process of 3D visualization of Gelnica Castle, gradually wrapping the architectural material on the old drawings shows Fig. 4. and the supposed appearance of Gelnica Castle before its destruction and abandonment in the first half of the 16th century illustrates Fig. 5.

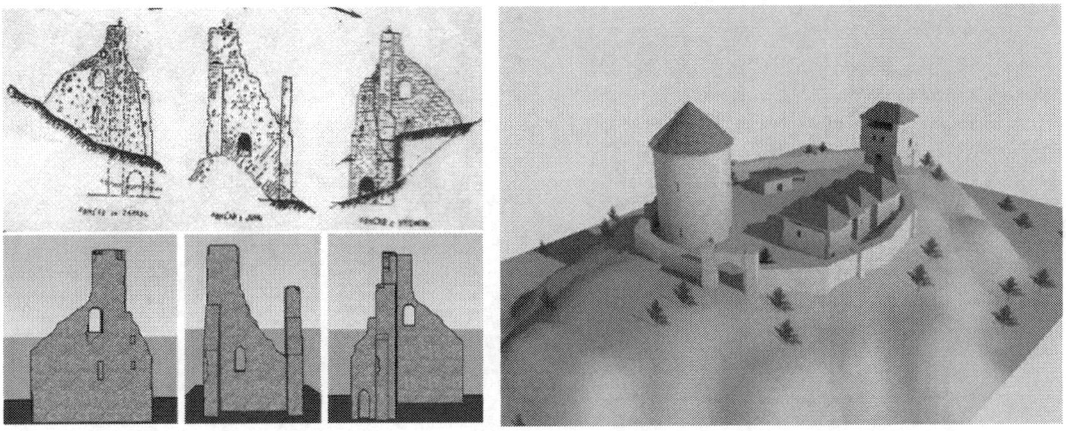

Fig. 4. The 3D visualization of Gelnica Castle

Source: [23]

Fig. 5. The supposed appearance of the Gelnica castle

Source: [23]

The research team from Technical University of Košice also dealt with 3D visualization of montaneous issues in Smolník. Here reconstructed a unique water management system used to supply water to the targeted excavated underground, where water was saturated with copper sulphate ($CuSO_4$). This water was subsequently extracted from the underground and copper was recovered on the surface by the cementation process in gutters. It is the second oldest economically used cementation process in the world, which was used already in the 14th century. It continued until the 20th century. Since the relics of the water management system have been partially preserved up to now, we present an example in the form of a non-existent specific montaneous underground excavated for the mining of enriched water. It is an underground under the Rotenberg Hill, how it was recorded by historical sources in the first half of the 19th century (Fig. 6.).

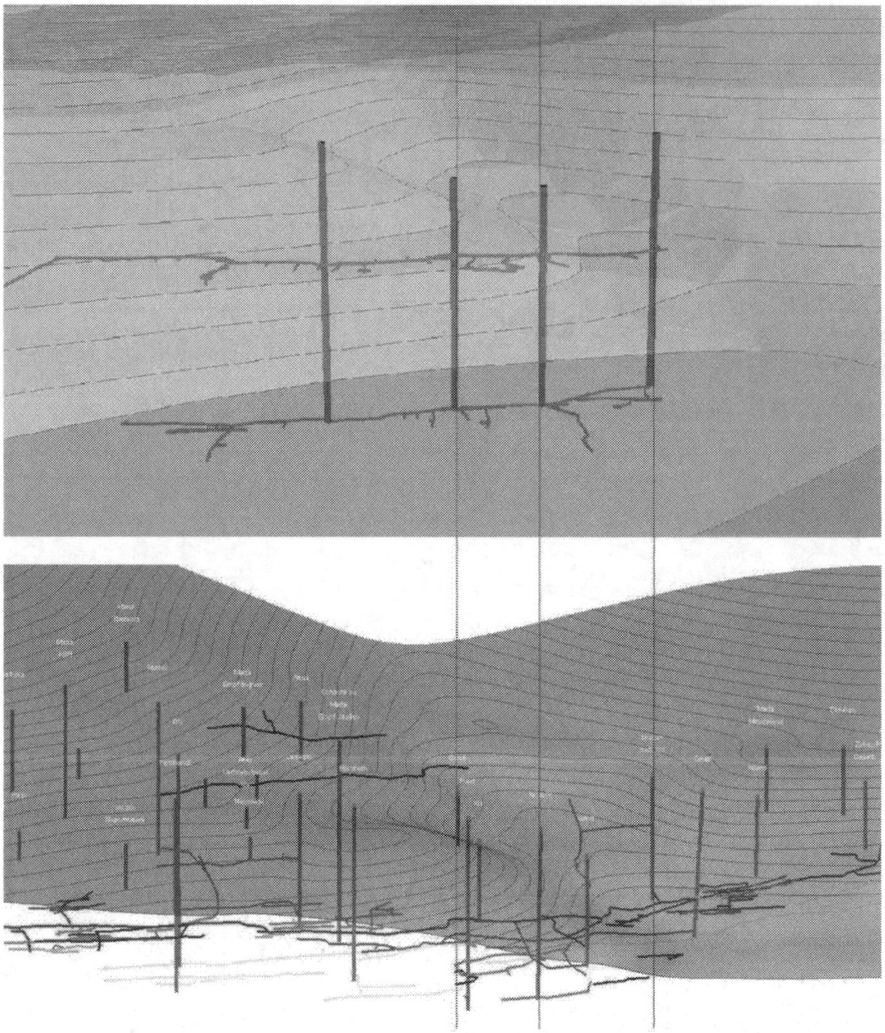

Fig. 6. Mining underground excavated for cementation in Smolník

Source: [19]

Conclusion

The phenomenon of the last decades to which we have become extremely accustomed is digitization. It has penetrated all spheres of our lives, including cultural and historical heritage. Classical forms of information are transformed into digital, what generally facilitates access to them, or modeling creates new, previously unknown information in history. It allows us to transfer virtually to places where we wouldn't otherwise be able to travel and see objects that otherwise cannot be seen. It is therefore clear that digital technologies have undoubtedly great application in the field of (mining) substantive, but also intangible heritage, not only in the visualization of existing objects, but especially in the creation of models of extinct objects. This process allows objects that no longer exist to be preserved for humanity in the form of a digital heritage.

3D models allow you to view montaneous objects from every side or angle, to move not only in the underground, but also to walk through entire mine from the surface to the deepest points. The models allow view the landscape (underground) or its elements (often no longer existing) from any point of view. Another advantage is that it is possible to look at the landscape, underground, or different object in different time horizons, of course, if the technical and content processing of the models allows it.

Another advantage of computer modeling of 3D models is that in addition to the primary, by modeling of visually obtained and directly visible and measurable parameters of the montaneous landscape and its objects, it is possible to derive from the digital model also secondary parameters that are invisible in historical materials. This process - research, however, requires expertise, so 3D models are also used in the scientific field, especially in montanistics, montaneous geomorphology, history, environmental history, etc.

References

1. H. J. Meyers. *Konversations-Lexikon: eine Encyklopädie des allgemeinen Wissens* (Bibliographisches Institut, Leipzig, 1890)
2. J. Špaňák. *Dictionarium Latino-Slovacum* (Aedes Paedagogicae Slovacae, Bratislava, 1987)
3. Kolektív. *Propedeutik der Geschichte, oder Historische Vorbereitungs- und Hulfswissenschaften* (Wien, 1821)
4. W. K. R. *Haidinger. Das Kaiserlich-Königliche montanistische Museum und die Freunde der Naturwissenschaften in Wien in den Jahren 1840-1850* (Braumüller, Wien, 1896)
5. Kolektív. *Ottův slovník naučný: illustrovaná encyklopaedie obecných vědomostí* (J. Otto, Praha, 1901)
6. K. Kavina. *Naučný slovník přírodních věd pro školu a dům.* (Elstner, Praha, 1942)
7. V. Teyssler, V. Kotyška. *Technický slovník naučný: ilustrovaná encyklopedie věd technických*. (Borský a Šulc, Praha, 1932)
8. P. Rybár, L. Hvizdák, M. Molokáč, J. Hvizdáková. *Information technologies in geotourism*. Geotourism and Montaneous tourism, 50-53 (2010)
9. P. Rybár, L. Hvizdák, M. Molokáč, J. Hvizdáková. *Reconstruction of historical mining teritory by information technology*. Pokročilé technológie v oblasti získavania a spracovania surovín .196-198 (2011)
10. L. Hvizdák, M. Molokáč. *Aplication of GIS to the model of the old mining objects and their exploitation in tourism.* SGEM, 965-972 (2012)
11. P. Rybár, L. Hvizdák, M. Molokáč, J. Hvizdáková. Acta Geoturistica **1**, 2 (2010)
12. P. Rybár, J. Engel, M. Molokáč. *Mediaval salt routes leading through the territory of Slovakia.* IRSE' 09, 7-19 (2009)
13. P. Rybár, J. Engel, M. Molokáč. *Medieval salt routes in Slovakia*. Mining in the Central Europe through history. 13-22 (2010)
14. M. Molokáč. *Salt routes in Slovakia in the middle ages*. (VŠB Technická univerzita, Ostrava, 2013)
15. P. Rybár, L. Hvizdák, P, Hronček, M. Jesenský. Acta Montanistica Slovaca. **22**, 3 (2017)
16. P. Hronček, P. Rybár. Acta Montanistica Slovaca. **21**, 1 (2016)
17. P. Hronček, P. Rybár. Acta Montanistica Slovaca. **21**, 4 (2016)

18. M. Bendárik, R. Putiška, I. Dostál, R, Tornai, K. Šilhán, F.Holzer, K. Weis, I. Ružek. *Multidisciplinary research of landslide at UNESCO site of Lower Hodruša mining water reservoir*. (Springer Berlin Heidelberg, Berlin, 2018)
19. K. Weis. P. Bednárik, M. Masný. Geografické informácie: časopis Katedry geografie a regionálneho rozvoja FPV UKF v Nitre. **20**, 2 (2016)
20. K. Weis, S. Jeleň, P. Bednárik. *3D visualisation of old mining works in Central Slovakia and their presentation on the internet* (Slovenské banské múzeum, Banská Štiavnica, 2015)
21. P. Hronček, K. Weis. *Terénny výskum a digitálne modelovanie reliktov starých banských štôlní*. Geografická revue. 10, 98-131 (2014).
22. K. Weis, P. Hronček. *Using historic postcards and photographs for the research of historic landscape in geography and the possibilities of their digital processing*. European journal of geography. **8**, 77-85 (2017)
23. L. Župčan, P. Hronček, M. Ďurčo. *3D model of selected cultural monument based on contemporary historical and graphical sources*. Geotur 2018. 27 – 27 (2018)
24. P. Hronček. Zobrazenie krajiny a sídla na sútoku Veľkej Michalovej a Rohoznej v starých mapách. 79-110 (2018)
25. P. Hronček, M. Jesenský, P. Jančura. Acta Regionalia. **2**, 1-2, (2017)

Macroeconomic indicators of raw material policy in Slovakia

Henrieta Pavolová[1], *Zuzana* Šimková[1,*], *Andrea* Seňová [1], and *Gabriel* Wittenberger[1]

[1] Technical University of Košice, Faculty of Mining, Ecology, Process Control and Geotechnologies, Institute of Earth Resources, Park Komenského 19, 042 00 Košice, Slovakia

Abstract. This paper points to the development tendencies of selected macroeconomic indicators of raw material policy in Slovakia, which forms an integral part of the national economic policy of the state. It describes in detail selected macroeconomic indicators of exploitation of raw materials, which form the basic platform of functioning of all national economic sectors in Slovakia. It also points out the benefits for the Slovak economy in the form of revenues from mining activities to the state budget, municipal budget and environmental fund. At the end the article summarizes the partial findings from the development tendencies of macroeconomic indicators of raw material policy, which is currently an integral part of the industrial policy of the European Union.

1 Introduction

Raw materials are important strategic resources not only for individual sectors of the economy, for the EU as a whole, but also for individual states, because they bring considerable revenues to the state budget and at the same time support socio-economic development at a lower level - at regional level. But they are also a negative form of human perception, because mining is often associated with negative phenomena in the country. Not ignoring the political risk, associated not only with legislative parameters, but also with the dissatisfaction of citizens who are active participants in politics through elections [1]. This is the reason why should raw material (RM) policy also take their needs into account as well as sustainable economic growth, which is conditioned by providing of reliable RM supply and proper protection of living environment [34].

RM policy has become an integral part of the European Union's industrial policy. At the Member State level it requires connection with industrial policy, innovation policy, resource efficiency, environment and economic competition. This integrated perspective will enable industry to use the necessary RMs in a smart and sustainable way, what is contributing to the strategy "European Union 2020". For these reasons raw materials policies at EU Member State level should include the following partial activities:

- to promote resource efficiency, namely primary energy and RM resources, so as to divide economic growth from growth in resource consumption,

* Corresponding author: zuzana.simkova@tuke.sk

- to strengthen research and development with regard to the possibility of replacing critical raw materials,
- to support existing or new RM extractions in Member States which are in-step with applicable environmental, social, health and safety legislation,
- to promote the supply of raw materials from domestic offer, these should be one of the pillars of all actions of RM policy,
- to maintain and increase employment in the European extractive sector by ensuring continuous education and training of the workforce, while the transition to more sustainable mining activities should be accompanied by social dialogue at all levels,
- to develop a coherent policy "urban mining" (extracting raw materials from municipal waste), which main goals is restoring and making available these resources of valuable raw materials,
- to promote the recycling of raw materials and the need to ensure the highest level of recycling where it is economically and technically feasible, including the exploitation of extractive wastes containing a significant amount and spectrum of different metals [2].

In addition to the above mentioned objectives of RM policy of EU in interaction with RM policy of Slovakia, we have analyzed guidelines, objectives, instruments, priorities, as well as national policies, plans, initiatives and measures with the impact and requirements for the efficient exploitation of raw materials as it is described in Table 1.

Table 1. Objectives of raw materials policy of Slovakia in the sense of the principles and principles of sustainable development.

EU policies, directives, goals, tools, priorities with impact on raw material industry
European raw material initiative
European innovation partnership and Knowledge innovation community KIC RM
Digital economy
European technological platform for sustainable raw material sources
EU policy in the area of raw materials: ERA-MIN
EU policy in the area of raw materials: Circular economy
EU tools: European Institute of Innovation and Technology (EIT)
EU tools: Horizon 2020 - SC5
Natura 2000
National policies, initiatives, and tools with impact on raw material policy
Government Program Statement SR
The strategy of permanent sustainable multispectral development and strengthening of competitive ability of Slovakia
Strategy for research and innovation for intelligent specialization SR
Slovakian economic policy
Energetic policy SR
Circular economy SR
Strategy Industry 4.0 SR
Digital economy – unified digital market
Environmental policy SR
Program for the waste economy in SR 2016-2020
The conception of geological research in Slovakia

As we mentioned above, raw materials can be considered as a basic platform of all production processes of industries of the Slovak Republic (SR), which determine the continuous development of its economy. The majority part of raw materials includes non-ore, construction and energy raw materials, while the production of larger part of non-ore and construction raw materials (limestone, dolomite, magnesite, gypsum, building stone, etc.) covers their domestic consumption in substantial rate.

2 Development of selected macroeconomic indicators of raw material policy in Slovakia

The exploitation of raw materials directly and indirectly affects also to its socio - economic area of sustainable development, which can be understood in accordance with § 6 of Act No. 17/1992 Coll. about the environment as „ *development that meets the needs of the present generation without compromising the ability of future generations to meet their own needs and also do not cut nature diverse and also save natural functions of ecosystems"*.

In view of the above, we decided to follow the development tendencies of selected macroeconomic indicators of socio-economic area of raw materials exploitation, among which we included:

- gross value added,
- foreign direct investments,
- Gross Domestic Product,
- development of state revenues from the raw materials exploitation.

2.1 Development of registered employment in the area of raw material policy of the SR

Employment in mining organizations recorded more or less decreasing trend of development, which connected logically with development of RM mining and using of technical and technological equipment during RM mining in Slovakia. During whole analyzed period higher employment had been recorded for employees, working at surface mining, which is again logically connected with prevailing surface way of RM exploitation in Slovakia. The highest employment had been recorded in 2000, when employment achieved level 13 968 employees (6 231 at deep mining and 7 737 at surface mining) and on the other hand the lowest level was achieved in 2016, when employment decreased to 5 476 employment (from which 1 666 in underground and 3 180 at surface mining. In following period total employment decreased to 45.5% of employment in 2000 with average number of employees 8 273 employees.year^{-1}.

In underground mining we recorded in analyzed period average employment - 3 169 employees.year^{-1} with its decrease in 2017 against 2000 by 62.3% (which means 3 884 persons) from original employment. At the surface mining decrease of employment was rather smooth, when recording decrease by 48.2% (which means 3 731 persons) from original employment in 2000, while averagely at surface mining 5 105 employees.year^{-1} had been employed (Fig. 1.).

 https://doi.org/10.1051/e3sconf/201913403017

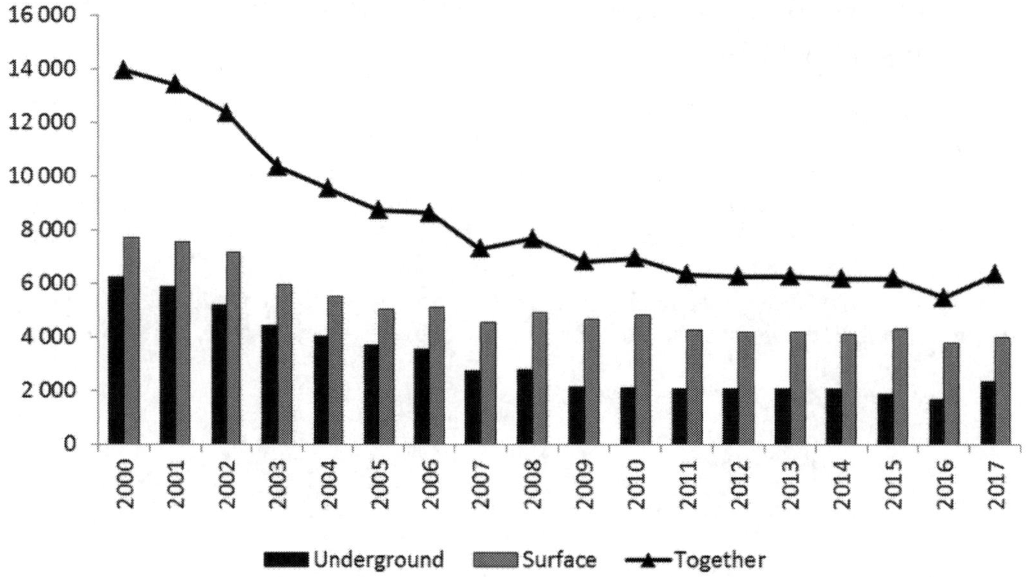

Fig. 1. Development of employment in mining organization of Slovakia [own processing according HBÚ].

2.2 Development of gross value added in the field of raw material policy

Exploitation of RM base contributes similarly to the creation of value added, which can be characterized as difference between production of commodities in basic prices and intermediate consumption of production factors in purchase prices. Production consists of products, produced in common accounting period and intermediate consumption consists of value of products and services, consumed in production process as inputs, excluding investment property, which consumption is reflecting in consumption of fixed capital. Also gross value added recorded in analyzed period fluctuating character of development with average value - 56 616 mil. Eur.year^{-1}, while the highest value was recorded in 2017, when this macroeconomic index achieved 73 602 mil. Eur, on the other hand the lowest value was in 2000, when value added recorded 37 698 mil. Eur (Table 2.). Similarly as employment development, also in the frame of gross value added creation in mining sector we recorded fluctuating tendency in analyzed period (Fig. 2.), while mining recorded average value added at level 295.81 mil. Eur.year^{-1}, which means 0.53% average rate. Generally we can state that the highest value added in mining had been recorded in 2009 with level 372.83 mil. Eur, on the other hand the lowest value was recorded in 2000, when mining recorded value added at level 186.65 mil. Eur. Comparing 2000 and 2017 we found gross value added, produced by mining, decreased in 2017 against 2000 by 144.56 mil. Eur. The highest rate of mining on value added was recorded in 2003 with level 0.64%. On the other hand the lowest rate of mining on value added decreased in 2017 against 2000 by 0.05% (Fig. 2.).

Table 2. Development of gross value added in Slovakia

Year	SR [mil. Eur]	Mining [mil. Eur]
2000	37 698,76	186,65
2001	39 486,60	187,48

2002	41 118,53	246,22
2003	42 973,05	275,94
2004	44 973,02	236,28
2005	47 492,91	249,22
2006	52 266,60	204,73
2007	58 021,92	328,49
2008	61 691,94	368,03
2009	58 314,00	372,83
2010	61 368,25	320,89
2011	62 898,59	344,24
2012	64 433,97	336,65
2013	65 167,74	337,71
2014	66 708,76	318,14
2015	69 295,49	352,96
2016	71 585,46	326,86
2017	73 602,75	331,21

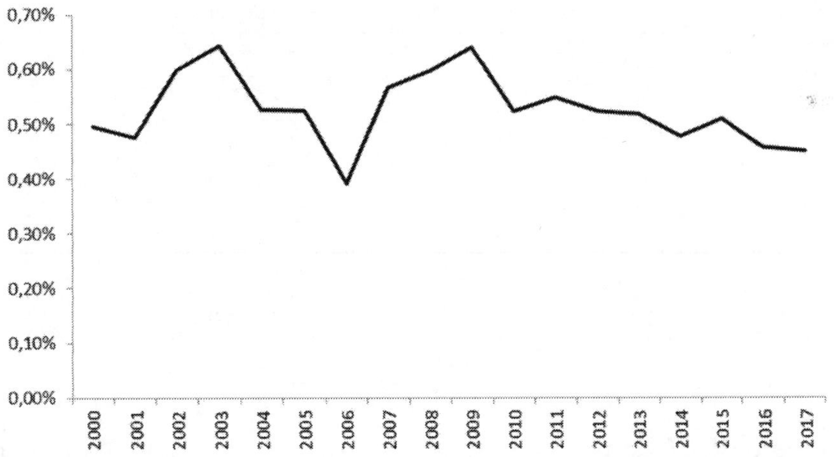

Fig. 2. Development of rate of mining on gross value added in Slovakia [own processing according SO SR]

2.3 Development of foreign direct investments in the field of raw material policy of the SR

Development of mining and connected single exploitation of RM base is determined also by domestic and foreign investments that support increasing of economic effectiveness. Mainly foreign direct investment (FDI) in Slovakia influences positively sustainable development, since they increase employment by creation of new working posts, GDP creation; they contribute to the implementation of innovative technologies or increasing of qualification of domestic human capital. FDI can be generally characterized as investment of money or other monetary values or rights with aim to base, obtain or extend permanent economic relations of domestic investor to business in abroad, resp. foreign investor investing to business in Slovakia. Mainly due to the mentioned FDI and rate of mining on FDI is considered as direct indicator of dynamic development of Slovakia in interaction with investment attractiveness of Slovakia for foreign investors. FDI development in Slovakia is characterizes more or less by increasing trend with slight decrease against 2013 by 2.62%. According available data we can state that average FDI level in analyzed periods was approximately 27 140 447 tis. Eur.year^{-1}, while exploitation of RM base recorded average FDI level 232 371 tis. Eur.year^{-1} (Table 3.), with average rate on total FDI in analyzed period at level 0.88%.

Table 3. FDI development in Slovakia and mining

Year	SR [tis. Eur]	Mining [tis. Eur]
2000	4 097 365,44	43 508,43
2001	5 376 306,25	65 094,73
2002	7 270 627,89	63 859,35
2003	12 490 727,63	71 948,77
2004	15 141 732,45	86 089,53
2005	19 700 181,02	102 308,47
2006	23 338 906,57	312 402,28
2007	29 075 292,32	300 263,44
2008	35 846 869,24	350 298,28
2009	36 469 023,00	396 317,98
2010	37 665 095,00	426 960,00
2011	40 173 448,00	476 802,00
2012	42 304 005,00	497 996,00
2013	42 071 874,94	211 726,00
2014	40 969 239,26	209 928,00
2015	42 256 452,00	102 437,00

2.4 Development of Gross Domestic Product in the field of raw material policy of the SR

Exploitation of RM base participates similarly as mentioned indexes of social and economic development also on the GDP development, which can be generally according Lisý et al., 2007 defined as "*summary of final goods and services, produced and provided during certain time period by production factors in Slovakia, without regard to their state correspondence*" [17]. GDP development had been ranked among basic indicators of economic effectiveness of Slovakia. Also this index, similarly as higher mentioned, recorded fluctuating trend. According available data we can state that in analyzed period 2000-2017 GDP recorded averagely 58 313.5 mil. Eur.year^{-1}, while the highest value had been recorded in 2015 at level 78 896.44 mil. Eur and on the other hand the lowest value was recorded in 2000 at level 31 601.29 mil. Eur. Exploitation of RM base in Slovakia participated at the GDP creation in analyzed period averagely by 58 313.5 mil. Eur.year^{-1}, while the highest level of RM base using on GDP in Slovakia had been recorded in 2008

(371.3 mil. Eur) and the lowest value was in 2002 (154.3 mil. Eur) (**Ошибка! Источник ссылки не найден.**.). At evaluation of rate of RM base on GDP creation we can state that this was directly determined by development of single GDP in interaction with development of mining sector that in analyzed period recorded average rate on single GDP creation in Slovakia 0.54% year^{-1}. Also rate of RM base using on GDP creation, similarly as higher mentioned analyzed indexes, recorded fluctuating trend, determined by GDP development in Slovakia, as well as by demand and offer development of single RM base. The highest rate of mining, which means using of RM base in Slovakia, was recorded in 2000, when the rate of mining on GDP creation in Slovakia achieved 0.79% and the lowest rate was in 2014, when the rate achieved only 0.42% (Fig. 3.).

Table 4. GDP development in Slovakia and mining

Year	SR [mil. Eur]	Mining [mil. Eur]
2000	31 601,286	161,2
2001	34 310,622	158,0
2002	37 279,842	154,3
2003	41 404,346	207,0
2004	46 101,527	160,7
2005	50 415,092	177,7
2006	56 272,653	204,2
2007	63 053,882	263,6
2008	68 491,623	371,3
2009	64 023,061	329,6
2010	67 577,288	358,1
2011	70 627,205	337,9
2012	72 703,513	330,3
2013	74 169,873	334,6
2014	76 087,789	319,8
2015	78 896,443	354,8

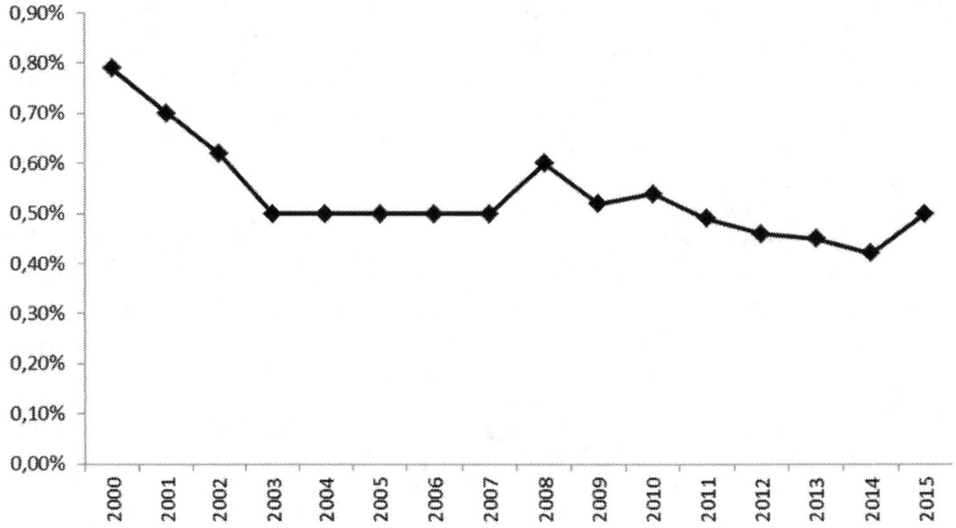

Fig. 3. Rate of mining on GDP in Slovakia [own processing according SGIDŠ]

2.5 Development of revenues from mining activities into the state budget

At higher mentioned development of employment in mining organization and development of average wage in SR we can state that income of state from payment from employees and employer by social security contributions presented averagely 2.78 mil. Eur.year^{-1}, when the highest incomes were in 2007, presenting approximately 2.56 mil. Eur, which connected logically with development of average wage and number of employees, since in 2007 average annual wage in SR was 718.91 Eur and number of employees was 7 312 and in 2017 average annual wage achieved 1095 Eur and number of employees in mining organizations was 6 353 (Fig. 4.). Income of state budget, budget of administrations and environmental fund presented also payment for mining area, mined minerals and stocking gases. In analyzed period the highest incomes was created in 2007, achieving approximately 5.56 mil. Eur, from which only 2.3% presented payment for mining area (approximately 126.55 tis. Eur), 20.0% payment for stocking gases (approximately 1.12 mil. Eur) and yet 77.8% (approximately 4.32 mil. Eur) presented payment for mined minerals (Fig. 5.). During analyzed period the highest rate of state budget income and administrations incomes presented payment for mined minerals, when its level is quantified according conditions amended by Government Decree No 50/2002 Coll. about payment for mining area, mined minerals and stocking of gases or liquids, as amended lately. Payment for mining area that presented lowest incomes of state and administrations is paid at the level 80% to administration, on which cadastral area the mining space is situated and 20% to state budget.

https://doi.org/10.1051/e3sconf/201913403017

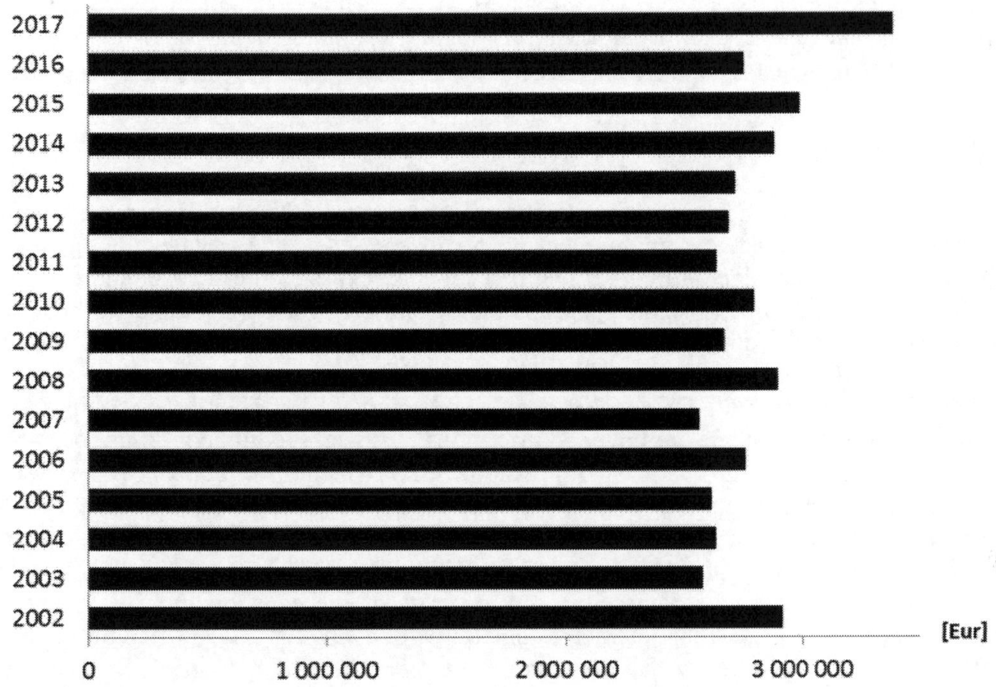

Fig. 4. State incomes from payment to social security of employees [own processing according HBÚ]

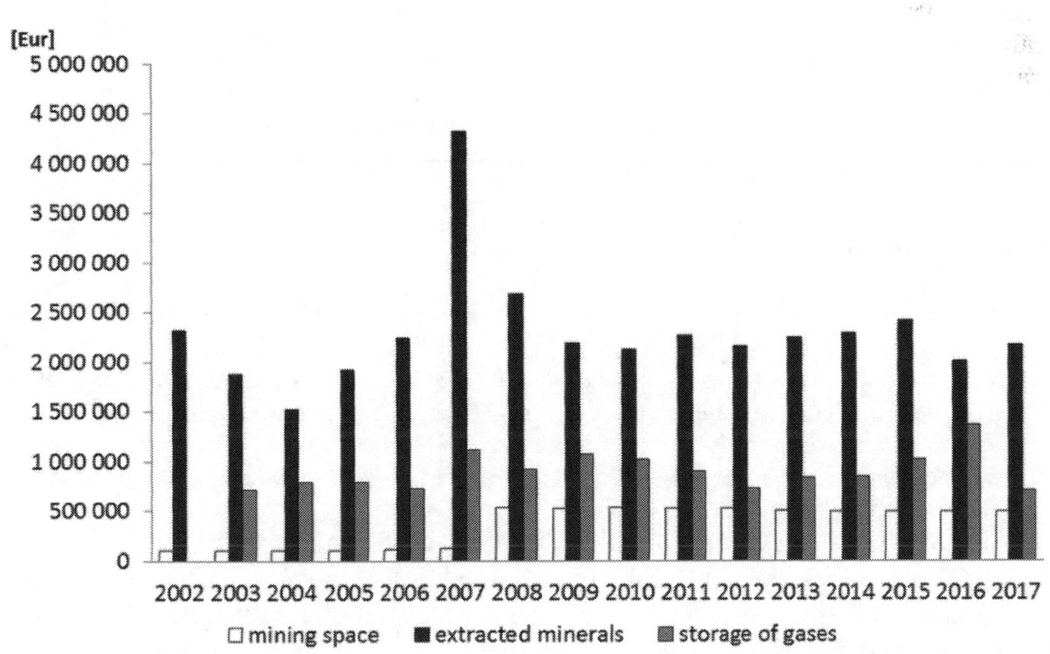

Fig. 5. Development of payment from RM exploitation in SR [own processing according HBÚ]

Due to the higher mentioned we can state that RM using is characterized by positive impacts to social and economic area of sustainable development in Slovakia, which is different from the negative impacts to the environmental area of sustainability.

3 Conclusion

Based on a detailed quantitative analysis of the above monitored macroeconomic indicators, namely registered employment in the surface and underground mining of raw materials, gross value added, foreign direct investments, Gross Domestic Product, we could conclude that in the time horizon 2000-2015 they showed fluctuating trends, which was logically related to the development of industrial sectors of the economy, in which the raw materials form the basic platform of all production processes.

The same character of the development was recorded also in the case of the state revenues from employee and employer contributions by social security contributions, which was logically related to the character of registered employment development in the case of raw materials exploitation. The revenues of the state budget, municipal budget and the environmental fund, which consisted of payments for mining space, extracted minerals and gas storage, also showed fluctuating trends in the years 2002-2017.

During the period under review, the largest part of the state budget and municipal budgets was paid as a fee of the extracted mineral, which amount is quantified on the basis of the conditions regulated by the Decree of the Government of the Slovak Republic No. 50/2002 Coll. about the reimbursement of the mining area, the reimbursement of the extracted minerals and the reimbursement for the storage of gases or liquids, as amended. Reimbursement for the mining area, which constituted the lowest revenues of the state and municipalities, is paid in the amount of 80% to the municipality in whose cadastral territory the mining area is situated and in the amount of 20% to the state budget.

4 Acknowledgements

Contribution is partial result of projects solving VEGA MŠVVaŠ SR 1/0515/18 „The decision-making model of process of evaluating raw material policy of regions".

References

1. M. Cehlár, L. Domaracká, I. Šimko, M. Puzder, *Mineral resource extraction and its political risks* (CRC Press/Balkema Leiden)
2. *Európsky a sociálny výbor. Stanovisko Európskeho hospodárskeho a sociálneho výboru na tému na tému „Oznámenie Komisie Európskemu parlamentu, Rade, Európskemu hospodárskemu a sociálnemu výboru a Výboru regiónov: Riešenie problémov na komoditných trhoch a trhoch s nerastnými surovinami*". URL: https://eur-lex.europa.eu/legal-content/SK/TXT/PDF/?uri=CELEX:52011AE1177&from=EN
3. *HBÚ. 2005. Správa o činnosti Hlavného banského úradu a obvodných banských úradov Slovenskej republiky za rok 2004.* URL: http://www.hbu.sk/sk/Vyrocna-a-rocna-sprava/Rocne-spravy.alej
4. *HBÚ. 2006. Správa o činnosti Hlavného banského úradu a obvodných banských úradov Slovenskej republiky za rok 2005.* URL: http://www.hbu.sk/sk/Vyrocna-a-rocna-sprava/Rocne-spravy.alej

5. *HBÚ. 2007. Správa o činnosti Hlavného banského úradu a obvodných banských úradov Slovenskej republiky za rok 2006.* URL: http://www.hbu.sk/sk/Vyrocna-a-rocna-sprava/Rocne-spravy.alej
6. *HBÚ. 2008. Správa o činnosti Hlavného banského úradu a obvodných banských úradov Slovenskej republiky za rok 2007.* URL: http://www.hbu.sk/sk/Vyrocna-a-rocna-sprava/Rocne-spravy.alej
7. *HBÚ. 2009. Správa o činnosti Hlavného banského úradu a obvodných banských úradov Slovenskej republiky za rok 2008.* URL: http://www.hbu.sk/sk/Vyrocna-a-rocna-sprava/Rocne-spravy.alej
8. *HBÚ. 2010. Správa o činnosti Hlavného banského úradu a obvodných banských úradov Slovenskej republiky za rok 2009.* URL: http://www.hbu.sk/files/documents/spravy/2009/spr%C3%A1va%20o%20%C4%8Dinnosti%20za%20rok%202009.pdf
9. *HBÚ. 2011. Správa o činnosti Hlavného banského úradu a obvodných banských úradov Slovenskej republiky za rok 2010.* URL: http://www.hbu.sk/files/documents/spravy/2009/spr%C3%A1va%20o%20%C4%8Dinnosti%20za%20rok%202009.pdf
10. *HBÚ. 2012. Správa o činnosti Hlavného banského úradu a obvodných banských úradov Slovenskej republiky za rok 2011.* URL: http://www.hbu.sk/files/documents/spravy/2011/spr%C3%A1va%20o%20%C4%8Dinnosti%20hb%C3%BA%20a%20ob%C3%BA%20sr%20za%20rok%202011.pdf
11. HBÚ. 2013. Správa o činnosti Hlavného banského úradu a obvodných banských úradov Slovenskej republiky za rok 2012. URL: <http://www.hbu.sk/files/documents/spravy/2012/spr%C3%A1va%20o%20%C4%8Dinnosti%20hb%C3%BA%20a%20ob%C3%BA%20sr%20za%20rok%202012.pdf
12. *HBÚ. 2014. Správa o činnosti Hlavného banského úradu a obvodných banských úradov Slovenskej republiky za rok 2013.* URL: http://www.hbu.sk/files/documents/spravy/2013/spr%C3%A1va%20o%20%C4%8Dinnosti%20hb%C3%BA%20a%20ob%C3%BA%20sr%20za%20rok%202013.pdf
13. *HBÚ. 2015. Správa o činnosti Hlavného banského úradu a obvodných banských úradov Slovenskej republiky za rok 2014.* URL: http://www.hbu.sk/files/documents/spravy/2014/spr%C3%A1va%20o%20%C4%8Dinnosti%20hb%C3%BA%20a%20ob%C3%BA%20sr%20za%20rok%202014.pdf
14. *HBÚ. 2016. Správa o činnosti Hlavného banského úradu a obvodných banských úradov Slovenskej republiky za rok 2015.* URL: http://www.hbu.sk/files/documents/spravy/2015/spr%C3%A1va%20o%20%C4%8Dinnosti%20hb%C3%BA%20a%20ob%C3%BA%20sr%20za%20rok%202015.pdf
15. *HBÚ. 2017. Správa o činnosti Hlavného banského úradu a obvodných banských úradov Slovenskej republiky za rok 2016.* URL: http://www.hbu.sk/files/documents/spravy/2016/sprava-o-cinnosti-hbu-a-obu-sr-za-rok-2016.pdf
16. *HBÚ. 2018. Správa o činnosti Hlavného banského úradu a obvodných banských úradov Slovenskej republiky za rok 2017.* URL: <http://www.hbu.sk/files/documents/spravy/2017/ro%C4%8Dn%C3%A1%20spr%C3%A1va%202017.pdf
17. J. Lisý, *Ekonómia v novej ekonomike* (Bratislava: JURAEDITION, 2007)
18. *Nariadenie vlády č. 50/2002 Z. z. o úhrade za dobývací priestor, úhrade za vydobyté nerasty a o úhrade za uskladňovanie plynov alebo kvapalín* (Bratislava, Law, 2018)

19. *Slovak minerals yearbook 2002* (Bratislava, SGIDŠ, 2002)
20. *Slovak minerals yearbook 2003* (Bratislava 2003, SGIDŠ)
21. *Slovak minerals yearbook 2005* (Bratislava 2005 SGIDŠ)
22. *Slovak minerals yearbook 2006* (Bratislava 2006, SGIDŠ)
23. *Slovak minerals yearbook 2007* (Bratislava 2007, SGIDŠ)
24. *Slovak minerals yearbook 2008* (Bratislava 2008, SGIDŠ)
25. *Slovak minerals yearbook 2009* (Bratislava 2009, SGIDŠ)
26. *Slovak minerals yearbook 2010* (Bratislava 2010, SGIDŠ)
27. *Slovak minerals yearbook 2011* (Bratislava 2011, SGIDŠ)
28. *Slovak minerals yearbook 2012* (Bratislava 2012, SGIDŠ)
29. *Slovak minerals yearbook 2013* (Bratislava 2013, SGIDŠ)
30. *Slovak minerals yearbook 2014* (Bratislava 2014, SGIDŠ)
31. *Slovak minerals yearbook 2015* (Bratislava 2015, SGIDŠ)
32. *Slovak minerals yearbook 2016* (Bratislava, SGIDŠ, 2016)
33. M. Taušová, E. Mihalíková, K. Čulková, B. Stehlíková, P. Tauš, D. Kudelas, L. Štrba, Sustainability, **11**:10, 1-16 (2019)
34. *Zákon č. 17/1992 Z.z. o životnom prostredí v znení neskorších predpisov* (Bratislava, Law, 1992)

Some features of rural economic diversification in mining regions

Tatyana Yakushnina[1]
[1]T.F. Gorbachev Kuzbass State Technical University, 650000 Kemerovo, 28 Vesennya st., Russian Federation

Abstract. Diversification is one of the main directions for improving the sustainability of rural development, as well as technological renewal of industries and certain sectors of the rural economy. The need to diversify the rural economy is due to the peculiarities of agricultural production, such as seasonality, low profitability of agricultural work, low cost of production. Diversification of the rural economy will make it possible to move from a predominantly monofile model of the rural economy to a multispectral one.

1 Introduction

Rural areas play an important economic role because of the large amount of both socio-demographic and natural resources in rural areas, which could contribute to economic growth in rural areas as well as in the economy as a whole. At present, however, one of the problems in rural areas is the uneven economic development of rural areas. While some regions are experiencing sustained economic growth, in contrast, production are declining, population is falling, which could cause their degradation.

In order to address these challenges, a number of countries, both developed and developing, are being encouraged to move towards a diversified rural economy model, which will address a number of problems in rural areas, namely rural employment, rural quality and standard of living, rural economic and food security, and nearby small towns. The need to diversify the rural economy, in our view, is also due to the feature of agricultural production such as seasonality, the low level of return on agricultural labour and the low cost of production.

Therefore, the "Concept of Sustainable Development of Rural Areas of the Russian Federation for the Period up to 2020" considers diversification as one of the main directions for improving the sustainability of rural development, as well as technological renewal of industries and certain sectors of the rural economy. However, the concept highlights the problem itself, but does not offer an integrated approach that could provide a basis for the systematic development of new directions for rural development.

2 Materials and Methods

Some foreign authors use a special term to reflect the combination of diverse activities in rural areas. Thus, Haggblade S. et al. [4] applies the concept of "non-agricultural economy" (rural areas). It includes many non-agricultural economic activities - manufacturing, mining, construction, trade, infrastructure, transportation, tourism and other services. The heterogeneity of the rural economy makes it necessary to consider the diversification of the rural economy as a complex multi-level and mutually agreed process, covering both the macro-level, the meso-level and the micro-level. Analysis of special literature has shown that there is no consensus among researchers as yet on what to consider diversification in relation to rural economies. The diversity of interpretations, in our view, is because the various works of researchers focus on different elements and levels of expansion of the necessary diversity. For example, a number of instruments on rural development programming address diversification primarily as a variety of individual farm activities that generate income for the household. At the same time, it was stressed that agricultural work on the farm is excluded from diversification, and special attention is paid to the fact that the direct use of the resources of the farm, i.e. land, buildings, equipment and equipment, is a prerequisite for the process of diversification of the population's farms [1].

The rural economy is "a system of industries and activities that includes, apart from agriculture and forestry, extractive and processing industries, rural services, and infrastructure."

The concept of infrastructure received a comprehensive study in foreign literature in the 40-50 yeas of the last century. P. Rosenstein-Rodan, in his work, emphasized the special importance of "social overhead" in industries such as energy, transport, communications, etc., to accelerate and increase the return on subsequent investments in the economy, which he called "framework" or "infrastructure" [15; 16]. Similarly, the early work of A. O. Hirschmann [5] emphasizes the high role of infrastructure as a driving force for economic development, but later, in exploring infrastructure projects, the author points out that it is wrong to view infrastructure formation as a key tool for economic development without corresponding demand, which can lead to significant damage [6]. However, infrastructure investment in the development of the territory and market has a major role to play today. With this approval, work on infrastructure investment research begins [3; 10; 12; 17; 19]. There is a symbiosis of technologies and infrastructure under the general term "infratechnology" [8], which means both methods of testing new technologies and the formation of a database for research on various types of activities. The efficiency of investments in infrastructure projects (including infratechnologies) is higher for the state than for the private sector [3; 10]. Current overseas research on infrastructure investment in the Territory aims to develop an integrated approach to developing a sustainable infrastructure investment system that takes into account both the economic, social, and environmental impacts of such investment. T. D. Foxon notes that infrastructure systems include physical and social networks, which mediate between flows of material resources and the provision of services for households and enterprises [12]. D. T. Scott emphasizes that there is a lack of investment in new technologies in a market that lacks appropriate infrastructure [17]. Research on the importance of infrastructure investment for the development of the economy at the macro, mesoia micro level is enshrined by a number of normative acts and is investigated by Russian scientists. The Urban Planning Code of the Russian Federation identifies the concept of engineering, transport and social infrastructures as tools for ensuring the sustainable development of the territory of N. A. Ivanov, considering the concept of "infrastructure," conducts a comparative characteristic of fifteen definitions and concludes that infrastructure is a set of industries designed to create and provide organizational, economic, social and legal conditions for the normal functioning of the economy and ensure the quality of life of people [8]. The analysis of the definitions of the concept under consideration showed that the main elements of infrastructure in the scientific literature include material and organizational and legal conditions, material and technical means, a set of enterprises, organizations and institutions, types and spheres of activity, as well as a set of interrelated service structures. These elements ensure the normal, free or (less frequently) effective functioning of the economy. The only thing that most definitions agree on is that the ultimate recipient of infrastructure development remains the person (population), and the degree of infrastructure involvement in the process may vary. Thus, social infrastructure, represented by housing and communal services, education, health care, social security, etc., has a direct impact on the processes of life of the population, while productive infrastructure affects indirectly through the development of the economic basis for improving the standard of living of the population.

3 Results and Discussion

The basic instrument for strategic planning of the diversification process is the integrated investment plan for the modernization of the Territory. The development of a comprehensive investment plan makes it possible to form an application for State funding for the Territory 's diversification and development programmer. However, these funds are exclusively for infrastructure projects. The latter determines the relevance of the formation of infrastructure for the diversification of the Territory 's economy.

At the present stage, the mono-relevant territories of Russia are experiencing a development crisis (economic, social and social-civil) caused by the outdated structure of local production, wear and tear of equipment, networks and systems, relatively low (compared to Western countries) qualification of the employed population, social passivity of most citizens, weakness of local budgets. The main challenges to be addressed are: 1) the high rate of unemployment moving into stagnant unemployment; 2) insufficient innovative activity of local enterprises; 3) limited functional orientation of local production; 4) increased degree of risks of the economy of the Territory; 5) immaturity of civil society relations.

At present, mono-profile territories are at the point of bifurcation, i.e. the choice of the path of their development. Two main options can be mentioned: the preservation of the current situation with all the resulting consequences or the launch of a mechanism of diversification and radical modernization, which will make mono city economically and socially different after a certain period of time [2].

Harmonization of interests of participants in the process of diversification of rural economy is one of the most important issues, as it affects almost all stages of development and implementation of diversification projects: Horizontal and vertical harmonization at the stage of setting goals and targets, as well as at the stages of determining development priorities, collecting potential ideas and attracting potential investors, selecting projects and developing a development program, implementing projects and a program of diversification of the territory, assessing the effectiveness of the achieved results (from the position of different actors will differ), monitoring, control, as well as developing corrective measures. The degree of harmonization of the interests of the participants in the diversification process has a direct impact on the effectiveness of the diversification process in rural areas. The study of approaches to the harmonization of interests of different groups of persons is based on the current theory of stakeholders developed by foreign authors [1-3] and actively applied to territorial development, interaction between industrial enterprises and universities [4], as well as other participants on the example of the process of diversification of the economy of the rural territory [5-7].

One of the first works to start studying the problems of mono industrial territory was the research of I. Harold [5] and P. Landis [9] in the 1930s, devoted to coal industrial territories, the need for modernization of which is overdue by the second half of the last century [10-12]. In Russia, the problems of development of mono-relevant territories in modern conditions are a pressing issue in all regions of the country. Currently, there are 66 single towns of various profiles in the Siberian Federal District. The undisputed leader in the number of single towns is Kemerovo Oblast, where out of 49 cities and urban settlements, 24 are single towns. At the same time, with the risk of deterioration of the socio-economic situation in the Kemerovo region - 16 territories.

At the same time, with all attention to the problems of single towns and taking into account all ways to solve them - from radical to the mildest and least painful - it is a matter of very distant perspective. There remain psychology of residents, traditions, inertia of previous stages of development of rural territory. In addition, significant financial resources and competent State and municipal policies are needed. Important policy decisions that directly determine the Territory 's future are often beyond the capacity of local governments to influence these decisions. Therefore, there is an objective need for an agreement (partnership, cooperation) in the system "federal power - regional power - municipal power - business" on concrete actions and measures of strategic importance for mono-specialized territory that should be taken. The system of public-private partnership is the key to solving many problems of mono-specialized territory. The federal authorities finance specific investment projects that create new knowledge-intensive jobs and diversify rural economies. Regional authorities can finance the development of infrastructure, in particular transport. The municipality creates an extensive system of services, increases the share of processing industries, finances social infrastructure, develops small entrepreneurship, determines points of economic and social development (growth), forms conditions for respect for work. The municipality is responsible for the establishment of a guarantee system for investment activities; Definition of clear and understandable procedures for opening a new enterprise; Establishment of clear rules for land use, removal of land for construction; Reasonable regulation of prices and tariffs for services; Granting tax incentives; Formation of attractive appearance (image) of the territory; Creation of a data bank that reveals investment and innovation potential; Information support of the project, etc. Business represented by corporations (city-forming enterprises) should first of all look for ways to increase innovation activity, determine sources of investments, justify opportunities of enterprise restructuring, solve problems of introduction of new technologies. At the same time, business managers should make greater use in practice of the provisions of the "Manifesto on the Moral Code of the Manager," adopted in Davos in 1973, which states: "The professional task of the management of the enterprise is to serve customers, employees, investors and society. Enterprise profit is a necessary means but not the ultimate goal of enterprise management. "

4 Conclusion

The main resource of mono profile territories is not the assets of the city-forming enterprise, but the cash population. However, at present in Russia as a whole there are no clearly developed methods of activation of this resource, disclosure of its potential. If programmers of reform of mono-relevant territories are launched without taking into account the intensification of human resources, especially the most creative groups of the population, the inefficient use of local resources can be predicted with a high degree of probability.

Thus, the importance of diversification, for the rural economy, is primarily that it will allow the transition from a predominantly monofile model of the rural economy to a multi-functional one, that is, a multi-sectoral one. This implies the creation in rural areas of as many economic forms and activities as possible to meet the employment needs of the population.

https://doi.org/10.1051/e3sconf/201913403018

Reference

1. Briefing European Parliamentary Research Service. Farm diversification in the EU. URL: http://www.europarl.europa.eu/RegData/etudes/
2. A. Bebbington, L. Hinojosa, D.H. Bebbington, M.L. Burneo, X. Warnaars, Development and Change, **39:6,** 887–914 (2008)
3. Emile A. Theoretical Approach of Public Infrastructures' Role in The Economic Growth // Indian Journal of Applied Research. 2016. Vol. 6, № 7. P. 464–470.
4. S. Haggblade, P.B., H. Reardon, *Transforming the rural non-farm economy: Opportunities and threats in the developing world* (Johns Hopkins University Press, Baltimore, 2007)
5. I. Harold, *Settlement and the Mining Frontier, in Settlement and the Forest Frontier in Eastern Canada* (Toronto, Macmillan of Canada, 1936)
6. A.O. Hirschman, *Development projects observed* (Brookings Institution Press, New York, 2011)
7. A.O. Hirshman, *The Strategy of Economic Development* (Yale University Press, Washington, 1958)
8. L.E. Karlsson, Evidence & Policy: a Journal of Research, Debate and Practice, **3,** 433-454 (2017)
9. P.H. Landis, *Three iron mining towns: a study in cultural change* (Edwards brothers, Ann Arbor, 1938)
10. A.N. Link, Scientometrics, **36**, 3, 325–342 (1996)
11. D. Littlewood, Journal of business ethics, **120:1,** 39-63 (2014)
12. T. J. Foxon, Infrastructure Complexity, **2:1,** 1-13 (2015)
13. S. Miles, Journal of Business Ethics, **142:3**, 437-459 (2017)
14. J. Moriarty, Business Ethics Journal Review, **4:8**, 47–52 (2016)
15. P.N. Rosenstein-Rodan, The Review of Economics and Statistics, **1,** 107-138 (1961)
16. P.N. Rosenstein-Rodan, *Notes on the theory of the 'big push'* (Palgrave Macmillan, London, 1961)
17. J.T. Scott, The Journal of Technology Transfer, **24:1,** 37–54 (1999)
18. M. Tonts, P. Plummer, M. Lawrie, Journal of Rural Studies, **28:3,** 288–301 (2012)
19. K. Young, J. W. Hall, Infrastructure Complexity, **2:1**, 1-18 (2015)

https://doi.org/10.1051/e3sconf/201913403019

Formation of competitive AIC in conditions of mining region sustainable development (on the example of Kuzbass)

Sergey Shelkovnikov[1], *Elmira* Lubkova[2], *Anna* Shilova[2], *Galina* Ermolaeva[2], and *Elena* Shustova[3]

[1] Novosibirsk State Agrarian University, 630039, 160 Dobrolyubova Street, Novosibirsk, Russian Federation
[2]T.F. Gorbachev Kuzbass State Technical University, 650000, 28 Vesennyaya Street, Kemerovo, Russian Federation
[3] Educational Institution "Kazakh Humanitarian Law innovation university", 070000, 94 Abay Street, Semey, Semey G.A., East Kazakhstan Region, Republic of Kazakhstan

Abstract. This article considers the particular features of the development of the agro-industrial complex (AIC) in the conditions of mining regions. The Kuzbass (Kemerovo region) has been chosen as the object of analysis. A conceptual approach to the food security of the industrial region is presented, in the framework of which the main goals and objectives, principles and directions of development of the agricultural sector of industrial regions are shown. The article systematizes the results of previous studies of the authors related to the conditions and development factors of the AIC of the Kemerovo region. It has been established that the Kemerovo region as an industrial region is characterized by specific features of the AIC development which have evolved under the influence of climate, the available natural resources and the results of local economic activity. The current state of the Kuzbass agro-industry suggests that the region cannot ensure high competitiveness in the main types of agricultural products and their processed products. In the current economic conditions, introduction of deep processing of agricultural products on an innovative basis, production of functional products, integration and cooperation of producers, and improvement of production technologies may become the promising areas for the development of the agro-industrial complex of the Kemerovo region. The development of these areas will make it possible to produce agricultural products and food products with higher added value, and therefore, to increase the competitiveness of the agricultural sector in the region.

1 Introduction

Sustainable development of mining regions implies the positive structural shift linking expanding processing sectors of economy with environmental problems long-term solving [1, 2]. It suggests relocating investments from coal mining to environmentally friendly industries, including agriculture [3].

A significant differentiation of natural and economic conditions in different regions of Russia determines the existence of regional features in various sectors of the Russian economy. The agro-industrial complex is characterized by significant regional imbalances that are primarily due to the high dependence of the distribution of agricultural enterprises as the main component of the agro-industry on the climatic and environmental conditions of specific territories.

The researchers are actively exploring issues of regional development and the functioning of the agro-industrial complex [5-9].

In industrial regions, unfavorable conditions for the development of agriculture are often observed. This is most typical of regions with developed extractive industries, in which a significant part of the territories is occupied by industrial enterprises and is not suitable for agriculture, and the ecological situation is very unfavorable.

The specific character of industrial regions in the development of their agribusiness and food supply was reflected in the works of G. Rachana, A. Joshi [8], V. Mazloev, M. Ozerova [9].

However, in order to provide the population of industrial regions with food, especially essentials, in such regions, agriculture should be developed taking into account their regional specific features. At the same time, it is important that the agribusiness of industrial regions could be competitive, because otherwise the enterprises of the complex will not be able to function in the long term without large-scale state support.

The relevance of the study is also due to the high importance of the agro-industrial complex for the sustainable development of industrial regions [10-11].

2 Research methodology

The object of observation of this study is the Kemerovo region. The conditions and factors of the AIC functioning in the industrial region are considered. The fundamental factors of increasing the competitiveness of agribusiness in the industrial region are detailed. In relation to the Kemerovo region, the promising areas for the development of the agro-industrial complex are determined taking into account the studied conditions and factors.

The article uses monographic, abstract logical, economic and statistical methods.

The empirical base of the study is represented by publications of scientists considering various aspects of food supply in the region, as well as by official statistics.

3 Results and discussion

A number of specific conditions affect the development of the agro-industrial complex of the Kemerovo region. The fundamental ones are natural climatic. The region is located in the temperate zone of the northern hemisphere. The climate of the Kemerovo Region is greatly influenced by the contrast of the relief structure, heterogeneity of the underlying surface and vegetation cover. The continental climate of the region, the high annual amplitude of temperature fluctuations, the unevenness in the amount of precipitation have a negative impact on agriculture. It should be noted separately that a high concentration of minerals is observed in the region. The main mineral is coal. In addition, in the region there are deposits of ore and non-metallic minerals. All this predetermined the development of industrial production in the region.

According to the Federal State Statistics Service, in the gross regional product (hereinafter - GRP) of the Kemerovo region, the share of industry in the period from 2000 to 2017 is in the range of 45-47% and tends to increase, while the share of agriculture is in the range of 3.6-5.7%, and there is a tendency to its decline. Such dynamics confirms that the economy of the industrial region is poorly diversified. It is worth noting that industrial regions are characterized by lower profitability of agricultural products as compared to the basic industries. In the Kemerovo region, the profitability of agricultural products in 2016 amounted to 3.3% compared to 16.4% profitability of extractive industries products.

The current environmental situation in the region reduces the possibility of producing quality products. In such conditions, the main task of developing the agribusiness of the industrial region is to maintain and develop the intensive agriculture and to expand production of deeply processed agricultural products on an innovative basis [11-12].

At the regional level, it is important to consider the environmental aspect of agricultural development [13-14]. For industrial regions, it takes on special significance.

Despite the fact that industry has an adverse effect on the environment, a high concentration of industrial production has led to the formation of a developed infrastructure in the region.

The human resources available in the region may become a possible growth point. Specialists for agriculture and the processing industry are trained at two universities in the region — the Kemerovo State Agricultural Academy and the Kemerovo State University, as well as in the Kemerovo Agricultural College.

The functioning of research centers and laboratories on the basis of regional universities specializing in the training of specialists for the agricultural sector suggests that the region has the scientific potential to develop areas and tools that will enable it to realize the growth factors in the competitiveness of the agricultural sector.

The priority development of industry in the region and the conditions considered in which the agro-industrial complex operates determine the existence of specific features of the regional agro-industrial complex.

According to the authors, the growth of food security of the population of the industrial region due to the development of competitiveness of agricultural products should become one of the most important goals of the regional agrarian policy at the present stage. At the same time, it is extremely important to rely on the principles of maintaining the unity of the market of agricultural products, raw materials and food and the priority of increasing the competitiveness of agricultural products.

Based on the goal, tasks and promising directions are set for the development of the agro-industrial complex of the industrial region (Figure 1).

Thus, since the competitiveness factor of the agricultural sector plays a high role, price and non-price factors of the competitiveness of the agricultural sector in industrial regions should be considered.

Speaking about the price competitiveness of the AIC agricultural products of industrial regions, it is worthwhile to understand that agricultural regions have a price advantage due to the existence of economies of scale. Large production volumes allow agricultural regions to reduce unit costs for agricultural production, so industrial regions should look for opportunities to build up non-price competitive advantages.

The advantages of industrial regions can be created due to high indicators of the efficiency and intensity of agriculture, as well as due to the technical and technological advantage over agricultural regions, which can be created through the investment of industrial enterprises in the development of agriculture and state support.

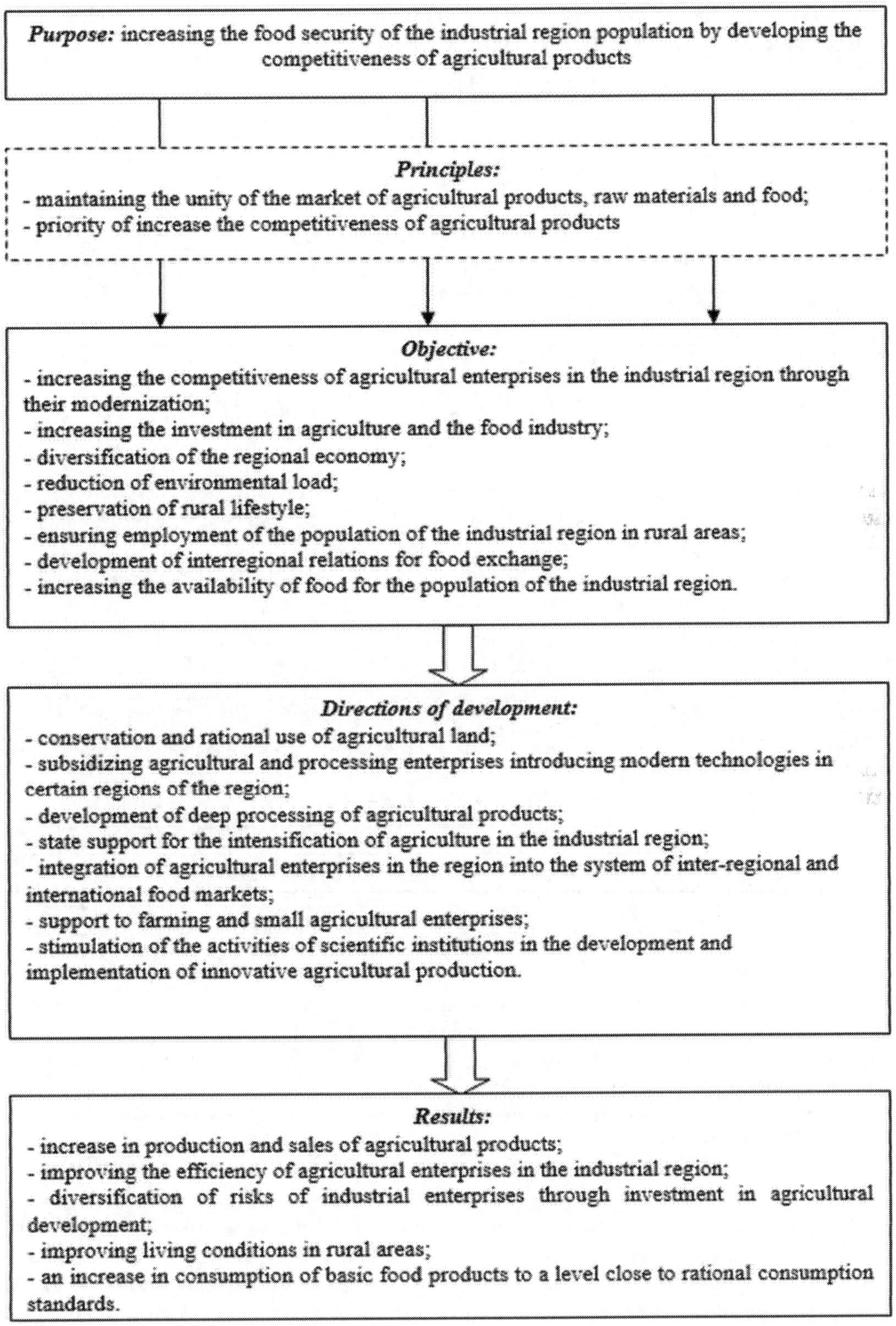

Fig. 1. A conceptual approach to the functioning and development of agribusiness in the industrial region.

 https://doi.org/10.1051/e3sconf/201913403019

Table 1 shows the indicators of intensity and efficiency of the agricultural production of the Kemerovo region examined by the authors.

Table 1. Indicators of production intensity and efficiency.

Indicators	**2000**	**2005**	**2010**	**2015**	**2016**	**2017**
Milk per 1 cow, t.	2693	3297	3886	4126	4168	4289
Yield of bovine offspring per 100 cows per year, heads	82	79	71	75.78	79	73
Yield of live calves per 100 cows and heifers	n/a	n/a	n/a	88.3	87.2	86.3
Yield of pig offspring per 100 sows per year, heads	891	n/a	2120	2687.07	3529	2932
The average egg production of laying hens in agricultural organizations per year, pcs	210	314	315	314.4	311	327
Yield of grain and leguminous crops in farms of all categories, dt / ha	14.6	15.0	17.3	17.1	15.6	18.2
Yield of potato in farms of all categories, dt / ha	134.5	104.9	145.7	143.3	152.1	158.7

In general, there is an increase in indicators of the effectiveness and intensity of agriculture in the Kemerovo region. However, yield indicators can be characterized as quite low compared to neighboring agricultural regions.

Table 2 shows the indicators of technical endowment of the development of agribusiness in the Kemerovo region.

Table 2. Indicators of technical endowment of the AIC development in the Kemerovo region.

Indicators	**2000**	**2005**	**2010**	**2015**	**2016**	**2017**
Seeding (planting) of the respective crops per one combine harvester, ha	273	340	494	476	504	515
Seeding (planting) of the respective crops per one potato harvester, ha	21	37	76	100	107	114
Number of tractors per 1000 ha of arable land	6.4	4.2	3.2	2.3	2.1	2
Arable land load per tractor	158	240	311	428	464	468
Tractor replacement coefficient, %	1.3	2.3	1.9	1.9	1.5	3.1
Harvester-thresher replacement coefficient, %	2.7	8.5	1.7	1.8	4.1	2.2
Potato harvester replacement coefficient, %	3.8	2.6	2.0	2.2	n/a	9.3

In the Kemerovo region in 2000-2017, an increase in the load on equipment is observed. The cultivated area per harvester increased significantly, and the arable load on the tractor almost tripled. At the same time, the growth in the replacement rate of agricultural machinery is a positive aspect of technical endowment of the development of the agro-industrial complex of the Kemerovo region.

According to the authors, the investment and personnel potential of industrial organizations should be aimed at optimization of the agro-industrial complex performance.

The conditions and factors of the AIC development and functioning in the Kemerovo region are shown in Figure 2.

Thus, the current orientation of the functioning of the agricultural sector in the Kemerovo region indicates the presence of specific conditions, and the state of price and non-price factors confirms the need to develop areas and tools that will increase the competitiveness of the AIC in the industrial region and produce products with high added value.

 https://doi.org/10.1051/e3sconf/201913403019

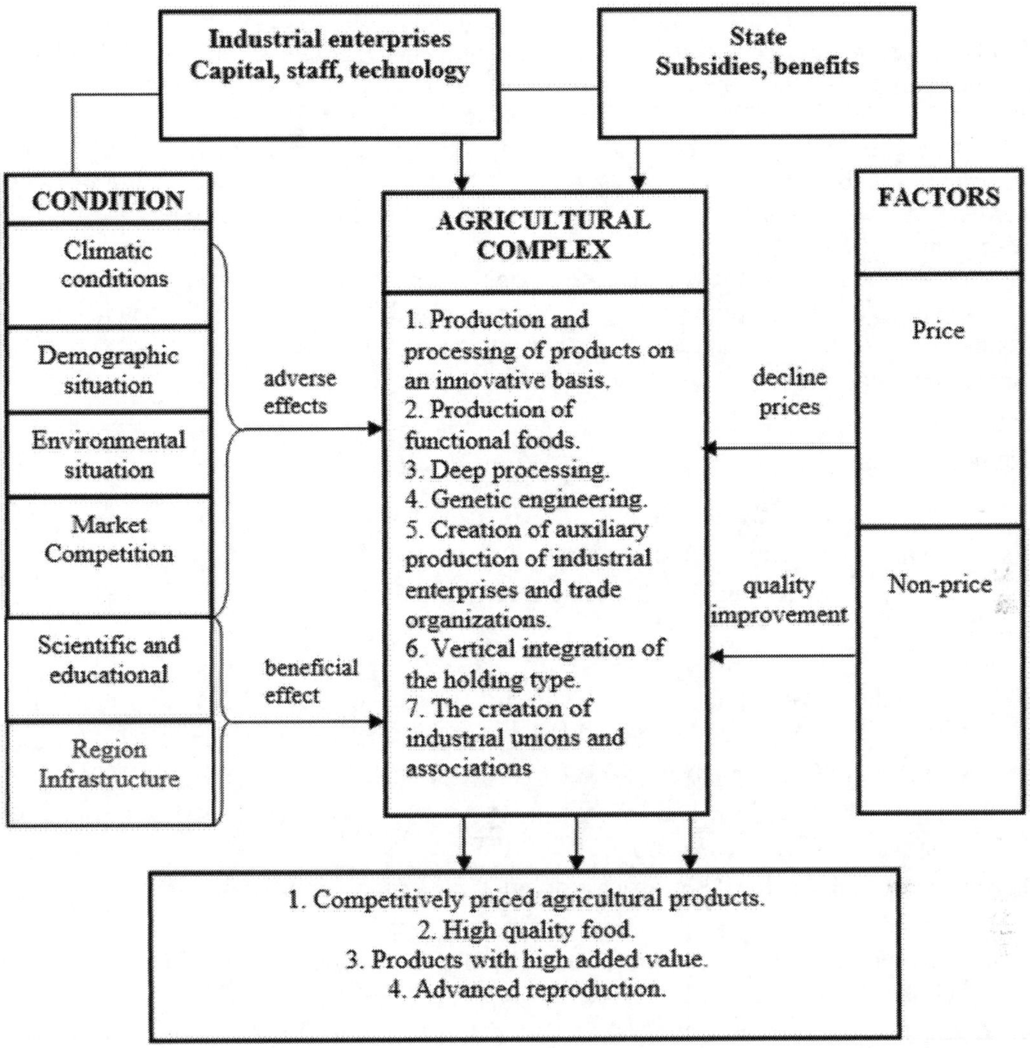

Fig. 2. Conditions and factors of development and functioning of the AIC of the Kemerovo region [15-16].

4 Conclusion

Since the development of agribusiness in industrial regions is influenced by various conditions, the impact of most of which is unfavorable, it is important for all participants in the food supply system of such regions to overcome them by creating growth points for the regional AIC on the basis of prices decrease and improvement of agricultural products quality.

The development of AIC in industrial regions is a task of state importance, as it allows us to solve the issue of food supply to the population.

Industrial enterprises interested in diversifying activities, creating subsidiary farms and having the potential to create high-quality food products and products with high added value can attribute to increasing the competitiveness of the agribusiness in such regions.

Industrial enterprises have the opportunity to create such growth factors. Despite the fact that in industrial regions the efficiency of the agro-industrial complex is significantly lower than that of the basic industries, industrial enterprises are interested in diversifying their own risks and creating subsidiary farms. In addition, the human potential and the ability to create products with high added value through integration and cooperation, participation in industry unions will make it possible to achieve a positive social and economic result and to maintain the rural territories of the industrial region.

Of the conditions reviewed in the article, the current infrastructure and the human resources available are positive for the Kemerovo region. Natural, climatic, environmental and demographic conditions, as well as competition in the food market adversely affect the development of the AIC.

The study suggests that today there is a need to develop and justify the promising areas for increasing the competitiveness of the agro-industrial complex in industrial regions, in particular, the Kemerovo region. The results obtained are of interest for other regions with a dominant share of extractive industries in GRP.

References

1. M.I. Agienko, E.P. Bondareva, G.V. Chistyakova, O.V. Zhironkina, O.I. Kalinina, IOP Conference Series: Earth and Environmental Science, **50:1**, 012022 (2017)
2. S. Zhironkin, O. Aleshina, V. Gorev, Y. Gunyakov, O. Zhironkina, E3S Web Conf., **105**, 04001 (2019)
3. S.A. Zhironkin, Ugol', 4, 29-31 (2001)
4. I.I. Doronina, V.N. Borobov, E.A. Ivanova, E.V. Gorynya, B.M. Zhukov, International Journal of Economics and Financial Issues, **6**, 295-299 (2016)
5. O. Tranchenko, Economic Annals-XXI. **9-10**, 7-10 (2013)
6. L. Shulgina, I. Chernyshova, A. Shulgin, IOP Conference Series: Earth and Environmental Science, **274:1**, 012097 (2019)
7. A.N. Chekavinskii, R.Y. Selimenkov, Modelling of food security in the region, **4**, 226 (2014)
8. G. Rachana, A. Joshi, International Journal of Recent Technology and Engineering, **7:6**, 276-282 (2019)
9. V. Mazloev, M. Ozerova, IOP Conference Series: Earth and Environmental Science, **274:1**, 012004 (2019)
10. P.C. Struik, T.W. Kuyper, Agronomy for Sustainable Development, **2**, 25 (2017)
11. S. Bereznev, O. Zonova, E. Lubkova, E3S Web Conf., **15**, 04002 (2017)
13. S.A. Shelkovnikov, E.M. Lubkova, A.E. Shilova, Economics, labor, management in agriculture, **7:52**, 51-56 (2019)
14. I.N. Merenkova, Ostrovsky Readings, **1**, 224-228 (2017)
15. A. Kotarev, IOP Conference Series: Earth and Environmental Science, **274:1**, 012035 (2019)
16. S.A. Shelkovnikov, E.M. Lubkova, A.E. Shilova, Economy of agricultural and processing enterprises, **6**, 22-27 (2019)

E3S Web of Conferences **134**, 03020 (2019) https://doi.org/10.1051/e3sconf/201913403020
SDEMR-2019

Energy-Saving Electronics Production in Viet Nam: Rise and Problems of Sustainable Development

Tran Thi Bich Ngoc[1,*], and *Dao Thanh* Binh[1]

[1]Hanoi University of Science and Technology, 1 Dai Co Viet, Hanoi, Viet Nam

Abstract. Electronic industry is the basis for the growth of industrial production in ASEAN countries, and represents the main lever of economic development when the region moves up to higher added value using modern technological solutions. At the same time, modern environmental imperatives require the electronics industry to make a significant contribution to reducing environmental stress and lowering energy consumption. This contribution is the production of energy-saving devices. Viet Nam's modern industry is increasing the production of electronic energy-saving devices, but still has problems, such as low added value, spontaneous structure, import dependence in production, lack of unified state administration, and the quality of training of labor resources. In the framework of this article, the authors investigated the state of the electronic industry in Viet Nam by analyzing its successes and shortcomings in recent years compared with the countries of Southeast Asia. The authors' proposals as the results of the study can be recommended for further development of a policy for the progress of energy-saving production in the electronic industry of Viet Nam and an import substitution program in the country's industry as a whole for a new period.

1 Introduction

Electronic industry is an industry for the production of electronic components and products and is one of the most important for any state, since it provides an electronic component base for almost all other sectors of the economy [1]. The release of energy-saving means differs from other industries in a high level of concentration, specialization and integration of production [2-4]. In addition, the very release of products for energy saving is characterized by a high level of international integration in production, the high technology level in its manufactured goods, short product life cycle and, therefore, high added value [5-6]. In the modern world, the state of electronic energy-saving means determines technological independence, state security and the pace of modernization of the country.

* Corresponding author: bichngoc2201@gmail.com

Viet Nam electronics industry was formed in the late 1980s and has been gradually improving since 2000. Its fast pace of development has gained since 2010, when foreign direct investment in the Viet Name economy is reviving impressively. Being a key sector of industrial production, electronic industry occupies an important place in the economy of Viet Nam and has an impact on the development of other industries. The development of the electronic industry also accelerates the process of industrialization in the country, stimulates the expansion of the service sector and provides the working population with jobs. The electronics industry in Viet Nam is characterized by rapid growth in production, expanding the range of products (semiconductors, integrated circuits, Printed Circuit Boards-PCBs, etc., as well as electronic energy-saving means (high-precision sensors, data integrators, equipment dispatch systems, components and accessories).

2 Materials and Methods

Since 2007, the Government of Viet Nam approved the "List of priority and key industries for the period from 2007 to 2010 with a vision to 2020.", according to which electronics, information and telecommunication technologies (ITT) were identified as key out of 10 priority industries. In accordance with this, the Government has established a number of measures to support their development.

The Decision number 879 / QĐ-TTg on June 9, 2014 approved the "Industrial development strategy of Viet Nam till 2025 with a vision to 2035", which re-emphasized that the production of energy-saving electronics and telecommunications is one of the three groups of highest priority branches of economy (manufacturing, electronics and telecommunications, new energy and renewable energy sources). By definition Strategy, the electronics industry will be a priority to develop a leading way compared with other industries; production of electronic industry represents a high technology and is widely used in energy-saving, creates the effect that the objective and prospective conditions for conversion into other areas of the national industry.

The data statistics shows the fast pace of development of the electronics industry. In 2000-2010 the growth of gross output of the electronic industry of Viet Nam has averaged 20-30% per year, especially increased by 96% in 2011. Industrial production index of the sector in 2014 - 2015 was respectively 138.2 and 137.0% [7].

In 2015, the value of exports of electronic products in Viet Nam amounted to $ 48.8 billion, beating Malaysia (about 43.64 billion) [8], and Thailand (32.2 billion) [9]. In 2016, according to the General Administration of Customs, the export volume of electronic products (including electronics for energy saving) reached the amount of US $ 56.231 billion, up 15.23% compared with 2015, contributes 32% of total export value of Viet Nam (US $ 176.63 billion in 2016). [7]

By the scale of production, according to official data of the Central Statistical Management of Viet Nam by the end of 2015, companies operating encounted up to1145 with the total number of people employed in the production in the electronics industry of 497037.

Viet Nam electronic industry uses preferential conditions for the development and transformation into a center of production of electronic energy-saving in the world:

1. The trend towards international economic integration of Viet Nam speeds up the process of rapprochement to other states in the region and the world in the field of economics, finance, distribution of labor. Participation in the global integration of electronic energy-saving production creates for Viet Name electronics industry the real opportunities to enter the world market and to enter more deeply into the global value chains. Accession to the WTO, ASEAN Economic Community and participation in many

free trade agreements with the European Union, the Eurasian Union shall contribute to the promotion of products to the regional and global market.

2. The inflow of global investments can turn Viet Nam into the second center of world production of electronics after China. According to experts, published by the American "Journal of Commerce", now a number of large multinational corporations of the world electronics industry from the United States, Japan, South Korea is gradually moving production from China to the countries of Southeast Asia. [10-11] Possible competitors of Viet Nam in the region for the production of energy-saving electronics and computer components are China, Malaysia and Thailand. However, due to the political stability, Viet Nam has become one of the areas where the moves of foreign capital make its inflow, which makes Viet Nam electronics industry the new impetus to the further development of the electronic energy-saving means. Large global corporations in the field of information technology and electronics continue to invest billions of dollars in projects in Viet Nam. The accumulated amount of direct foreign investment in the electronics industry to date has reached $ 20 billion, including the Samsung – 14,8 billion, Intel – 1 billion, LG – 1.5 billion, Microsoft – 302 million, Japanese manufacturer of optical sensors and micromotors (Nidec) – 1 billion, Foxconn Group (Taiwan – the production of electronic components) – 1 billion). [12]

3. The attractiveness of the investment environment is one of the advantageous conditions for attracting foreign direct investment. According to Viet Namese law number 32/2013 / QH13 of June 19, 2013 on amendments and additions to the articles of the Law on Corporate Income Tax and other legal instruments of Viet Nam, all organizations and enterprises use the benefits of electronic industry and support from the State, established for subjects high-tech production, namely:

- exemption from the payment of land rent for 11 years for investment projects in areas with special socio-economic difficulties; projects related to the list of specific sectors for investment in areas with socio-economic difficulties; or 15 years, if the investment is implemented in areas with special socio-economic difficulties;
- the provision of benefits in relation to corporate income tax rate of 10% applied for 15 consecutive years since the introduction of the taxable income, including: exemption from corporate income tax for 4 years; 50% reduction for not more than 9 years of follow-up;
- tax exemption on the import of equipment, materials, vehicles and other goods for investment projects in Viet Nam.

These incentives are considered to be more attractive than those in Thailand, where the tax benefits from the government included an exemption from corporate income tax for 8 years, and a decline of 50% over the next 5 years; tax exemption on equipment imports. In the region of corporate income tax rate in Viet Nam, Thailand accounted for 20%, higher than in Singapore (17%) and lower than in Malaysia and Indonesia (25%). [13].

Despite the successes and large-scale development, production of electronic energy-saving means in the electronics industry of Viet Nam is evaluated as a new sub-sector, which is facing problems directly affecting its efficiency and competitiveness.

1. The low value added and underdeveloped supporting industries. In recent years, Viet Nam electronic industry in general is gaining significant output growth (an average of 37% per year) and reaches a large volume of exports, but it is not enough to earn the added value in proportion.

2. The tendency to assembly of products for multinational corporations; lack of own brands and dependence on foreign investors in the choice of technologies and products. The energy-saving production in the electronic industry in Viet Nam has some restrictions as the assembly of products of famous brands and manufacture of parts and accessories. The reasons for this are as follows: (i) the pressure from a competitor with other producers in the market; (ii) weak market conditions and a lack of knowledge about the market and

consumers from Viet Nam producers; (iii) high dependence on imported components, equipment and software. Such import dependence cannot be solved in one day. In our opinion, it can be added to the main reason as follows: the weak R & D, dependence on imports of important components due to lack of domestic production of critical parts. For this reason, the production of Viet Nam electronics industry depends on the well-known corporations such as like Samsung, LG, Intel.

3. The lack of a unified plan for the development of Viet Nam electronics industry and a uniform control for a long time. Viet Nam electronics industry since the advent existed almost spontaneous manner without a development plan, and the attention of the competent authorities for 20 years. Only in 2007 for the first time, "General electronics industry development plan of Viet Nam for the period until 2010 with a vision to 2020" were approved by the Government decision number 75/2007 / QĐ-TTg dated May 28, 2007. Under the plan, the state management of Viet Nam electronics industry is the responsibility of the Ministry of Information and Communications. In our opinion, there is a number of problems: it is necessary to solve the basic issues relating to investment, production and sales of electronics for energy saving, financial support for the activities and training. But these functions are respectively implemented by the Ministry of Planning and Investment, Ministry of Industry and Trade, the Ministry of Finance and the Ministry of Education. At the same time, other ministries carry out only a part of them relating to problems associated with the production of energy-saving electronics, which reduces to the duplication of functions, untimely responses and slow decision-making problems, and finally, the inhibition of the development of the entire electronics industry.

4. The business environment and competitiveness issues. In the ranking of Doing Business 2017, published recently by the World Bank, Viet Nam occupies the 82th place of 190 classified countries and economies, while Malaysia – the 23rd, Thailand – the 46th, China – the 78th and Indonesia – the 91st. Although much has been done, but on the Global Competitiveness Index of the World Economic Forum (World Economic Forum / The Global Competitiveness Report 2016–2017) with an index of 4.3 / 7 rating Viet Nam occupies the 60th place out of 138 economies studied. Average ratings and the ratings given by 12 targets, determining national competitiveness, namely the quality of institutions (index and rating) – 3,8 and 82; infrastructure 3.9 and 79; macro stability – 4.5 and 77; health and primary education – 5,8 and 65; higher education and training – 4,1 and 83; efficiency of the market for goods and services – 4,2 and 81; labor market efficiency – 4,3 and 63; development of the financial market – the 3.9 and 78; level of technological development 3,5 and 92; the size of the domestic market – 4.8 and 32; the competitiveness of companies – 3.6 and 96; innovation potential – 3.3 and 73 [14].

Consequently, in spite of the open investment policy and measures to improve the business environment, Viet Nam still is in a group with the threat of the backlog, and should improve the performance of the release of energy-saving electronics and, thereby, increase the competitiveness of the electronics industry at the regional and global market.

3 Results and Discussion

Studying the state of Viet Nam electronics industry, the authors believe that to solve the problems in the production of energy-saving electronics the approach is required, which includes the following comprehensive measures aimed at its development:

1. Measures to strengthen state regulation:

- transfer function of state regulation of the electronics industry from the Ministry of Communications to the Ministry of Industry and Trade as well as the implementation of management of the industry in general is a function of Industry and Trade, or create by the

Government a special body for an overall settlement of the electronics industry and coordination of related ministries and agencies;

- to develop measures to control the phenomenon of transfer pricing of transnational corporations and their businesses in Viet Nam;

- improve the business environment for all indicators in determining national competitiveness.

2. Measures for the development of supporting industries for the production of energy-saving electronics:

- develop further measures to encourage and attract investment in supporting industries and provide incentives and financial support, respectively, the Program of development of supporting industries for the period 2016-2025 years, with attention given to the development of specific efforts and activities to address the problems of import substitution;

- to develop a range of priority of energy-saving electronic products (individual components and parts and components) and create the conditions for the needs of the domestic assembly plants and external market;

- expand and improve conditions for the energy-saving electronic other areas, such as banking, transport services.

3. Measures for the formation of industrial clusters:

- to create an e-based clusters of industries with industrial parks in Ho Chi Minh City and Hanoi, in Haiphong, Bac Ninh, and Thaynguen Bình Dương Province, with the focus assembly plants and companies supporting industries;

- provide incentives and financial support to the local and foreign investors and to any businesses and individuals which wish to conduct their business and services in the cluster, with the highlight the attention and support to small and medium-sized enterprises according to the established by the government for their orders.

4. Measures to improve the quality of labor resources:

- to draw the national educational institutions onto the training of manpower for all sectors of the economy, including the labor in electronics industry, create research centers and institutes, while enhancing the cooperation on the type of public-private partnerships; focusing attention and efforts on training of specialists and workers;

- to take into account the implementation of the reforms carried out by the Ministry of Education for the restructuring of the educational system in order to optimize the relationship between higher, secondary and primary education in order to meet the increasing demand of the labor market;

- to develop measures to strengthen the material-and-technical base of educational and training institutions, to improve the teaching staff;

- to form a close mutual cooperation between educational institutions and enterprises – employers – to examine their demand for specialists in the field of energy-saving electronics and implement targeted training principle.

5) Measures of Innovation Policy:

- to identify the main types of advanced electronics for energy saving; enhance innovation in R & D in research institutions, universities and enterprises, at the same time orient their efforts on the research and development of new priority kinds of electronic products provided by the decision of the Prime Minister of the number 49/2010 / QĐ-TTg of July 19, 2010 on the nomenclature of priority high-tech and knowledge-intensive energy-saving goods;

- to develop a policy to promote the commercialization and transfer of modern technologies in high-tech manufacturing processes for the production of energy-saving electronics.

Conclusion

In conclusion, it should be noted that the producing of energy saving electronics as a fusion of modern scientific and technological achievements and technology, gains a paramount importance today. Virtually there is no area of human activity, where production of energy-saving electronics has not found its application. In Viet Nam, the electronics industry is recognized as a basic material production sector of the national economy. Its development has a direct impact on other industries and plays an important role in dealing with the process of industrialization and modernization in the country, as well with the growing worldwide trend of responsible environmental management.

References

1. *Viet Nam General Statistical Office (2016). Review of the socio-economic situation in 2015 - index of industrial production.* URL: https://www.gso.gov.vn/default.aspx?tabid=621&ItemID=15507
2. S. Zhironkin, S. Demchenko, G. Kayachev, M. Ryzhkova, O. Zhironkina, E3S Web Conf., **41**, 04011 (2018)
3. S. Zhironkin, S. Demchenko, G. Kayachev, E. Taran, O. Zhironkina, E3S Web of Conferences, **105**, 03008 (2019)
4. S.A. Zhironkin, K.A. Kolotov, A.E. Genin, F.V. Agafonov, S.A. Kovalevsky, IOP Conf. Ser.: Earth Environ. Sci., **50:1,** 012011 (2017)
5. S. Zhironkin, M. Gasanov, G. Barysheva, K. Kolotov, O. Zhironkina, E3S Web Conf., **15**, 03012 (2017)
6. E.A. Gasanov, T.S. Boyko, N.S. Frolova, Economics and Innovation Management, **3**, 4-8 (2018) DOI: 10.26730/2587-5574-2018-3-4-8
7. General Statistics Office *Statistical Yearbook of Viet Nam* (Statistics, Hanoi, 2017)
8. Rabobank in Malaysia. *Foreign Trade in Figures. 2017.* URL: https://rabowereldwijd.nl/malaysia-economic-and-political-outline/c/2245
9. OIE The Office of Industrial Economics, Thailand. *Industrial Economic Conditions in 2015 and Outlook for 2016.* (OIE, Bangkok, 2015)
10. G. Knowler, *Factory shift from China to Viet Nam accelerates, Barclays data shows. The latest news from the Industry.* URL: http://www.aslglobal.com/en/news-detail?id=73
11. S. Zhironkin, D. Khloptsov, N. Skrylnikova, I. Petinenko, O. Zhironkina, E3S Web Conf., **41**, 04010 (2018)
12. Gary Rynhart, Jae-Hee Chang, Phu Huynh, *ASEAN in transformation: electrical and electronics: on and off the grid. Working Paper No.13 ILO.* URL: http://www.ilo.org/public/english/dialogue/actemp/downloads/publications/2016/asean_in_transf_2016_r5_electr.pdf
13. Hirunya Suchinai, *Thailand Investment environment and BOI Investment Promotion Policy. Thailand Board of Investment* (South Korea, Changwon, 2003)
14. *2016 JETRO Survey on Business Conditions of Japanese Companies in Asia and Oceania.* URL: https://www.jetro.go.jp/ext_images/en/reports/survey/pdf/2013_05_01_biz.pdf

AUTHOR INDEX

Abramovich Alexander 1
Alama Faruz 82
Alitkina Oksana 86
Andryushkov Alexey 191
Anyona Seroni 51
Bátorová Kristína 186
Beer Martin 127
Begunov Artyom 146
Bel'kov Anton 231
Binh Dao Thanh 345
Bizyaev Aleksey 23
Blištan Peter 186
Brodny Jaroslaw 15
Brzózka Marcin 15
Bykov Anatoly 158
Cehlar Michal 291
Cherkasova Elizaveta 191
Cherkasova Tatyana 191, 196
Cupov Michail 23
Demidov Anton 175
Demirel Nuray 110
Diederichs Ryszard 15
Dormeshkin Oleg 191
Drozdenko Yuriy 30
Dyagileva Anna 51
Egorova Natalia 252
Èitbajová Jana 302
Eltsova Elena 181
Ermolaeva Galina 338
Falaleev Albert 296
Fridman Yuri 212
Garafonova Olga 275, 283
Gazizullin Rafail 71
Gerike Boris 30
Gerike Pavel 30
Goryunov Sergey 86
Grigashkina Svetlana 275
Grigoryeva Natalya 86
Ibragimova Flyura 93
Ivanov Vadim 51
Jacko Stanislav 186
Janocko Juraj 165
Janovská Kamila 302
Kalinina Oksana 223
Kandinsky Vladimir 146
Karasev Andrei 36
Kassikhina Elena 98
Katanov Igor 105
Kayachev Gennady 296
Kharitonov Aleksey 252
Khoreshok Aleksey 86
Kipczak Piotr 7
Kolechkina Irina 181
Kolmakov Nikolay 196
Kondratyev Sergey 105
Kopytov Aleksandr 56
Korchagina Tatiana 158
Kosinskiy Piotr 252, 259
Kosolapov Andrey 93
Kovacs Péter 267
Kozel Roman 302
Kozureva Olena 275
Kozyreva Marina 231
Kretschmann Jurgen 1
Kudrevatykh Natalia 115, 243
Kurgansky Sergey 296
Kusurgasheva Ludmila 217
Kutáè Josef 302
Lapinskas Arunas 217
Lazarenko Yuliia 275
Loginova Ekaterina 212
Lubkova Elmira 338
Lubyagina Lydia 217
Lyubimov Oleg 267
Lyukshin Vladimir 153
Mamzina Tatiana 243
Marasová Daniela 121
Marhasova Vyktoriia 275
Medvedev Aleksey 259
Merkuriev Vladimir 259
Meshkov Anatoly 137
Mihailov Vladimir 236
Mikhailov Vladimir 115
Mikhalchenko Vadim 203
Nesterov Valery 71
Ngoc Tran Thi Bich 345
Nguyen Cao Khai 64
Nguyen Phi Hung 64
Nikolaev Alexander 71
Oliinyk Oksana 283
Pavolová Henrieta 312, 321
Pershin Vladimir 56, 98
Petráš Miloš 312
Petrov Vadim 77
Petrova Mariana 181
Pham Duc Thang 64
Piecha Marian 302
Pimonov Alexander 45, 236
Plotnikova Larisa 146
Popov Anton 137
Preis Elena 86
Prokopenko Eugeniya 175
Radchenko Yuriy 196
Raevskaya Elena 236
Rechko Galina 212
Rubanik Yuriy 203
Rybár Radim 127
Samorodova Lyudmila 267

AUTHOR INDEX

Savchenko Andrey 23
Schastlivcev Evgeny 158
Sedina Natalia 231
Šèerbáková Barbora 186
Selyukov Alexei 110
Semenov Yuri 121
Semenova Olga 121
Seòová Andrea 321
Shabaev Sergey N. 82
Shabarov Arcady 137
Shalamanov Victor A. 82
Shatko Dmitry 153
Shelkovnikov Sergey 338
Sheveleva Oksana 243
Shilova Anna 338
Shustova Elena 338
Shut'ko Larisa 267
Šimková Zuzana 312, 321
Slesarenko Ekaterina 243
Snegireva Tatyana 296
Solodov Vyacheslav 196
Sotnikov Igor 175
Stefanek Pawel 217
Stepanov Yuri 1
Strelnikov Pavel 153
Subbotin Sergey 196
Sysoyev Andrey 105
Szurgacz Dawid 15
Ta Van Kien 64
Tailakova Anna 45
Tan Yunliang 98
Tihomirova Anastasia 191
Tyuleneva Tatiana 115
Udovitsky Vladimir 146
Verchagina Irina 181
Verezomska Irina 283
Vilamová Šárka 302
Vlasenko Andrey 175
Voronets Dmytro 283
Voronkina Natalia 23
Weber Dieter 36
Wittenberger Gabriel 321
Wladzielczyk Krzysztof 7
Yakunina Yulia 267
Yakushnina Tatyana 333
Yazevich Mariya 223
Yutiaev Evgeny 137
Zacharov Aleksandr 30
Zaruba Natalia 252
Zhdanov Vyacheslav 93
Zhironkin Sergey 291
Zhironkin Vitaly 165, 291
Zhironkina Olga 223
Zolkin Denis 77
Zolotukhin Vladimir 231
Zolotukhina Natalia 231
Zvarych Evgeny 36
Zykov Pyotr 36